ENCOUNTERS WITH THE PARANORMAL

ENCOUNTERS WITH THE PARANORMAL

Science,

Knowledge,

and Belief

EDITED BY
KENDRICK FRAZIER

Prometheus Books
59 John Glenn Drive
Amherst, New York 14228-2197

Published 1998 by Prometheus Books

01 00 99 98 97   5 4 3 2 1

Library of Congress Cataloging-in-Publication Data

Encounters with the paranormal : science, knowledge, and belief / edited by
    Kendrick Frazier.
        p.   cm.
    Includes bibliographical references (p.   ) and index.
    ISBN 1-57392–203–X (alk. paper)
    1. Parapsychology and science. 2. Science—Philosophy. I. Frazier,
Kendrick.
BF1045.S33E53   1998
133—dc21   97–51468
                                                                             CIP

Printed in Canada on acid-free paper.

CONTENTS

PART THREE: SCIENCE, PSEUDOSCIENCE, AND PATHOLOGICAL SCIENCE

PART FOUR: PSEUDOSCIENCE AND PATHOLOGICAL CULTS: HEAVEN'S GATE

PART FIVE: SCIENCE AND THE PSYCHOLOGY OF BELIEF

Contents

PART EIGHT: SOCIAL DYNAMICS AND BELIEF

PART NINE: THE MALLEABILITY OF MEMORY

INTRODUCTION

What is the relationship between science and wonder? What is the role of open-minded skepticism in science, and how do you balance that with no-holds-barred scientific exploration? Does the scientific quest for knowledge diminish or increase the aesthetic experience? How do we know what we know? How do beliefs form, anyway, and what gives them so much power, sometimes in face of logic and rationality? If eyewitnesses are so much sought in courtroom testimony, why is eyewitness testimony so suspect when it comes to some scientific claims?

Do polygraph tests really detect lies? Do honesty tests work? Can memories be repressed, and, if so, can they be recovered? Can false memories be implanted? Is subliminal perception a reality or an illusion? What is a lucid dream? Why do conspiracy theories have such strong appeal to many? What are the lessons of the Heaven's Gate cult suicides? Have there been outbreaks of ritualistic satanic cult crimes? Do we have evidence for psychic functioning? Can trained psychics really help intelligence agencies locate secret enemy facilities? How can we improve secondary science education? If knowledge always increases, does ignorance do so as well?

Encounters with the Paranormal: Science, Knowledge, and Belief explores these and scores of other questions. They all arise one way or another along the tumultuous interfaces between science, knowledge, and belief. The forty-five noted authors (including two Nobel laureates and many other distinguished scholars) represent the physical and biological sciences, psychology, philosophy, the social sciences, and forensic science. Their insights are not just for people interested in, or concerned about, the "paranormal," although I think they are worthwhile reading for everyone who fits those categories. The emphasis on the articles selected for this volume is on topics that have much wider implications. The forty-six chapters bring scientific perspective to issues and controversies that continually make news, influence public policy, affect what we think we know, and shape our perceptions and misperceptions of reality and the natural world.

These chapters, many of them updated specially for this volume, originally appeared as articles in the *Skeptical Inquirer: The Magazine for Science and Reason,* the journal of the Committee for the Scientific Investigation of Claims of the Paranormal (CSICOP). CSICOP is an independent nonprofit scientific and educational organization established in 1976 to encourage the critical investigation of paranormal and fringe-science claims from a responsible scientific point of view and to disseminate the results of such inquiries to the scientific community and the public. It encourages science, critical thinking, science education, and open-minded, critical inquiry. The idea is to use both the formal processes of science and our innate common sense and skepticism to try to examine the validity of all manner of claims to truth that lie along or just outside the borders of science.

The general feeling of the philosophers, scientists, psychologists, and others who helped found CSICOP was that by ignoring the many kinds of unusual claims and topics that have great popular appeal, scientists were missing an opportunity. For one thing, they were allowing many kinds of dubious or unsupported assertions to go unchallenged, and as a result a distressing amount misinformation was being spread about and taken for fact. At the same time, they were missing a marvelous chance to help show students, the public, and the media some of the wonderful processes, experiences, and discoveries of science. Likewise, there were enormous opportunities to educate about the differences between real science and those activities that adopt some of the language and superficial trappings of science but fail to use scientific methods to sift fact from fancy, sense from nonsense, and fact-based scientific speculation from sensationalism and pseudoscience.

All the evidence indicates that almost all of us need some help in this regard. Even scholars sometimes have trouble sifting through the factual assertions, arguments, and issues that fall outside their own specialties. Among the general public, there is a very natural and understandable fascination with all these topics, and very little of the information readily available about them comes via a scientific perspective. CSICOP and the *Skeptical Inquirer* have helped fill that void.

Many distinguished scientists and scholars have contributed to this effort over the years. Philosopher Paul Kurtz is the founder and chairman of CSICOP. Among those core people involved from the beginning were author/critic Martin Gardner, psychologist Ray Hyman, and magician/investigator James Randi. Five Nobel laureate scientists (Francis Crick, Murray Gell-Mann, Leon Lederman, Glenn T. Seaborg, and Steven Weinberg) are now among CSICOP Fellows. Some of the other distinguished Fellows (just to give a few examples) include the evolutionary scientists Richard Dawkins and Stephen Jay Gould, philosophers W. V. Quine and Stephen Toulmin, anthropologists Eugenie Scott and Thomas Sebeok, cognitive scientists Douglas Hofstadter and Marvin Minsky, physicists Gerald Holton and Richard A. Muller, astronomers E. C. Krupp and Jill Tarter, space scientists David Morrison and Cornelis de Jager, medical scientists Wallace Sampson and Stephen Barrett, and science editors John Maddox and Leon Jaroff. Other fields represented by CSICOP Fellows and Scientific Consultants include mathematics, statistics, sociology, folklore, engineering, materials science, the earth sciences, public health, forensics, conjuring, and science journalism. (The late Carl

Sagan, Isaac Asimov, B. F. Skinner, and Sydney Hook were also among the most prominent Fellows.)

However, if there is any one discipline that is most central to understanding and elucidating the entire range of concerns of the CSICOP and the *Skeptical Inquirer,* it is psychology. And CSICOP's roster of Fellows appropriately includes a host of distinguished psychologists. Among them are James Alcock, Robert Baker, Barry Beyerstein, Susan Blackmore, Thomas Gilovich, C. E. M. Hansel, the previously mentioned Ray Hyman, Elizabeth Loftus, David Marks, Loren Pankratz, Milton Rosenberg, and Carol Tavris. Also, at least eight of CSICOP's scientific consultants are psychologists.

Psychologists bring expertise in the psychology of belief, in the influence of our expectations on our perceptions, and in the nature of mind and memory. They understand how to carry out complex experimental studies. They know how best to test claims of unusual human abilities. They can provide people natural explanations for their powerful and extraordinary personal experiences that may otherwise be considered mystical or supernatural. They also bring a deep understanding of how the natural workings of our minds can—even while functioning totally normally—sometimes lead us astray.

Having emphasized some of the most distinguished scholars associated with CSICOP, I must quickly point out that the effort to promote science and bring the methods of science to examining borderland claims is a very democratic activity. It is open to anyone committed to using science and reason. The authors who publish in the *Skeptical Inquirer* need have no affiliation with the organization itself, and many do not. Acceptance of articles is based solely on interest, clarity, significance, relevance, authority, and topicality.

Forty-six of the best of those articles published in the past six years in the *Skeptical Inquirer* now await you. I have organized them into nine sections that seem to me to flow fairly naturally from what real science is all about, to concerns about antiscience, pseudoscience, and pathological science, and then to the many psychological and social aspects of these topics. These sections are: (1) Science, Imagination, and Responsibility; (2) Science and Antiscience; (3) Science, Pseudoscience, and Pathological Science; (4) Pseudoscience and Pathological Cults: Heaven's Gate; (5) Science and the Psychology of Belief; (6) Psychology and the Claims of Psi; (7) Psychology and the Anomalous Experience; (8) Social Dynamics and Belief; and (9) The Malleability of Memory.

This is the fourth general anthology of *Skeptical Inquirer* articles.* The previous three are *The Hundredth Monkey and Other Paradigms of the Paranormal* (1991), *Science Confronts the Paranormal* (1986), and *Paranormal Borderlands of Science* (1981). All are still in print with Prometheus Books. This one is, by necessity, the most selective. Even more so than with the earlier anthologies, many excellent articles

---

*The first single-subject anthology of *Skeptical Inquirer* articles, *The UFO Invasion* (K. Frazier, B. Karr, and J. Nickell, editors), was published by Prometheus in 1997. It includes virtually all articles published on UFOs, the Roswell crashed-saucer claim, alien abductions, crop circles, and the search for extraterrestrial intelligence.

had to be omitted. For one reason, this book covers a six-year period (1991–1997), where the others covered five. For another, beginning with the first issue of 1995, we increased the frequency of the *Skeptical Inquirer* from quarterly to bimonthly, and so more material than ever has been available to chose from.

I am very grateful to all those who contribute to the *Skeptical Inquirer*. They are its strength. Likewise the readership. The readers come from all walks of life, not just science and academia, and they are intelligent, curious, discerning, demanding, and knowledgeable. This volume is dedicated to our contributors and our readers.

PART ONE

SCIENCE, IMAGINATION, AND RESPONSIBILITY

1

WONDER AND SKEPTICISM
Carl Sagan

I was a child in a time of hope. I grew up when the expectations for science were very high: in the thirties and forties. I went to college in the early fifties, got my Ph.D. in 1960. There was a sense of optimism about science and the future. I dreamt of being able to do science. I grew up in Brooklyn, New York, and I was a street kid. I came from a nice nuclear family, but I spent a lot of time in the streets, as kids did then. I knew every bush and hedge, streetlight and stoop and theater wall for playing Chinese handball. But there was one aspect of that environment that, for some reason, struck me as different, and that was the stars.

Even with an early bedtime in winter you could see the stars. What were they? They weren't like hedges or even streetlights; they were different. So I asked my friends what they were. They said, "They're lights in the sky, kid." I could tell they were lights in the sky, but that wasn't an explanation. I mean, what were they? Little electric bulbs on long black wires, so you couldn't see what they were held up by? What were they?

Not only could nobody tell me, but nobody even had the sense that it was an interesting question. They looked at me funny. I asked my parents; I asked my parents' friends; I asked other adults. None of them knew.

My mother said to me, "Look, we've just got you a library card. Take it, get on the streetcar, go to the New Utrecht branch of the New York Public Library, get out a book and find the answer."

That seemed to me a fantastically clever idea. I made the journey. I asked the librarian for a book on stars. (I was very small; I can still remember looking up at her, and she was sitting down.) She was gone a few minutes, brought one back, and gave it to me. Eagerly I sat down and opened the pages. But it was about Jean Harlow and Clark Gable, I think, a terrible disappointment. And so I went back to her, explained (it wasn't easy for me to do) that that wasn't what I had in mind at all, that what I

This article first appeared in *Skeptical Inquirer* 19, no. 1 (January/February 1995): 24–30. Copyright © 1994 by Carl Sagan.

wanted was a book about real stars. She thought this was funny, which embarrassed me further. But anyway, she went and got another book, the right kind of book. I took it and opened it and slowly turned the pages, until I came to the answer.

It was in there. It was stunning. The answer was that the Sun was a star, except very far away. The stars were suns; if you were close to them, they would look just like our sun. I tried to imagine how far away from the Sun you'd have to be for it to be as dim as a star. Of course I didn't know the inverse square law of light propagation; I hadn't a ghost of a chance of figuring it out. But it was clear to me that you'd have to be very far away. Farther away, probably, than New Jersey. The dazzling idea of a universe vast beyond imagining swept over me. It has stayed with me ever since.

I sensed awe. And later on (it took me several years to find this), I realized that we were on a planet—a little, non-self-luminous world going around our star. And so all those other stars might have planets going around them. If planets, then life, intelligence, other Brooklyns—who knew? The diversity of those possible worlds struck me. They didn't have to be exactly like ours, I was sure of it.

It seemed the most exciting thing to study. I didn't realize that you could be a professional scientist; I had the idea that I'd have to be, I don't know, a salesman (my father said that was better than the manufacturing end of things), and do science on weekends and evenings. It wasn't until my sophomore year in high school that my biology teacher revealed to me that there was such a thing as a professional scientist, who got paid to do it; so you could spend all your time learning about the universe. It was a glorious day.

It's been my enormous good luck—I was born at just the right time—to have had, to some extent, those childhood ambitions satisfied. I've been involved in the exploration of the solar system, in the most amazing parallel to the science fiction of my childhood. We actually send spacecraft to other worlds. We fly by them; we orbit them; we land on them. We design and control the robots: Tell it to dig, and it digs. Tell it to determine the chemistry of a soil sample, and it determines the chemistry. For me the continuum from childhood wonder and early science fiction to professional reality has been almost seamless. It's never been, "Oh, gee, this is nothing like what I had imagined." Just the opposite: It's exactly like what I imagined. And so I feel enormously fortunate.

Science is still one of my chief joys. The popularization of science that Isaac Asimov did so well—the communication not just of the findings but of the methods of science—seems to me as natural as breathing. After all, when you're in love, you want to tell the world. The idea that scientists shouldn't talk about their science to the public seems to me bizarre.

There's another reason I think popularizing science is important, why I try to do it. It's a foreboding I have—maybe ill-placed—of an America in my children's generation, or my grandchildren's generation, when all the manufacturing industries have slipped away to other countries; when we're a service and information-processing economy; when awesome technological powers are in the hands of a very few, and no one representing the public interest even grasps the issues; when the people (by "the people" I mean the broad population in a democracy) have lost the ability to set their own agendas, or even to knowledgeably question those who do set the agendas; when there is no practice in questioning those in authority; when, clutching our crystals and religiously consulting our horoscopes, our critical faculties in steep decline, unable to distinguish between what's true and what feels good, we slide, almost without noticing, into superstition and darkness. CSICOP plays a sometimes lonely but still—and in this case the word may be right—heroic role in trying to counter some of those trends.

We have a civilization based on science and technology, and we've cleverly arranged things so that almost nobody understands science and technology. That is as clear a prescription for disaster as you can imagine. While we might get away with this combustible mixture of ignorance and power for a while, sooner or later it's going to blow up in our faces. The powers of modern technology are so formidable that it's insufficient just to say, "Well, those in charge, I'm sure, are doing a good job." This is a democracy, and for us to make sure that the powers of science and technology are used properly and prudently, we ourselves must understand science and technology. We must be involved in the decision-making process.

The predictive powers of some areas, at least, of science are phenomenal. They are the clearest counterargument I can imagine to those who say, "Oh, science is situational; science is just the current fashion; science is the promotion of the self-interests of those in power." Surely there is some of that. Surely if there's any powerful tool, those in power will try to use it, or even monopolize it. Surely scientists, being people, grow up in a society and reflect the prejudices of that society. How could it be otherwise? Some scientists have been nationalists; some have been racists; some have been sexists. But that doesn't undermine the validity of science. It's just a consequence of being human.

So, imagine—there are so many areas we could think of—imagine you want to know the sex of your unborn child. There are several approaches. You could, for example, do what the late film star who Annie and I admire greatly—Cary Grant—did before he was an actor: In a carnival or fair or consulting room, you suspend a watch or a plumb bob above the abdomen of the expectant mother; if it swings left-right it's a boy, and if it swings forward-back it's a girl. The method works one time in two. Of course he was out of there before the baby was born, so he never heard from customers who complained he got it wrong. Being right one chance in two—that's not so bad. It's better than, say, Kremlinologists used to do. But if you really want to know, then you go to amniocentesis, or to sonograms; and there your chance of being right is 99 out of 100. It's not perfect, but it's a whole lot better than one out of two. If you really want to know, you go to science.

Or suppose you wanted to know when the next eclipse of the sun is. Science does something really astonishing: It can tell you a century in advance where the eclipse is going to be on Earth and when, say, totality will be, to the second. Think of the predictive power this implies. Think of how much you must understand to be able to say when and where there's going to be an eclipse so far in the future.

Or (the same physics exactly) imagine launching a spacecraft from Earth, like the Voyager spacecraft in 1977; 12 years later Voyager 1 arrives at Neptune within 100 kilometers or something of where it was supposed to be—not having to use some of the mid-course corrections that were available; 12 years, 5 billion kilometers, on target!

So if you want to really be able to predict the future—not in everything, but in some areas—there's only one regime of human scholarship, of human claims to knowledge, that really delivers the goods, and that's science. Religions would give their eyeteeth to be able to predict anything like that well. Think of how much mileage they would make if they ever could do predictions comparably unambiguous and precise.

Now how does it work? Why is it so successful?

Science has built-in error-correcting mechanisms—because science recognizes that scientists, like everybody else, are fallible, that we make mistakes, that we're driven by the same prejudices as everybody else. There are no forbidden questions. Arguments from authority are worthless. Claims must be demonstrated. Ad hominem arguments—arguments about the personality of somebody who disagrees with you—are irrelevant; they can be sleazeballs and be right, and you can be a pillar of the community and be wrong.

If you take a look at science in its everyday function, of course you find that scientists run the gamut of human emotions and personalities and character and so on. But there's one thing that is really striking to the outsider, and that is the gauntlet of criticism that is considered acceptable or even desirable. The poor graduate student at his or her Ph.D. oral exam is subjected to a withering crossfire of questions that sometimes seem hostile or contemptuous; this from the professors who have the candidate's future in their grasp. The students naturally are nervous; who wouldn't be? True, they've prepared for it for years. But they understand that at that critical moment they really have to be able to answer questions. So in preparing to defend their theses, they must anticipate questions; they have to think, "Where in my thesis is there a weakness that someone else might find—because I sure better find it before they do, because if they find it and I'm not prepared, I'm in deep trouble."

You take a look at contentious scientific meetings. You find university colloquia in which the speaker has hardly gotten 30 seconds into presenting what she or he is saying, and suddenly there are interruptions, maybe withering questions, from the audience. You take a look at the publication conventions in which you submit a scientific paper to a journal, and it goes out to anonymous referees whose job it is to think, Did you do anything stupid? If you didn't do anything stupid, is there anything in here that is sufficiently interesting to be published? What are the deficiencies of this paper? Has it been done by anybody else? Is the argument adequate, or should you resubmit the paper after you've actually demonstrated what you're speculating on? And so on. And it's anonymous: You don't know who your critics are. You have to rely on the editor to send it out to real experts who are not overtly malicious. This is the everyday expectation in the scientific community. And those who don't expect it—even good scientists who just can't hold up under criticism—have difficult careers.

Why do we put up with it? Do we like to be criticized? No, no scientist likes to be criticized. Every scientist feels an affection for his or her ideas and scientific results. You feel protective of them. But you don't reply to critics: "Wait a minute, wait a minute; this is a really good idea. I'm very fond of it. It's done you no harm.

Please don't attack it." That's not the way it goes. The hard but just rule is that if the ideas don't work, you must throw them away. Don't waste any neurons on what doesn't work. Devote those neurons to new ideas that better explain the data. Valid criticism is doing you a favor.

There is a reward structure in science that is very interesting: Our highest honors go to those who disprove the findings of the most revered among us. So Einstein is revered not just because he made so many fundamental contributions to science, but because he found an imperfection in the fundamental contribution of Isaac Newton. (Isaac Newton was surely the greatest physicist before Albert Einstein.)

Now think of what other areas of human society have such a reward structure, in which we revere those who prove that the fundamental doctrines that we have adopted are wrong. Think of it in politics, or in economics, or in religion; think of it in how we organize our society. Often, it's exactly the opposite: There we reward those who reassure us that what we've been told is right, that we need not concern ourselves about it. This difference, I believe, is at least a basic reason why we've made so much progress in science, and so little in some other areas.

We are fallible. We cannot expect to foist our wishes on the universe. So another key aspect of science is experiment. Scientists do not trust what is intuitively obvious, because intuitively obvious gets you nowhere. That the Earth is flat was once obvious. I mean, really obvious; obvious! Go out in a flat field and take a look: Is it round or flat? Don't listen to me; go prove it to yourself. That heavier bodies fall faster than light ones was once obvious. That blood-sucking leeches cure disease was once obvious. That some people are naturally and by divine right slaves was once obvious. That the Earth is at the center of the universe was once obvious. You're skeptical? Go out, take a look: Stars rise in the east, set in the west; here we are, stationary (do you feel the Earth whirling?); we see them going around us. We are at the center; they go around us.

The truth may be puzzling. It may take some work to grapple with. It may be counterintuitive. It may contradict deeply held prejudices. It may not be consonant with what we desperately want to be true. But our preferences do not determine what's true. We have a method, and that method helps us to reach not absolute truth, only asymptotic approaches to the truth—never there, just closer and closer, always finding vast new oceans of undiscovered possibilities. Cleverly designed experiments are the key.

In the 1920s, there was a dinner at which the physicist Robert W. Wood was asked to respond to a toast. This was a time when people stood up, made a toast, and then selected someone to respond. Nobody knew what toast they'd be asked to reply to, so it was a challenge for the quick-witted. In this case the toast was: "To physics and metaphysics." Now by metaphysics was meant something like philosophy—truths that you could get to just by thinking about them. Wood took a second, glanced about him, and answered along these lines: The physicist has an idea, he said. The more he thinks it through, the more sense it makes to him. He goes to the scientific literature, and the more he reads, the more promising the idea seems. Thus prepared, he devises an experiment to test the idea. The experiment is painstaking. Many possibilities are eliminated or taken into account; the accuracy of the measurement is refined. At the end of all this work, the experiment is completed and . . . the idea is shown to be worthless. The physicist then discards the idea, frees his mind (as I was saying a moment ago) from the clutter of error, and moves on to something else.

The difference between physics and metaphysics, Wood concluded, is that the metaphysicist has no laboratory.

✦ ✦ ✦

Why is it so important to have widely distributed understanding of science and technology? For one thing, it's the golden road out of poverty for developing nations. And developing nations understand that, because you have only to look at modern American graduate schools—in mathematics, in engineering, in physics—to find, in case after case, that more than half the students are from other countries. This is something America is doing for the world. But it conveys a clear sense that the developing nations understand what is essential for their future. What worries me is that Americans may not be equally clear on the subject.

Let me touch on the dangers of technology. Almost every astronaut who has visited Earth orbit has made this point: I was up there, they say, and I looked toward the horizon, and there was this thin, blue band that's the Earth's atmosphere. I had been told we live in an ocean of air. But there it was, so fragile, such a delicate blue: I was worried for it.

In fact, the thickness of the Earth's atmosphere, compared with the size of the Earth, is in about the same ratio as the thickness of a coat of shellac on a schoolroom globe is to the diameter of the globe. That's the air that nurtures us and almost all other life on Earth, that protects us from deadly ultraviolet light from the sun, that through the greenhouse effect brings the surface temperature above the freezing point. (Without the greenhouse effect, the entire Earth would plunge below the freezing point of water and we'd all be dead.) Now that atmosphere, so thin and fragile, is under assault by our technology. We are pumping all kinds of stuff into it. You know about the concern that chlorofluorocarbons are depleting the ozone layer; and that carbon dioxide and methane and other greenhouse gases are producing global warming, a steady trend amidst fluctuations produced by volcanic eruptions and other sources. Who knows what other challenges we are posing to this vulnerable layer of air that we haven't been wise enough to foresee?

The inadvertent side effects of technology can challenge the environment on which our very lives depend. That means that we must understand science and technology; we must anticipate longterm consequences in a very clever way—not just the bottom line on the profit-and-loss column for the corporation for this year, but the consequences for the nation and the species 10, 20, 50, 100 years in the future. If we absolutely stop all chlorofluorocarbon and allied chemical production right now (as we're in fact doing), the ozonosphere will heal itself in about a hundred years. Therefore our children, our grandchildren, our great-grandchildren must suffer through the mistakes that we've made. That's a second reason for science education: the dangers of technology. We must understand them better.

A third reason: origins. Every human culture has devoted some of its intellectual, moral, and material resources to trying to understand where everything comes from—our nation, our species, our planet, our star, our galaxy, our universe. Stop someone on the street and ask about it. You will not find many people who never thought about it, who are incurious about their ultimate origins.

I hold there's a kind of Gresham's Law that applies in the confrontation of science and pseudoscience: In the popular imagination, at least, the bad science drives out the good. What I mean is this: If you are awash in lost continents and channeling and UFOs and all the long litany of claims so well exposed in the *Skeptical Inquirer*, you may not have intellectual room for the findings of science. You're sated with wonder. Our culture in one way produces the fantastic findings of science, and then in another way cuts them off before they reach the average person. So people who are curious, intelligent, dedicated to understanding the world, may nevertheless be (in our view) enmired in superstition and pseudoscience. You could say, Well, they ought to know better, they ought to be more critical, and so on; but that's too harsh. It's not very much their fault, I say. It's the fault of a society that preferentially propagates the baloney and holds back the ambrosia.

The least effective way for skeptics to get the attention of these bright, curious, interested people is to belittle, or condescend, or show arrogance toward their beliefs. They may be credulous, but they're not stupid. If we bear in mind human frailty and fallibility, we will understand their plight.

For example: I've lately been thinking about alien abductions, and false claims of childhood sexual abuse, and stories of satanic ritual abuse in the context of recovered memories. There are interesting similarities among those classes of cases. I think if we are to understand any of them, we must understand all of them. But there's a maddening tendency of the skeptics, when addressing invented stories of childhood sexual abuse, to forget that real and appalling abuse happens. It is not true that all these claims of childhood sexual abuse are silly and pumped up by unethical therapists. Yesterday's paper reported that a survey of 13 states found that one-sixth of all the rape victims reported to police are under the age of 12. And this is a category of rape that is preferentially under-reported to police, for obvious reasons. Of these girls, one-fifth were raped by their fathers. That's a lot of people, and a lot of betrayal. We must bear that in mind when we consider patients who, say, because they have an eating disorder, have suppressed childhood sexual abuse diagnosed by their psychiatrists.

People are not stupid. They believe things for reasons. Let us not dismiss pseudoscience or even superstition with contempt.

In the nineteenth century it was mediums: You'd go to the séance, and you'd be put in touch with dead relatives. These days it's a little different; it's called channeling. What both are basically about is the human fear of dying. I don't know about you; I find the idea of dying unpleasant. If I had a choice, at least for a while, I would just as soon not die. Twice in my life I came very close to doing so. (I did not have a near-death experience, I'm sorry to say.) I can understand anxiety about dying.

About 14 years ago both my parents died. We had a very good relationship. I was very close to them. I still miss them terribly. I wouldn't ask much: I would like five minutes a year with them; to tell them how their kids and their grandchildren are doing, and how Annie and I are doing. I know it sounds stupid, but I'd like to ask them, "Is everything all right with you?" Just a little contact. So I don't guffaw at women who go to their husbands' tombstones and chat them up every now and then. That's not hard to understand. And if we have difficulties on the ontological status of who it is they're talking to, that's all right. That's not what this is about. This is humans being human.

In the alien-abduction context, I've been trying to understand the fact that humans hallucinate—that it's a human commonplace—yes, under conditions of sensory deprivation or drugs or deprival of REM sleep, but also just in the ordinary course of existence. I have, maybe a dozen times since my parents died, heard one of them say my name: just the single word, "Carl." I miss them; they called me by my first name so much during the time they were alive; I was in the practice of responding instantly when I was called; it has deep psychic roots. So my brain plays it back every now and then. This doesn't surprise me at all; I sort of like it. But it's a hallucination. If I were a little less skeptical, though, I could see how easy it would be to say, "They're around somewhere. I can hear them."

Raymond Moody, who is an M.D., I think, an author who writes innumerable books on life after death, actually quoted me in the first chapter of his latest book, saying that I heard my parents calling me Carl, and so, look, even he believes in life after death. This badly misses my point. If this is one of the arguments from chapter 1 of the latest book of a principal exponent of life after death, I suspect that despite our most fervent wishes, the case is weak.

But still, suppose I wasn't steeped in the virtues of scientific skepticism and felt as I do about my parents, and along comes someone who says, "I can put you in touch with them." Suppose he's clever, and found out something about my parents in the past, and is good at faking voices, and so on—a darkened room and incense and all of that. I could see being swept away emotionally.

Would you think less of me if I fell for it? Imagine I was never educated about skepticism, had no idea that it's a virtue, but instead believed that it was grumpy and negative and rejecting of everything that's humane. Couldn't you understand my openness to being conned by a medium or a channeler?

The chief deficiency I see in the skeptical movement is its polarization: Us vs. Them—the sense that we have a monopoly on the truth; that those other people who believe in all these stupid doctrines are morons; that if you're sensible, you'll listen to us; and if not, to hell with you. This is nonconstructive. It does not get our message across. It condemns us to permanent minority status. Whereas, an approach that from the beginning acknowledges the human roots of pseudo-science and superstition, that recognizes that the society has arranged things so that skepticism is not well taught, might be much more widely accepted.*

If we understand this, then of course we have compassion for the abductees and those who come upon crop circles and believe they're supernatural, or at least

---

*If skeptical habits of thought are widely distributed and prized, then who is the skepticism going to be mainly applied to? To those in power. Those in power, therefore, do not have a vested interest in everybody being able to ask searching questions.

of extraterrestrial manufacture. This is key to making science and the scientific method more attractive, especially to the young, because it's a battle for the future.

Science involves a seemingly self-contradictory mix of attitudes: On the one hand, it requires an almost complete openness to all ideas, no matter how bizarre and weird they sound, a propensity to wonder. As I walk along, my time slows down; I shrink in the direction of motion, and I get more massive. That's crazy! On the scale of the very small, the molecule can be in this position, in that position, but it is prohibited from being in any intermediate position. That's wild! But the first is a statement of special relativity, and the second is a consequence of quantum mechanics. Like it or not, that's the way the world is. If you insist that it's ridiculous, you will be forever closed to the major findings of science. But at the same time, science requires the most vigorous and uncompromising skepticism, because the vast majority of ideas are simply wrong, and the only way you can distinguish the right from the wrong, the wheat from the chaff, is by critical experiment and analysis.

Too much openness and you accept every notion, idea, and hypothesis—which is tantamount to knowing nothing. Too much skepticism—especially rejection of new ideas before they are adequately tested—and you're not only unpleasantly grumpy, but also closed to the advance of science. A judicious mix is what we need.

It's no fun, as I said at the beginning, to be on the receiving end of skeptical questioning. But it's the affordable price we pay for having the benefits of so powerful a tool as science.

UPDATE BY KENDRICK FRAZIER

Carl Sagan's death on December 20, 1996, of complications of the bone marrow disease myelodisplasia set off expressions of shock and grief rarely accorded a scientist's passing. He was only sixty-two. He had been battling the disease for two years and had led an active intellectual and literary life to the end.

His book *The Demon-Haunted World: Science as a Candle in the Dark,* about some of the themes in the foregoing article—the value of scientific thinking and skepticism and the dangers of pseudoscience, superstition, and the "siren song of unreason"—had been published earlier that year. Another book, *Billions & Billions,* was essentially completed and was published in 1997. It was subtitled *Thoughts on Life and Death at the Brink of the Millennium.* Several scientific papers he had coauthored on planetary science had been accepted by journals and were published posthumously.

His death was an international news story. Scientists praised his contributions to planetary science and his great ability, as National Science Foundation Director Neal Lane put it, "to bring science into American living rooms, to show its relevance to our everyday lives, and to share the excitement and discovery." His Cornell astronomy colleague Yervant Terzian called Sagan "the best teacher of science in the world."

The *Skeptical Inquirer* invited tributes. They were published as "The Darkened Cosmos: A Tribute to Carl Sagan," and there were so many they had to be spread over three 1997 issues, March/April, May/June, and July/August. Richard Dawkins said Sagan was "one of the great literary stylists of our age, and he did it by giving proper weight to the poetry of science." Arthur C. Clarke called him a "superstar of popular science communication" and praised his taking on "ably and daringly"—the pseudoscience and nonscience that rots the American mind." Fellow planetary astronomer David Morrison said, "Carl Sagan did more than any other individual in this century to bring science to the public in a way that was consistently honest, thoughtful, and humane." James Randi said he was devastated and added, "The burden now falls on us to provide as much light as we can generate, to banish the darkness [of pseudoscience, superstition, and unreason] and make sure it does not triumph over us." Nobel laureate physicist Leon Lederman called Sagan's efforts in the public understanding of science "unique and heroic" and "an awesome example for the rest of us." Paul Kurtz noted that his death was a profound loss to CSICOP and the entire skeptical movement.

Tributes came from ordinary people as well as scientific luminaries. Some wrote poems and sonnets. Many expressed a great *personal* sense of loss. "I'm just a regular guy who likes to read and think," said one. "But the passing of Carl Sagan is a horrible loss.... He never knew me but in every sense he was my teacher." Said another: "I will miss the articulate exuberance with which he celebrated the wonders of our cosmos and his stern but sympathetic admonitions against those of us who would succumb to easy preconceptions about ourselves and our world." "My loss seemed so personal," said another. "Carl Sagan wasn't 'just' a hero; he was a rare and precious sort of hero." Another reader wrote movingly: "He never knew me, but he saved my life, and I never got to thank him." He described how he had been a "screwed-up kid in the late '70s on the fast road to nowhere. Then came [Sagan's TV series] 'Cosmos.'... He gave me back my sense of wonder at the mystery of the universe...something I wasn't even aware I'd lost." This reader changed his way of living, began reading and studying science, and now has a degree in physics and a web page devoted to the advancement of science. "This man, who never knew me, but saved my life, and whom I never got to thank. Well, Carl Sagan, thank you.... You were my greatest teacher. There may be billions of stars in the universe, but few have shined as brightly as you."

In an epilogue to *Billions & Billions*, Ann Druyan, Sagan's wife and longtime collaborator, wrote of his final days and the worldwide response to his death. "I sit surrounded by cartons of mail from people all over the planet who mourn Carl's loss. Many of them credit him with their awakenings. Some of them say that Carl's example has inspired them to work for science and reason and against the forces of superstition and fundamentalism. These thoughts comfort me and lift me out of my heartache. They allow me to feel, without resorting to the supernatural, that Carl lives."

2

A STRATEGY FOR SAVING SCIENCE
Leon M. Lederman

(T) he support of scientific (and engineering) research in the U.S. has been steadily eroding for the past decade. There are three institutions that conduct research and all are in retreat. Industry, once the source of basic research with vast technological implications, has been retreating, with once-great laboratories now only shadows of their former splendor. Universities, both public and private, are losing money and are, through hiring freezes and general austerity, reducing their own contribution to match the reduced overhead collected from federal contracts. Opportunities for young faculty are thereby restricted. And the U.S. government, by far the largest contributor to basic research, has not kept pace with the consumer price index, not to mention the increases needed to continue research with ever more profound, challenging, and therefore more expensive measurements. Some of the results are the worsening of morale, the loss of young scientific recruits, and the loss of new science as senior investigators spend increasing amounts of time searching for funding to keep their experiments going and their graduate students fed.

Superposed upon this state of gracelessness is the emergence of antiscience. Perhaps it has always been there but our explosive growth in communications changes things. More likely, antiscience waxes and wanes over the decades and we are in a waxing phase. And the antiscience comes in many flavors. "Parascience" or "junk science" includes astrologers, mystics, psychics, clairvoyants, fortune tellers, soothsayers, pyramidalists, pendulum dowsers, spiritualists, Israeli magicians, faith healers, UFO witnesses, ESP wizards. Whew! I suspect that most of them simply want to earn a dishonest dollar. I'm personally protected from astrology because I am a Gemini, and Geminis don't believe in astrology.

At a more sophisticated but equally dishonest level, in my opinion, are the late-night talk show liars, the now mostly jailed crowd of TV evangelists ("In the

---

This article first appeared in *Skeptical Inquirer* 20, no. 6 (November/December 1996): 23–28.

name of God" send money), and the TV networks (NBC is clearly in the lead here!) that prey on gullible viewers with their sensational "science-like" programs such as "Sightings," "Paranormal Borderlands," and the NBC special "The Mysterious Origins of Man." There are also credentialed scientists who, for a buck or a fat book advance, will provide a mathematical proof that God exists (there was an error in the proof!) or that life after death is wonderful enough to discourage you from exercising and eating oat bran. Included here are also the political gumshoes, the modern-day McCarthy sleuths, Congressional investigators like Congressman John Dingell. Then there were Senator William Proxmire's obscene "Golden Fleece" awards—cheap, ignorant publicity devices disguised as protection of the public purse. Let's not forget the self-appointed, tattletale peeping Toms disguised as whistle-blowers, who make their reputations as St. Georges, slaying the dragon of guilty-until-proven-innocent scientists.

Antiscience includes fundamentalists, creationists, cultists, the Religious Wrong—people who have a desperate need to believe; that's okay, but they have an equally desperate need to have you believe, too.

On another plane are the academicians who know no science, are proud of their ignorance, or who, mortified by this disconnectedness, generate contempt for the scientific enterprise. Others, swept into the antimodernist, end-of-objectivity collective, hold the goals of science to improve human society as not only unrealizable but harmful. Surely the seemingly intractable problems of poverty, urban decay, crime, hatred, and general inequities seem irrelevant to how much we learn about the world. Certainly some of the more exotic views of this group, often at the edge of environmental, feminist, or animal rights movements, are rooted in a deep ignorance of the nature of science—what it does successfully and what it cannot do. If only we could get into their heads, expose them to the luminescent, emerging worldview of the physical and biological universes and the scientific magic that allows arrogant, passionate, revolutionary, imperfect humans—scientists—to create so magnificent an edifice as our ever-tentative but ever-evolving scientific knowledge base!

Now scientists are too often genetically infected optimists. We believe that problems, even seemingly intractable ones, can be addressed and solved. In any case, it is too late to retreat—the world is stuck with science and technology. The problem will be to design a strategy that maximizes the possibilities of using science and technology for the advance of humankind.

Now we have recently been reassured by the biennial NSF survey (National Science Board, *Science and Engineering Indicators—1996.* Washington, D.C.: U.S. Government Printing Office, 1996. [NSB 96–21]), which tells us that 70 percent of the American public loves us, thinks that scientists are among the most honorable of professionals, and believes that scientific research is a public good. And this number has been about the same for decades! So why the paranoia? Perhaps because antiscience has grown from 1 percent to 10 percent in the past decade or so. Perhaps because the 1 percent or 10 percent are congressmen, TV anchors, newspaper and magazine editors. Perhaps because many of the favorable 70 percent also watch dopey TV programs and listen to Rush Limbaugh. I don't know.

But, given this background, let me address a much deeper problem that underlies the success of antiscience in all of its forms and then suggest a strategy.

The deeper problem already well identified today is the problem of illiteracy, scientific illiteracy. If, as our experts tell us, the U.S. public is 93.8 percent scientifically illiterate (or 97.3 percent, depending on how one measures), then small wonder that our population is helpless before the onslaught of antiscience. Look at its advantages: antiscience is positive, authoritative, a haven for people who need safety and assurance, whereas science is hesitant, skeptical, even of—*especially of*—its own heritage. Not too much comfort there. And science demands some effort, a thought process. Antiscience says: Don't think! Believe! Trust us! We know! Science says: This is the best we can do here, the most we can say, note the error bars in our statements. . . . Not a fair fight.

Some nasty critics out there will point out that, much like religious organizations, science has its high priests who speak to each other in an esoteric and incomprehensible jargon and, when pressed, often invoke their gods with exotic names like Heisenberg, Schrödinger, Pasteur, and Einstein. However, the truth is that scientific authority exists to be overthrown, usually by very young revolutionaries, wielding logical bombs but perhaps not too unlike the caricature of the old Latin American country with a revolution every two weeks. The dramatic distinction, without penetrating to the details, is that scientific revolutions leave behind an improved knowledge base, a more comprehensive synthesis of what the laboratory has revealed.

Science has a four-hundred-year track record of progress, and this is measured in many ways: by the ever-widening domain in space, time, and conditions over which we can describe nature and make predictions. All the antiscience armies combined could not tell you the date of arrival of Halley's comet, whereas science can give you the year, day, hour, and minute. Science's only competition in the prediction business are the long-haired, barefoot sandwich board carriers upon whose boards doomsday is spelled out to the microsecond.

It is science that has converted night to day, extended human longevity, cured many dread diseases, enabled people of very modest means to drive across continents, fly over oceans, and surf webs. Following the rules of antiscience (collectively) would condemn the vast majority of humans to extremes of poverty, starvation, and early death, allowing the priests and kings to inhabit their drafty castles, monasteries, and rectories. Science makes available to many the best of what mankind produces—jazz, rock, Beethoven's late quartets; Shakespeare, Leonardo da Vinci, and Michael Jordan. Antiscience sells some comfort in community and the dubious promise of eternity in a better place or potions that guarantee all kinds of joy through strength. However, the ultimate argument for not abandoning science to the dark forces of superstition, ignorance, and rigid belief systems is that the planet will not survive a population of upwards of ten billion people (by the year 2050?) without significant increases in our knowledge base, without new forms of energy, food production, and mechanisms for raising the standard of living of the poorest people.

But of course there are more serious tensions between science, with its associated technology, and society; and these problems are also well known: the distribution of scientific knowledge is uneven, and the benefits are far from uniformly spread. Yes, there is trickle down, but it's no fun to be poor in an affluent society even if you have a car and an average life span of sixty or more years. It is even less fun if you are trapped in a ghetto housing project, with your children in dysfunctional schools on dangerous streets; or if you live in a dirt-floor cabin in the Amazon but you can watch the good life on television. Even the most successful systems for harvesting the new wealth that science generates are still full of glaring inequities and filled with damages inflicted by greed and ignorance.

Like art, music, and literature, science creates (or reveals) beauty and wonder, but science and technology also create wealth and empowerment; control passes to government, commerce, and ultimately, in principle, to the citizens of democratic society. The challenge to problem solvers is how to prepare these elements to better manage this truly unimaginable power. So the thoughtful citizen, whether science illiterate or not, asks good questions. Isn't it science that generates the greenhouse gases, the acid rain, the new weapons, nuclear waste dumps? And doesn't technology mostly benefit the rich?

As we watch with bemusement or astonishment the incredible changes being brought about by developments of technology, we see the following major issues:

• A growth in short-termism with its consequent turning away from such investments as research and education.

• A prevailing ignorance about how science works.

• Excesses of science and technology deployments: industrial wastes, huge oil spills, toxic and radioactive debris, production of CO_2 . . .

• An exponential world population growth, a product of the spread of our understanding of sanitation and health care.

• A failed educational system that has settled so long for low standards and lack of priority.

• Failed cities, crowded jails—as Roosevelt said in his famous 1932 presidential campaign speech, "One third of a nation is still ill-housed, ill-clothed, ill-fed."

• A commercial-TV wasteland that, for its easy profits, encourages mindlessness, violence, vacuous entertainment, and rampant consumerism. The exceptions only compound the frustration of what could be but isn't.

And now we can contemplate the wave of antiscience that pushes and pulls at our four-hundred-year-old commitment to science and rationality.

Almost like the drunk clutching the lamp post, scientists clutch *education* as a long-term solution to the problem. People have long placed faith in education—from Thales of Miletus to Richard Feynman of Brighton Beach, from the ancient prophets of biblical times to the profits of McGraw Hill, from King Solomon, who was the education king, to George Bush, the education president. Education is my approach, too. (It is now known as ".edu".) This almost unreasoned belief in education is our article of faith. Of course, there are a depressing number of examples to tell us that education does not inevitably produce ethical, virtuous, or even wise human beings. Creationists count heavily on their dozen or so Ph.D.s, ignoring the fact that in a Ph.D.-counting contest they'd lose by over a thousand to one.

Today we have the Unabomber example, and whereas one can dismiss him as one individual having gone "over the edge," we also have the example of the Nazi scientists cheerfully participating in World War II concentration camp experiments; we have pillars of our society and graduates of our best colleges, the tobacco company executives whose products continue to generate more casualties than Hitler, Stalin, Ghengis Khan, and Napoleon put together. Articles in the *New York Times* and *Time* magazine on alternative medicines are discouraging. Facing the attacking antiscience forces, if I had the Unabomber on the left and the new Republican congressmen on the right, I would attack left. Why? The Unabomber only wanted to kill scientists and engineers, and he blew you up cleanly and humanely, whereas these others will starve and torture science to a slow and agonizing decay while their leaders murmur in your ear about how much they love basic research.

Still, we must not lose faith in education. It has to be done better, and the ambitious reforms I will review need to be organized and coordinated in some kind of new strategy, perhaps a general high-command center. We are still stuck with the age-old problem of how to teach ethics, morality, and social responsibility. However, "they are working on it" and I really do not know how else to proceed. Education must be the antidote to superstition, victimization, totalitarianism, bigotry. If it fails here and there we must make it better. We must work together—scientists, educators, psychologists, neuroscientists, linguists, and anthropologists—to make it better. The urgency of doing better and the common peril should make this easier. But education spans a K–16, or better, what I call a "K–100" domain—in other words, education spans a lifetime; it does not stop with formal, conventional education. The strategic vision is that if an ever-increasing number of our citizens could be taught to think scientifically, to understand the critical methods that have allowed scientists and engineers to create so much wealth, these citizens, in the democratic context, would be intolerant of sound bites and baloney, would insist on the proper allocation of national resources, would insist on a balance between operation and investment, would insist that the products of science and technology be deployed for the long-term benefit of the many, and would understand the role of *knowledge* in social, economic, and cultural contexts. They would be shielded from the philosophical con men and women and snake-oil purveyors. They would surely understand that education must count almost as much as deficit reduction in the future well-being of their children. Whereas in an earlier time, public understanding of science and technology was a cultural plus, in today's and tomorrow's world the stakes are much higher— nothing less than the preservation of our four-hundred-year-old commitment to a rational worldview.

There is, in our activities, the belief, perhaps born of despair, that the universal absorption of the scientific tradition, leavened by the liberal arts, will produce a new citizenship insistent on the application of science to fulfill the promise of progress.

So let's look at our educational system with pithy comments on what is going on or should be going on in each of the conventional phases of education. It is here that I will restrict myself to my personal experiences and activities. I do, however, make the disclaimer that whereas I address science and math education, I realize that the social sciences, language, the arts, and literature are also in trouble and need urgent attention. I believe that the pioneering new teaching styles in math and science can have a strong influence on the rest of the liberal arts curriculum if we can only give teachers time to talk and learn from one another. In fact, I believe we must be allies and work together for a renaissance of education across all subjects.

I like to think about education as a kind of circle [see figure p. 33]. We learn in elementary electricity that if the circuit is not complete, but is cut somewhere, that no current will flow. Education is a circle because it starts, for example, with pre-school children, circles up to grades K through 8, continues to grades 9 through 12, whereupon some threads separate out—those who leave school—and others continue on to grades 13 through 16 or further; but they all circle around as part of the "general public," with the majority returning to the circle as parents of the pre-school children. So we have: preschool, K–8, 9–12, 13–16, and general public, which of course includes media professionals, politicians, doctors, lawyers, voters, and parents, and so around we go.

It is clear that we need a coordinating strategy that does not now exist. We must cover all arcs of the circle. We can improve the schools only if we get to parents, teachers, school boards, the media. We need a coordinated battle plan—I've already mentioned a central headquarters, perhaps festooned with electronic screens and nattily outfitted *Star Trek* crews—with reports coming in from the front lines on where there is resistance, with reinforcements to dispatch to the places of need, with public information releases so that progress can be followed, with CNN coverage of frontline action in the war on ignorance, and with yellow ribbons tied outside the homes of teachers! *Most of all, we need a national priority for education.*

My personal opinions as we go around this education circle are dominated by the big IF. We can succeed *if* we motivate our teaching of science by the desired outcome that is all-important: we want our educated people to have incorporated scientific thinking—the blend of curiosity and skepticism, the habit of critical questioning—into their very nature. This should reduce the vulnerability of our citizens to quackery and misdirection. This should raise the level of debate where science, technology, and skepticism have a role. Of course, grade K–12 students must have a basic grasp of science content: Earth is one of nine or so planets; the solar system is one of billions in our galaxy; there are chemical elements and chemical compounds; the seed is father of the plant; there are cells, membranes, nerves, and muscles; there is a molecular basis of genes; and so on.

Many of these details will be forgotten in the adult, but in a coherent, hopefully seamless K–12 experience, the student should permanently absorb a grasp of general scientific principles and a sense of why such principles—for example, conservation of energy—play so vital a role in science. The basic requirements of the educational system have been written down in some detail in new national standards that have been developed such as the Benchmarks issued by the American Association for the Advancement of Science in 1992 and the National Science Standard issued by the National Academy of Sciences in 1995. So we do have a consensus on what students should master. The new standards aim to: ". . . emphasize a new way of teaching and learning about science that reflects how science itself is done, emphasizing inquiry as a way of achieving knowledge and understanding about the world . . ." and ". . . strengthen many of the skills that people use everyday, like solving problems creatively, thinking critically, working cooperatively, in teams. . . ."

We obviously do not teach science and mathematics very well to children (K–8). The situation is so bad that we are indeed "a nation at risk." Incidentally, in this case, we are not much worse than most other nations of the world. The problem is easily traced to the poor training of primary school teachers, but also to constraints on supplies, preparation time, and above all, professional development. Although this seems to be an international invariant, the reasons why we do so poorly in the U.S. can be catalogued. Teachers have lost social status in the post–World War II era. Teacher training institutions have failed and largely continue to fail to produce elementary school teachers who can teach math and science. What makes this so tragic is that the new pedagogies for teaching math and science are so brilliant and so engaging.

Yet the situation in primary schools is far from hopeless. The buzzwords of hands-on, inquiry-based, cooperative, constructivist teaching of math and science are spreading from sea to shining sea. The problem is that professional development of teachers has been grossly neglected by the school systems. Teacher colleges must change. The new pedagogy is nothing less than a cultural change, and it requires patience and persistence. Acquiring scientific literacy must become a central aspect of K–8 education to match the central aspect that science and technology increasingly play in our society.

Consider grades 9–12, high schools. The science curriculum in 99 percent of U.S. high schools begins with biology, continues to chemistry and, for about 20 percent of U.S. students, ends with physics. This has been the way for one hundred years, and it is obviously backwards. As the science and math standards become part of state and local school policy, it will be clear that *all* students, whatever their futures, will need at least three years of math and three years of science. We have learned that to begin to understand chemistry, one needs to understand the construction and working of atoms and molecules. We have learned that DNA is the bridge between physics, chemistry, and biology. A few cities, including New York and Chicago, have adopted the policy of installing a three-year requirement. We must see to and encourage this for all schools, and there is some optimism here. This would be the time to rethink the high school curriculum: Science I, II, and III, a coherent, integrated sequence which, at every stage, makes continuous use of what has been learned, to enlarge the scope. Several promising programs are around, and one I am most interested in is project ARISE (American Renaissance in Science Education), a loose coalition of scientists and educators from the National Academy of Sciences, the American Association for the Advancement of Science, and the National Science Teachers Association, and a few government agencies and universities. It seems to me that it makes sense to start with ninth-grade conceptual physics (very light on math, heavy on concepts), which would culminate in the development a model of the structure of atoms. Science II, mostly chemistry, would take over here and hand over students with a command of chemical reactions and molecule formation, ready for modern biology.

In such a sequence, we would include some history of science, some science in history and society, the interplay of science and technology, and the structure of such multidisciplinary subjects as earth science and ecology.

If our fifteen thousand high schools would adopt one or another sequence of three years for *all* students (subject to the national standards), in a decade we should have substantially raised the level of behavior of our general public. The impact on grades K–8 and the use of community colleges, trade schools, as well as the offerings in liberal arts colleges, would all have to change. Here again my assessment is optimistic. The resistance to change in the high schools must not be underestimated, but the virtues of reform are overwhelming. On our side we have the as-yet unorganized support of the science establishment. We have the customers for high school graduates in industry and government. And we have the support of perceptive educators. Give it ten years!

I will skip grades 13–16 except to make a plea, guaranteed to fall on resistant ears, that a two-year science requirement for nonscience and nonengineering students is really minimum. Certainly the standard "Rocks for Jocks" will no longer work for our ARISE generation high school graduates. It is here that critical examinations of antiscience can be made. It is also here that the moral and ethical obligations on science and technology can be discussed at a much deeper level.

Finally, let me comment on the "grade 17–100" students—John Q. Public in the wide spectrum of our citizens as they exist today: 97.3 percent scientifically illiterate; addicted to their PCs, CD-ROMs, digitized toothbrushes, Superbowls, and mindless sitcoms; and filled with a love for their children. We must reach them because they vote now, because they are the parents and school board members, because they produce NBC documentaries, because national and local and personal decisions increasingly involve science and technology. Here are some typical questions out of today's news:

- Should we ban cigarettes?
- Are silicone breast implants dangerous?
- Can peach pits cure cancer?
- Is there any good data on the efficacy of alternative medicines?
- Should our military continue to employ psychics?
- Are aliens kidnapping and molesting U.S. citizens?
- Should we put a tax on carbon emissions?
- What *is* the population crisis problem?
- Are humans influencing global climate change?
- Should we decriminalize drugs?
- How do we understand and control the information revolution?

Perhaps it is too idealistic to hope that a science-savvy public will be able to follow the scientific debates and to reasonably weigh the pros and cons of public

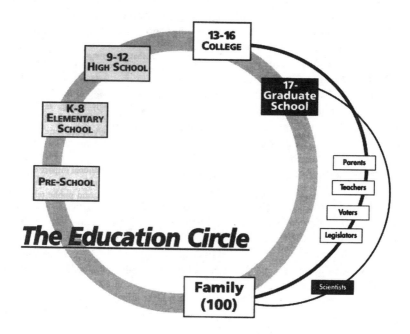

policy decisions. Any approach to raising the level of the public's science savvy must be saturated with reality. Where do we start? Like Zsa Zsa Gabor's seventh husband, we ask: How do we make it interesting? Science needs all the help it can get from media professionals. Just consider the power of a prime time TV dramatic series that glues the viewers in their couches and teaches them some science. Such a program, its episodes cycling over the variety of research disciplines, could show scientists as humans (I know it's hard to believe), as often young, occasionally and increasingly female, as spanning the range of human qualities but as addressing scientific and technological problems (entertainingly apocalyptic) in the scientific spirit, demonstrating the essential qualities of skepticism and curiosity, insight, and imagination. Were it to catch on and become highly popular, its educational impact could be huge.

But approaches to public understanding have many more avenues. There is cable, and public television and radio (for example, Ira Flatow's *Science Friday* on NPR). There are op-ed opportunities like the corner of the *New York Times* that is usually rented by Mobil Oil. Occasionally, Henry Kendall's Union of Concerned Scientists writes good science stories as does *Scientific American*. And, as they said when a cruise ship full of lawyers sank in Lake Erie, "It's a start!" Let me challenge you to write a 600- to 800-word essay on your favorite science story. Make it lively and readable. Mail it to me, and I'll try hard to get it published.

In all of this, we should pay close attention to the professionals who have studied the history and the complexities of "public understanding."

My penultimate remark may not be necessary for this audience: raising the level of public understanding of science has clear objectives. The future is like navigating a sea with islands of disaster, islands of human fulfillment, and islands not yet explored. The steering of the ship cannot be left to captains who can't read the maps, nor should it be left to scientists or any special priesthood. History and our love and respect for democracy favors the selection of able representatives by all the literate and knowledgeable passengers, sensitized to the scientific spirit. The captain must also be wise, compassionate, visionary, and managerial (but, of course, not perfect!).

Finally, let's be clear on the difference between education and marketing. That a science-literate public would favor increasing science funding is not at all clear. Honest education may generate, will generate, as many critics as admirers of science. This is the kind of criticism that science needs to keep it sharp. Ethical and social responsibility at the highest level is essential to our future. The skepticism we so desperately want to instill in citizens may turn to skepticism about the usefulness or value of science. But I do believe most of us are willing to take our chances.

3

THE NEW SKEPTICISM
Paul Kurtz

(**S**) kepticism, like all things, is good if used in moderation. It is essential for the healthy mind; but if taken to excess, it can lead to overweening doubt. Skepticism, if properly understood, is not a metaphysical picture of the unknowability of "ultimate reality"; it does not lead to an inevitable epistemological impasse; it need not culminate in existential despair or nihilism. Rather it should be considered as an essential methodological rule guiding us to examine critically all claims to knowledge and affirmations of value. Without it, we are apt to slip into complacent self-deception and dogmatism; with it, if prudently used, we can effectively advance the frontiers of inquiry and knowledge, and also apply it to practical life, ethics, and politics.

Briefly stated, a skeptic is one who is willing to question any claim to truth, asking for clarity in definition, consistency in logic, and adequacy of evidence. The use of skepticism is thus an essential part of objective scientific inquiry and the search for reliable knowledge.

Skepticism has deep roots in the philosophical tradition. The term derives from the Greek word *skeptikos*, which means "to consider, examine." It is akin to the Greek *skepsis*, which means "inquiry" and "doubt."

Skepticism provides powerful tools of criticism in science, philosophy, religion, morality, politics, and society. It is thought to be exceedingly difficult to apply it to ordinary life or to live consistently with its principles. For human beings seek certitudes to guide them, and the skeptical mode is often viewed with alarm by those who hunger for faith and conviction. Skepticism is the intractable foe of pretentious belief systems. When people demand definite answers to their queries, skepticism always seems to give them further questions to ponder. Yet in a profound sense, skepticism is an essential ingredient of all reflective conduct and

This article first appeared in *Skeptical Inquirer* 18, no. 2 (Winter 1994): 134–41. It has been adapted from Paul Kurtz's *The New Skepticism: Inquiry and Reliable Knowledge* (Amherst, N.Y.: Prometheus Books, 1992).

an enduring characteristic of the educated mind. Still, skeptics are considered dangerous because they question the reigning orthodoxies, the shibboleths and hosannas of any age. Although the skeptical attitude is an indelible part of reflective inquiry, can a person get beyond the skeptical orientation to develop positive directions and commitments in belief and behavior, and will skepticism enable one to do so?

Skeptics always bid those overwhelmed by Absolute Truth or Special Virtue to pause. They ask, "What do you *mean*?"—seeking clarification and definition— and "*Why* do you believe what you do?"—demanding reasons, evidence, justification, or proof. Like natives of Missouri, they say, "Show me." All too often probing skeptics discover that the unquestioned beliefs and many cherished values of the day rest on quixotic sands and that, by digging at their tottering foundations, they may hasten their collapse. Skeptics are able to detect contradictions within belief systems; they discover hypocrisies, double standards, disparities between what people profess and what they actually do; they point to the paucity of evidence for most of humankind's revered belief systems.

Skeptics are viewed as dissenters, heretics, radicals, subversive rogues, or worse, and they are bitterly castigated by the entrenched establishments who fear them. Revolutionary reformers are also wont to turn their wrath on skeptical doubters who question their passionate commitment to ill-conceived programs of social reconstruction. Skeptics wish to examine all sides of a question; and for every argument in favor of a thesis, they usually can find one or more arguments opposed to it. Extreme skepticism cannot consistently serve our practical interests, for insofar as it sires doubt, it inhibits actions. All parties to a controversy may revile skeptics because they usually resist being swept away by the dominant fervor of the day.

Nevertheless, skepticism is *essential* to the quest for knowledge, for it is in the seedbed of puzzlement that genuine inquiry takes root. Without skepticism, we may remain mired in unexamined belief systems that are accepted as sacrosanct yet have no factual basis in reality. With it, we allow for some free play for the generation of new ideas and the growth of knowledge. Although the skeptical outlook may not be sufficient unto itself for any philosophy of the practical life, it provides a necessary condition for the reflective approach to life. Must skepticism leave us floundering in a quagmire of indecision? Or does it permit us to go further, and to discover some probabilities by which we can live? Will it allow us to achieve reliable knowledge? Or must all new discoveries in turn give way to the probing criticisms of the skeptic's scalpel? The answer to these questions depends in part on what one means by skepticism, for there are various kinds that can be distinguished.

NIHILISM

Total Negative Skepticism. The first kind of skepticism that may be identified is nihilism. Its most extreme form is total negative skepticism. Here I am referring to skepticism as a complete rejection of all claims to truth or value. This kind of skepticism is mired in unlimited doubt, from which it never emerges. Knowledge is not possible, total skeptics aver. There is no certainty, no reliable basis for convictions, no truth at all. All that we encounter are appearances, impressions, sensations; and we have no guarantee that these correspond to anything in external reality. Indeed, we have no assurance that we are properly describing external objects "in themselves." Our senses, which lie at the heart of our experiential world, may deceive us. Our sense organs act as visors, shielding and limiting our perceptions, which vary from individual to individual and from species to species. Similar pitfalls await those who seek to root knowledge in the cognitive intuitions or deductive inferences of mathematics or logic, claim total skeptics. Meanings are irreducibly subjective and untranslatable into intersubjective or objective referents. Purely formal conceptual systems tell us more about the language we are using than about the nature of ultimate reality itself. In any case, human beings are prone to err. For every proof in favor of a thesis, one may pose a counterproof. Like the web that is spun by a spider, the entire structure may collapse when we disturb the glue that holds the threads together.

Not only is epistemic certainty impossible, maintain total skeptics, the very criteria by which we judge whether something is true or false are questionable. Knowledge is based upon the methods by which we evaluate claims to truth—whether empirical or rational. But these are simply assumed, they insist, and cannot be used to validate themselves without begging the question. Thus we can never get beyond the first stage of inquiry. Total skeptics end up in utter subjectivity, solipsists imprisoned in their own worlds, confused about the nature of knowledge. This is the total skeptics' approach to science, philosophy, and religion.

Nihilistic skepticism has also been used in ethics with devastating results. Here the total skeptic is a complete relativist, subjectivist, and emotivist. What is "good" or "bad," "right" or "wrong," varies among individuals and societies. There are no discernible normative standards other than taste and feeling, and there is no basis for objective moral judgment. We cannot discern principles that are universal or obligatory for morality. Complete cultural relativity is the only option for this kind of skepticism. Principles of justice are simply related to power or the social contract; there are no normative standards common to all social systems. In the face of moral controversy, total skeptics may become extreme doubters; all standards are equally untenable. They may thus become conservative traditionalists. If there are no reliable guides to moral conduct, then the only recourse is to follow custom. Ours is not to reason why, ours is but to do or die, for there are no reasons. Or total skeptics may become cynical amoralists for whom "anything goes." Who is to say that one thing is better or worse than anything else? they ask, for if there are no standards of justice discoverable in the nature of things, political morality in the last analysis is a question of force, custom, or passion, not of reason or of evidence.

This kind of total skepticism is self-contradictory; for in affirming that no knowledge is possible, these skeptics have already made an assertion. In denying that we can know reality, they often presuppose a phenomenalistic or subjectivistic metaphysics in which sense impressions or ideas are the constitutive blocks out of which our knowledge of the world, however fragmented, is constructed. In asserting that there are no normative standards of ethics and politics, total skeptics sometimes advise us either to be tolerant of individual idiosyncrasies and respect cultural relativity, or to be courageous and follow our own quest to satisfy ambition or appetite. But this imperceptibly masks underlying value judgments that skeptics cherish. This kind of skepticism may be labeled "dogmatism"; for in resolutely rejecting the very possibility of knowledge or value, such skeptics are themselves introducing their own questionable assertions.

NEUTRAL SKEPTICISM

One form of nihilistic skepticism that seeks to avoid dogmatism does so by assuming a completely neutral stance. Here skeptics will neither affirm nor deny anything. They are unwilling to utter any pronouncements, such as that sense perception or formal reasoning is unreliable. They reject any type of skepticism that masks a theory of knowledge or reality in epistemology, metaphysics, ethics, and politics. Neutralists claim to have no such theory. They simply make personal statements and do not ask anyone to accept or reject them or be convinced or persuaded by their arguments. These are merely their own private expressions they are uttering, and they are not generalizable to others. For every argument in favor of a thesis, they can discover a counterargument. The only option for neutral skeptics is thus to suspend judgment entirely. Here agnosticism rules the roost. They are unable in epistemology to discover any criteria of knowledge; in metaphysics, a theory of reality; in religion, a basis for belief or disbelief in God; in ethics and politics, any standards of virtue, value, or social justice.

The ancient pre-Socratic Greek philosopher Cratylus (fifth-fourth century B.C.E.) was overwhelmed by the fact that everything is changing, including our own phenomenological worlds of experience; and he therefore concluded that it is impossible to communicate knowledge or to fully understand anyone. According to legend, Cratylus refused to discuss anything with anyone and, since it was pointless to reply, only wiggled his finger when asked a question. The neutral state of suspension of belief, now known as Pyrrhonism, was defended by Pyrrho of Elis and had a great impact on the subsequent development of skepticism. It applied primarily to philosophical and metaphysical questions, where one is uncertain about what is ultimately true about reality, but it put aside questions of ordinary life, where convention and custom prevail. This form of skepticism also degenerates into nihilism, for in denying any form of knowledge it can lead to despair.

MITIGATED SKEPTICISM

There is a fundamental difficulty with the forms of skepticism outlined above, for they are contrary to the demands of life. We need to function in the world—whatever its ultimate reality—and we need to develop some beliefs by which we may live and act. Perhaps our beliefs rest ultimately upon probabilities; nevertheless we need to develop knowledge as a pragmatic requirement of living and acting in the world. A modified form of skepticism was called *mitigated skepticism* by David Hume, the great eighteenth-century Scottish philosopher. It is a position that was also defended by the Greek philosopher Carneades of the second century B.C.E. Mitigated skeptics have confronted the black hole of nothingness and are skeptical about the ultimate reliability of knowledge claims. They are convinced that the foundations of knowledge and value are ephemeral and that it is impossible to establish ultimate truths about reality with any certainty. Nonetheless, we are forced by the exigencies of practical life to develop viable generalizations and to make choices, even though we can give no ultimate justification for them. One cannot find any secure basis for causal inferences about nature, other than the fact that there are regularities encountered within experience, on the basis of which we make predictions that the future will be like the past. But we have no ultimate foundation for this postulate of induction. Similarly, one cannot deduce what *ought* to be the case from what is. Morality is contingent on the sentiments of men and women who agree to abide by social convention in order to satisfy their multifarious desires as best they can.

Mitigated skepticism is not total, but only partial and limited, forced upon us by the exigencies of living. It would be total if we were to follow philosophy to the end of the trail, to irremediable indecision and doubt. Fortunately, we take detours in life, and thus we live and act *as if* we had knowledge. Our generalizations are based upon experience and practice, and the inferences that we make on the basis of habit and custom serve as our guide.

UNBELIEF

The term *skepticism* has sometimes been used as synonymous with *unbelief* or *disbelief* in any domain of knowledge. Actually there are two aspects to this—one is the *reflective* conviction that certain claims are unfounded or untrue, and hence not believable, and this seems a reasonable posture to take; the other is the negative a priori rejection of a belief without a careful examination of the grounds for that belief. Critics call this latter form of skepticism "dogmatism." The word *unbelief* in both of these senses is usually taken to apply to religion, theology, the paranormal, and the occult.

In religion the unbeliever is usually an atheist—not simply a neutral agnostic —for this kind of skepticism rejects the claims of theists. The atheist denies the basic premises of theism: that God exists, that there is some ultimate purpose to the universe, that men and women have immortal souls, and that they can be saved by divine grace.

Reflective unbelievers find the language of transcendence basically unintelligible, even meaningless, and that is why they say they are skeptics. Or, more pointedly, if they have examined the arguments adduced historically to prove the existence of God, they find them invalid, hence unconvincing. They find the so-called appeals to experience unwarranted: neither mysticism nor the appeal to miracles or revelation establishes the existence of transcendental realities. Moreover, they maintain that morality is possible without a belief in God. Unbelievers are critics of supernaturalistic claims, which they consider superstition. Indeed, they consider the God hypothesis to be without merit, a fanciful creation of human imagination that does not deserve careful examination by emancipated men and women. Many classical atheists (Baron d'Holbach, Diderot, Marx, Engels) fit into this category, for they were materialists first, and their religious skepticism and unbelief followed from their materialistic metaphysics. Such skeptics are dogmatic only if their unbelief is a form of doctrinaire faith and not based on rational grounds.

In the paranormal field, unbelievers similarly deny the reality of psi phenomena. They maintain that ESP, clairvoyance, precognition, psychokinesis, and the existence of discarnate souls are without sufficient evidence and contrary to our knowledge of how the material universe operates. Some skeptics deny paranormal phenomena on a priori grounds, i.e., they are to be rejected because they violate well-established physical laws. They can be considered dogmatists only if they refuse to examine the evidence brought by the proponents of the paranormal, or if they consider the level of science that has been reached on any one day to be its final formulation. Insofar as this kind of unbelief masks a closed mind, it is an illegitimate form of skepticism. If those who say that they are skeptics simply mean that they deny the existence of the paranormal realm, they are aparanormalists. The question to be asked of them always is, *Why?* For much as the believers can be judged to hold certain convictions on the basis of inadequate evidence or faith, so the dogmatic unbelievers may reject such new claims because these violate their own preconceptions about the universe. This latter kind of skepticism has many faults and is in my judgment illegitimate. These skeptics are no longer open-minded inquirers, but debunkers. They are convinced that they have the Non-Truth, which they affirm resolutely, and in doing so they may slam shut the door to further discoveries.

SKEPTICAL INQUIRY

There is yet another kind of skepticism, which differs from the kinds of skepticism encountered above. Indeed, this form of skepticism strongly criticizes nihilism, total and neutral; mitigated skepticism; and dogmatic unbelief—although it has learned something from each of them. This kind of skepticism I label "skeptical inquiry," with inquiry rather than doubt as the motivation. I call it the *new skepticism*, although it has emerged in the contemporary world as an outgrowth of pragmatism. A key difference between this and earlier forms of skepticism is that it is *positive* and *constructive*. It involves the transformation of the negative critical analysis of the claims to knowledge into a positive contribution to the growth and development of skeptical inquiry. It is basically a form of *methodological* skepticism, for here skepticism is an essential phase of the process of inquiry; but it does not and need not lead to unbelief, despair, or hopelessness. This skepticism is not total, but is limited to the context under inquiry. Hence we may call it *selective* or *contextual* skepticism, for one need not doubt everything at the same time, but only certain questions in the limited context of investigation. It is not neutral, because it believes that we do develop knowledge about the world. Accordingly, not only is human knowledge possible, but it can be found to be reliable; and we can in the normative realm act on the best evidence and reasons available. Knowledge is not simply limited to the descriptive or the formal sciences, but is discoverable in the normative fields of ethics and politics. Although this is a modified form of skepticism, it goes far beyond the mitigated skepticism of Hume; for it does not face an abyss of ultimate uncertainty, but is impressed by the ability of the human mind to understand and control nature.

The new skepticism is not dogmatic, for it holds that we should never by a priori rejection close the door to any kind of responsible investigation. Thus it is skeptical of dogmatic or narrow-minded atheism and aparanormalism. Nonetheless, it is willing to assert reflective *unbelief* about some claims that it finds lack adequate justification. It is willing to assert that some claims are unproved, improbable, or false.

Skepticism, as a method of doubt that demands evidence and reasons for hypotheses, is essential to the process of scientific research, philosophical dialogue, and critical intelligence. It is also vital in ordinary life, where the demands of common sense are always a challenge to us to develop and act upon the most reliable hypotheses and beliefs available. It is the foe of absolute certainty and dogmatic finality. It appreciates the snares and pitfalls of all kinds of human knowledge and the importance of the principles of fallibilism and probabilism in regard to the degrees of certainty of our knowledge. This differs sharply from the skepticisms of old, and it can contribute substantially to the advancement of human knowledge and the moral progress of humankind. It has important implications for our knowledge of the universe and our moral and social life. Skepticism in this sense provides a positive and constructive eupraxophy that can assist us in interpreting the cosmos in which we live and in achieving some wisdom in conduct.

The new skepticism is more in tune with the demands of everyday knowledge than with speculative philosophy. Traditional skepticism has had all too little to say about the evident achievements of constructive skeptical inquiry. And derisive skeptical jabs hurled from the wings of the theater of life are not always appreciated, especially if they inhibit life from proceeding without interruption.

Skeptical inquiry is essential in any quest for knowledge or deliberative valuational judgment. But it is limited and focused, selective, and positive, and it is part and parcel of a genuine process of inquiry. This form of modified skepticism is formulated in the light of the following considerations:

There has already been an enormous advance in the sciences, both theoretically and technologically. This applies to the natural, biological, social, and behavioral sciences. The forms of classical skepticism of the ancient world that reemerged in the early modern period were unaware of the tremendous potential of scientific research. Pyrrhonistic skepticism is today invalidated, because there now exists a considerable body of reliable knowledge. Accordingly, it is meaningless to cast all claims to truth into a state of utter doubt. The same considerations apply to postmodernist subjectivism and Richard Rorty's pragmatic skepticism, which I believe are likewise mistaken.

Contrary to traditional skeptical doubts, there are methodological criteria by which we are able to test claims to knowledge: (*a*) empirical tests based upon observation, (*b*) logical standards of coherence and consistency, and (*c*) experimental tests in which ideas are judged by their consequences. All of this is related to the proposition that it is possible to develop and use objective methods of inquiry in order to achieve reliable knowledge.

We can apply skeptical inquiry to many areas. Thoroughgoing investigations of paranormal claims can only be made by means of careful scientific procedures. Religious claims, using biblical criticism, the sciences of archaeology, linguistics, and history, have today given us a basis for skeptical criticism of the appeals to reveration and theories of special creation.

We have long since transcended cultural relativism in values and norms and are beginning to see the emergence of a global society. Thus extreme cultural subjectivity is no longer valid, for there is a basis for transcultural values. There is also a body of tested prima facie ethical principles and rules that may be generalizable to all human communities.

Therefore, the methods of skeptical inquiry can be applied to the political and economic domain in which we frame judgments of practice. Indeed, it is possible to develop a eupraxophy, based on the most reliable knowledge of the day, to provide a generalized interpretation of the cosmos and some conceptions of the good life.

Doubt plays a vital role in the context of ongoing inquiry. It should, however, be selective, not unlimited, and contextual, not universal. The principle of fallibilism is relevant. We should not make absolute assertions, but be willing to admit that we may be mistaken. Our knowledge is based upon probabilities, which are reliable, not ultimate certainties or finalities.

Finally, skeptical inquirers should always be open-minded about new possibilities, unexpected departures in thought. They should always be willing to question or overturn even the most well-established principles in the light of further inquiry. The key principle of skeptical inquiry is to seek, when feasible, adequate evidence and reasonable grounds for any claim to truth in any context.

4

THE CRISIS IN PRE-COLLEGE SCIENCE AND MATH EDUCATION

Glenn T. Seaborg

Our *nation is at risk. Our once unchallenged preeminence in commerce, industry, science and technological innovation is being overtaken by competitors throughout the world.... If an unfriendly power had attempted to impose on America the mediocre educational performance that exists today, we might well have viewed it as an act of war. As it stands, we have allowed this to happen to ourselves.... We have, in effect, been committing an act of unthinking, unilateral educational disarmament.*

These are the dramatic opening lines of the report "A Nation at Risk: The Imperative for Educational Reform" that the members of our National Commission on Excellence in Education handed to President Ronald Reagan in a ceremony at the White House in April 1983.

The Commission was created in August 1981 by then-Secretary of Education Terrel H. Bell, whom the Seaborg Center is very fortunate to have as chairman of its National Advisory Council. The Commission was charged with reporting on the quality of education in our country and making positive recommendations for remedying our deficiencies. What we learned in the course of our 20-month study and at public hearings across the country was so appalling that we decided to make our report as dramatic as possible to draw attention to these serious problems and to reach maximum readership.

We succeeded in drawing almost unprecedented attention from educators, parents, public, and press. It is now apparent that the educational crisis and the urgent need for reform are broadly perceived as being a top priority. In the last presidential election George Bush made education an important campaign issue, thus putting the need for educational reform at the top of the national agenda. Last year President Bush reaffirmed his commitment to improving education when he visited the exhibits of the 40 Westinghouse Science Talent Search win-

---

The article first appeared in *Skeptical Inquirer* 14, no. 3 (Spring 1990): 270–75.

ners. This was the first time in the 48-year history of this competition that the president of the United States made a tour of the students' exhibits. In his remarks President Bush encouraged the efforts of programs like the Science Talent Search and said that he believed that these efforts, aided by the federal government, would result in significant changes in our educational system. He said that "as a nation, we have no natural resource more precious than our intellectual resources" and that he wanted to make science education one of the most important investments for the future of our nation.

So, how bad is the crisis in education in this country?

Since the "Nation at Risk" report, there have been dozens of other reports, by a wide spectrum of American organizations, emphasizing and deploring the state of precollege education in science and math in the United States today. (See box p. 45.) These reports indicate that, while some progress has been made, there is still much work to be done to resolve the crisis in education.

In a recent survey conducted in six nations, 13-year-old American school children placed last in mathematics, behind South Korea, Spain, Great Britain, Ireland, and Canada. Moreover, the high-ranking Koreans surpassed the Americans by a wide margin: 78 percent of Korean 13-year-old students have the ability to use intermediate math skills to solve two-step problems but only half as many Americans could do so. Moreover, 40 percent of the Korean children can understand and apply mathematical concepts, but less than 10 percent of American schoolchildren can do so.

While most American 13-year-old schoolchildren can add, subtract, multiply, and divide, they are seriously lacking in cognitive skills, such as reasoning, investigating, and estimating. However, many entry-level jobs today are demanding workers with high-order, more sophisticated skills for which these students are not being adequately prepared.

In another report, released in September 1988, results showed that, compared with students from 13 other countries, twelfth-grade American students scored in the lower range on mathematics achievement tests. In geometry, American students did only slightly better than those in Hungary and Thailand; in algebra, only Thailand was worse; and the United States ranked last in calculus. Overall the American students obtained only half as many points on tests as those from Hong Kong, the highest-ranking group.

In science, the statistics are just as grim. In the same six-nation survey of 13-year-old schoolchildren, Americans again scored below average. Ironically, the United States came in behind Spain, which does not have a reputation for being in the forefront in science and technology. Again, as in math, when the 13-year-olds were tested for conceptual understanding, interpretation, and application of what they had learned, the performance of the American schoolchildren dropped dismally. At the highest level, 33 percent of Koreans had the skills to apply scientific principles, while only 12 percent of Americans could do so.

In a 13-nation survey of twelfth-graders, American students scored thirteenth in biology, eleventh in chemistry, and ninth in physics.

In report after report, American students consistently score in the lower end of the scale. It is evident that too many students are leaving our schools without adequate skills to be full participants in our increasingly technologically oriented society. By 1993, those 13-year-old students, who today are performing so poorly in math and science, will be voting and entering the work force—the scientific and industrial leadership of our nation will be in their hands.

Already it is evident that the industrial supremacy of our nation is being seriously threatened. Currently only about 6 percent of American adults are scientifically literate. This greatly diminishes their productivity in a more technologically demanding workplace.

Scientific illiteracy also affects the ability to function effectively as citizens and the ability to play an informed role in social and political decision-making on issues with scientific or technological content, such as those involving nuclear power, acid rain, the ozone layer, genetic engineering, chemical warfare, and so forth. The vitality of a democracy assumes a certain "core of knowledge," shared by everyone, that serves as a unifying force.

There can be no doubt that scientific literacy, a solid understanding of science and mathematics, is now more important than ever before—there is irrefutable evidence that the skills of our youth are not only failing to keep up with the increasing demands—but actually are deteriorating at an alarming rate.

This country cannot afford another generation of students who are unpre-

---

A Litany of Concern

Here are some of the recent reports that have documented and deplored the state of precollege science and math education in the United States today.

Everybody Counts: A Report to the Nation of the Future of Mathematics Education
National Research Council, National Academy Press, Washington, D.C. (1989)

The Forgotten Half: Pathways to Success for America's Youth and Young Families
The William T. Grant Foundation Commission on Work, Family and Citizenship (November 1988)

1988 Education Indicators
Joyce D. Stern, editor; Marjorie O. Chandler, associate editor. U.S. Department of Education (1988)

Moral Education and Character
Ivor Pritchard, editor. U.S. Department of Education (September 1988)

American Education: Making It Work
"A Report to the President and the American People," William J. Bennett (April 1988). This is a followup report assessing the progress that has been made in education since 1983, when "A Nation at Risk" was published.

The Forgotten Half: Non-College Youth in America
The William T. Grant Foundation Commission on Work, Family and Citizenship (January 1988)

Women and Minorities in Science and Engineering
National Science Foundation (January 1988)

The Condition of Education: Postsecondary Education
Joyce D. Stern, editor; Marjorie O. Chandler, associate editor. U.S. Department of Education (1988)

Undergraduate Science, Mathematics and Engineering Education
National Science Board, NSB Task Committee on Undergraduate Science and Engineering Education (March 1986)

---

pared to respond to the worldwide rapid growth of scientific knowledge and technological power. The nation's future depends on them. Therefore, we must improve general science and math education for all our young.

The task of guiding the intellectual (and often social) development of our young is all-important. We must begin to recognize teachers' contributions not only by adequately compensating them for their service, but also by giving them the due respect that would motivate them to refine their skills and expand their knowledge to meet future challenges. While some teachers are eminently qualified, a significant number of them have little background in science or mathematics or have not had any involvement with these subjects in many years and have simply lost touch with changes in their fields.

According to a 1985–1986 national survey supported by the National Science Foundation and released in 1988, many science and math teachers feel they lack adequate training and are not qualified to teach. Indeed, the study showed that only about one in three elementary science teachers has taken a college chemistry course and only one in five teachers has taken a college physics course. While 82 percent of them felt they were very well qualified to teach reading, only 66 percent felt they could teach math. In the science disciplines, fewer than one-third of the teachers felt very well qualified to teach the life sciences and only 15 percent felt very well qualified to teach the physical or earth and space sciences.

At the high school level, less than one in three teachers had included earth and space sciences in their undergraduate curriculum. The report also states that "more than half of all secondary science teachers have never had a college computer-science course and almost half have had no college calculus."

However, most science and math teachers feel that they would enjoy teaching these subjects if they had adequate preparation. Whatever their situation, these teachers need opportunities to upgrade their math and science teaching skills. Thus there is not only a need to increase the available pool of qualified math and science teachers, but there is also a need to enhance the capabilities of those now teaching these subjects.

The Lawrence Hall of Science is an institution committed to improving the quality of mathematics and science instruction for precollegiate students. For more than two decades the Hall has dedicated its superior resources as part of the University of California to the continuing battle against educational mediocrity.

The Lawrence Hall of Science, which I serve as chairman, was conceived in 1958 and built in 1968 as a memorial to Ernest O. Lawrence, the University of California's first Nobel Laureate and inventor of the cyclotron. As a dynamic research and educational institution, the Hall continues today, 22 years after its dedication in 1968, to focus its efforts on three main objectives:

• To improve the quality of mathematics and science instruction for the benefit of precollegiate students through the development of innovative math and science courses and accompanying curriculum materials and teacher training services.

• To augment the mathematics and science instruction provided by our schools by offering special mathematics and science courses at the Hall.

• To enhance the knowledge, appreciation, and enjoyment of mathematics and science for the general public by providing the community with a math and science center.

In its efforts to improve instruction in math and science the Hall has developed programs for students and teachers in their own schools. Two particular programs that have been recognized for their excellence are CHEMStudy, a comprehensive high school chemistry curriculum, and the Science Curriculum Improvement Study (SCIS), an activity-oriented science program for children K-6 that was developed in the late 1960s and is now used in more than 20 percent of the nation's elementary schools. Through its in-school programs, the Hall reaches more than 122,000 children every year in the San Francisco Bay Area.

The Hall sets as a priority the development of programs that deal with the issue of attracting and retaining underrepresented students—such as young women and minorities—in mathematics and computer education.

For more than two decades, the Hall has also provided innovative leadership in precollegiate math and science education through the publication of major curricula. These learning materials are utilized by millions of students in the United States and around the world. Curricula and exhibits developed by the Hall are currently used by schools and science centers in more than 30 countries. Each year, more than 700 educators from around the world visit the Hall to learn new techniques to improve science and mathematics instruction.

The Lawrence Hall of Science has achieved national and international prominence as a result of its innovative and effective programs. However, despite the Hall's numerous discoveries and noted accomplishments, much more work still needs to be done.

There is a need for more institutions like the Hall. Therefore, I was deeply honored to participate recently in the dedication ceremonies of another institution committed to improving the education of our young—the Glenn T. Seaborg Center for Teaching and Learning Science and Mathematics. As part of Northern Michigan University (NMU), the Seaborg Center is in a unique position to provide, through its services, resources, and programs, quality education for students and teachers of the Upper Peninsula.

The Seaborg Center, while still young, is rapidly gaining a national reputation that has brought funding agencies to the Center with specific requests. The Lawrence Hall of Science at the University of California at Berkeley, the Kellogg Foundation, and the Public Service Satellite Consortium have all sought out the Seaborg Center and Northern Michigan University to direct or participate in programs that will impact not only the Upper Peninsula but teachers and students throughout the United States.

The Lawrence Hall of Science and the Seaborg Center can join forces to help achieve the goal of educational reform so urgently needed in our country today. We all have a vested interest in education and we must all work together, employing all our resources, to reform and improve our educational system and ensure a prosperous future for our nation. Whatever the expense of improving education, it is an investment in the future we must make. Excellence costs. But in the long run mediocrity costs far more.

5

SCIENCE VS. BEAUTY?
Martin Gardner

N owhere is the gulf that too often divides the culture of science from the culture of the liberal arts more noticeable than in the lines of certain poets who believe that a knowledge of science somehow destroys one's awareness of the wonders and beauty of nature.

Over the decades I've collected some examples:

> The moon shines down with borrowed light,
> So savants say—I do not doubt it.
> Suffice its silver trance my sight,
> That's all I want to know about it.
> A fig for science. . . .
> > —Robert Service

> The goose that laid the golden egg
> Died looking up its crotch
> To find out how its sphincter worked.
> Would you lay well? Don't watch.
> > —X. J. Kennedy, "Ars Poetica"

> I'd rather learn from one bird how to sing
> Than teach ten thousand stars how not to dance.
> > —e. e. cummings

This article is reprinted from Martin Gardner, *Weird Water and Fuzzy Logic: More Notes of a Fringe Watcher* (Amherst, N.Y.: Prometheus Books, 1996). Copyright © 1996 by Martin Gardner. Reprinted by permission of the author and publisher. It first appeared under the title "Science and Beauty" in *Skeptical Inquirer* 19, no. 2 (March/April 1995): 14–16, 55.

While you and i have lips and voice which
are for kissing and to sing with
who cares if some one-eyed son of a bitch
invents an instrument to measure Spring with?

—e. e. cummings

... Do not all charms fly
At the mere touch of cold philosophy?
There was an awful rainbow once in heaven:
We know her woof, her texture; she is given
In the dull catalogue of common things.
Philosophy will clip an Angel's wings,
Conquer all mysteries by rule and line,
Empty the haunted air and gnomed mine—
Unweave a rainbow. . . .

—John Keats, "Lamia"

"Arcturus" is his other name—
I'd rather call him "Star."
It's very mean of Science
To go and interfere!

I pull a flower from the woods—
A monster with a glass
Computes the stamens in a breath
And has her in a "class"!

—Emily Dickinson

A Color stands abroad
On Solitary Fields
That Science cannot overtake
But Human Nature feels.

It waits upon the Lawn,
It shows the furthest Tree
Upon the furthest Slope you know
It almost speaks to you.

Then as Horizons step
Or Noons report away
Without the Formula of sound
It passes and we stay—

A quality of loss
Affecting our Content
As Trade had suddenly encroached
Upon a Sacrament.

—Emily Dickinson

Sweet is the lore which Nature brings;
Our meddling intellect
Mis-shapes the beauteous forms of things:—
We murder to dissect.

Enough of Science and of Art;
Close up those barren leaves;
Come forth, and bring with you a heart
That watches and receives.

       —Wordsworth, "The Tables Turned"

Today we breathe a commonplace,
Polemic, scientific air;
We strip illusion of her veil;
We vivisect the nightingale
To probe the secret of his note
The Muse in alien ways remote
Goes wandering.

       —Thomas Bailey Aldrich

Science! true daughter of Old Time thou art!
   Who alterest all things with thy peering eyes.
Why preyest thou thus upon the poet's heart,
   Vulture, whose wings are dull realities?

       —Edgar Allan Poe

There's machinery in the butterfly,
There's a mainspring to the bee.
There's hydraulics to a daisy
And contraptions to a tree.

If we could see the birdie
That makes the chirping sound
With psycho-analytic eyes,
With X-ray, scientific eyes,
We could see the wheels go round.

And I hope all men
Who think like this
Will soon lie underground.

       —Vachel Lindsay

Similar sentiments have been expressed in prose. Here are a few: Coleridge: "The real antithesis of poetry is not prose but science." Billy Rose: "I wish the engineers would keep their slide rules out of the bits of fairyland left in this bollixed up world." Nietzsche: "They [scientists] have cold, withered eyes before which all birds are unplumed."

There is something to be said for such sentiment, though not much. It is possible for scientists to become so wrapped up in their work that they lose all sense of nature's beauty and mystery. "When you understand all about the sun and all about the atmosphere and all about the rotation of the earth," wrote Alfred North Whitehead, a philosopher who stood astride the two cultures, "you may still miss the radiance of the sunset."

G. K. Chesterton made the same point in his amusing story "The Unthinkable Theory of Professor Green," in *Tales of the Long Bow*. Green is an astronomer who forgot about the world around him until one day when he fell in love with a farmer's daughter. He announces a lecture on his discovery of a new planet. The auditorium is packed with colleagues while he describes one of the planet's strange creatures. Slowly it dawns on Green's listeners that he is describing a cow.

I suppose that scientists like Professor Green, before he discovered the earth, exist, but if so, I have yet to encounter one. On the contrary, almost all scientists believe that as their knowledge increases, their sense of wonder also grows. The scientist sees a flower, said physicist John Tyndall, "with a wonder superadded."

Professor Green's unthinkable theory reminds me of stanza xcii from the first canto of Byron's *Don Juan*:

> He thought about himself, and the whole earth,
>   Of man the wonderful, and of the stars,
> And how the deuce they ever could have birth;
>   And then he thought of earthquakes, and of wars,
> How many miles the moon might have in girth,
>   Of air-balloons, and of the many bars
> To perfect knowledge of the boundless skies;—
> And then he thought of Donna Julia's eyes.

The Laputans, in Gulliver's Travels, describe a woman's beauty by "rhombs, circles, parallelograms, ellipses, and other geometrical terms." Arthur S. Eddington, writing on "Science and Mysticism" in *The Nature of the Physical World*, quotes from a page on winds and waves in a textbook on hydrodynamics. He then compares this with the aesthetic experience of watching actual sea waves "dancing in the sunshine."

That knowledge of science adds to one's appreciation of the mystery and splendor of the cosmos has nowhere been more vigorously expressed than by the late physicist Richard Feynman in Christopher Syke's *No Ordinary Genius* (W. W. Norton, 1994). He described an artist friend who would hold up a flower and say: "I, as an artist, can see how beautiful a flower is. But you, as a scientist, take it all apart and it becomes dull."

"I think he's kind of nutty," says Feynman, and he adds:

First of all, the beauty he sees is available to other people—and to me too. Although I might not be quite as refined aesthetically as he is, I can appreciate the beauty of a flower.

At the same time, I see much more about the flower than he sees. I could imagine the cells in there, the complicated actions inside, which also have a beauty. I mean, it's not just beauty at this dimension of one centimeter: there is also beauty at a smaller dimension—the inner structure. The fact that the colors in the flower are evolved in order to attract insects to pollinate it is interesting—it means that the insects can see the color. It adds a question: does this aesthetic sense also exist in the lower forms? Why is it aesthetic? All kinds of interesting questions which a science knowledge only adds to the excitement and mystery and the awe of a flower. It only adds. I don't understand how it subtracts.

Does it make any less of a beautiful smell of violets to know that it's molecules? To find out, for example, that the smell of violets is very similar to the chemical that's used by a certain butterfly (I don't know whether it's true, like my father's stories!), a butterfly that lets out this chemical to attract all its mates? It turns out that this chemical is exactly the smell of violets with a small change of a few molecules. The different kinds of smells and the different kinds of chemicals, the great variety of chemicals and colors and dyes and so on in the plants and everywhere else, are all very closely related, with very small changes, and the efficiency of life is not always to make a new thing, but to modify only slightly something that's already there, and make its function entirely different, so that the smell of violets is related to the smell of earth. . . . These are all additional facts, additional discoveries. It doesn't take away that it can't answer questions of what, ultimately, does the smell of violets really feel like when you smell it. That's only if you expected science to give the answers to every possible question. But the idea that science takes away is something I don't understand.

It's true that technology can have an effect on art that might be a kind of subtraction. For example, in the early days painting was to make pictures when pictures were unavailable, that was one reason: it was very useful to give people pictures to look at, to help them think about God, or the Annunciation, or whatever. When photography came as a result of technology, which itself was the result of scientific knowledge, then that made pictures very much more available. The care and effort needed to make something by hand which looked exactly like nature and which was once such a delight to see now became mundane in a way (although of course there's a new art—the art of taking good pictures). So yes, technology can have an effect on art, but the idea that it takes away mystery or awe or wonder in nature is wrong. It's quite the opposite. It's much more wonderful to know what something's really like than to sit there and just simply, in ignorance, say, "Oooh, isn't it wonderful!"

A famous poem by Walt Whitman tells how he listened to a "learn'd astronomer" lecture about the heavens until he (Walt) became "unaccountably tired and sick." He walks out of the lecture room into the "mystical moist night air" so he can look up "in perfect silence at the stars."

Here is how Feynman, in his Lectures on Physics, reacted to the notion that astronomical knowledge dulls one's sense of awe toward the cosmos:

"The stars are made of the same atoms as the earth." I usually pick one small topic like this to give a lecture on. Poets say science takes away from the beauty of the stars—mere gobs of gas atoms. Nothing is "mere." I too can see the stars on a desert night, and feel them. But do I see less or more? The vastness of the heavens stretches my imagination—stuck on this carousel my little eye can catch one-million-year-old light. A vast pattern—of which I am a part—perhaps my stuff was belched from some forgotten star, as one is belching there. Or see them with the greater eye of Palomar, rushing all apart from some common starting point when they were perhaps all together. What is the pattern, or the meaning, or the *why?* It does not do harm to the mystery to know a little about it. For far more marvelous is the truth than any artists of the past imagined! Why do the poets of the present not speak of it? What men are poets who can speak of Jupiter if he were like a man, but if he is an immense spinning sphere of methane and ammonia must be silent?

Isaac Asimov, writing on "Science and Beauty" in *The Roving Mind,* quotes Whitman's poem. "The trouble is that Whitman is talking through his hat," says Asimov. Of course the night sky is beautiful, but is there not a deeper, added beauty provided by astronomy? Asimov continues with lyrical paragraphs about the "weird and unearthly beauty" of our sister planets, as recently disclosed by space probes, about the awesome wonders of the stars, of the billions of galaxies each containing billions of suns, of clusters of galaxies, and superclusters fleeing from each other as the universe expands from its incredible origin in the explosion of a tiny point some 15 billion years ago.

And all of this vision—far beyond the scale of human imaginings—was made possible by the works of hundreds of learn'd astronomers. All of it; *all* of it was discovered after the death of Whitman in 1892, and most of it in the past twenty-five years, so that the poor poet never knew what a stultified and limited beauty he observed when he look'd up in perfect silence at the stars.

Nor can we know or imagine now the limitless beauty yet to be revealed in the future—by science.

ADDENDUM

The following three letters appeared in the *Skeptical Inquirer* (July/August 1995):

As always, Martin Gardner makes an excellent point in his piece "Science and Beauty." Knowledge of how the universe works certainly can and should deeply increase one's sense of its loveliness and, yes, mysteriousness.

However, perhaps Gardner overlooks an important source of much emotional reaction against the scientific worldview. Increasingly, scientific findings and theories extend the domain of physical causality into humankind itself. The more thoroughly our thoughts, feelings, and actions are shown to be matters of chemistry and structure, the less room is left for traditional free will, not to mention an immortal soul or a God.

It doesn't really do much good to to say that these questions are not subject to empirical test and therefore lie beyond the purview of science. At best, this simply sweeps them under the rug. At worst, it is dishonest. The fact is, most of them are integrally involved with material reality. For example, St. Paul pointed out quite some time ago that Christianity stands or falls by whether or not the resurrection of Christ is a historical fact, i.e., a testable proposition. Moral and ethical standards depend largely on what the nature of the human organism is.

And so on. Whether or not they articulate this to themselves, most people are quite able to perceive it, and to many it is a bitter pill to swallow.

Poul Anderson, Orinda, Calif.

In asserting a conflict between "the culture of science" and "the culture of liberal arts," all Martin Gardner accomplishes is to reinforce the fortress mentality I perceive in the pages of *Skeptical Inquirer*.

The liberal arts are not competitors with science, but rather serve us all as observers of and commentators on the human zoo. Both areas are complementary in creating and maintaining the well-functioning human being, and neither stands independent of the society in which it is found.

Jack Miller, Port Clements, B.C., Canada

Disparaging Emily Dickinson, Wordsworth, Poe, et al. in the name of scientific skepticism? I can't think of a more foolproof way to alienate the romantics among us.

Surely we can allow the poet enchanted by nature to shun the skeptical lens for a few moments of reverie. I'm a nature poet myself, and I shudder to think that some of my lyrical lines could be skewered by the hardened skeptic as anti-scientific.

Please, have some room in your scientific universe for the value of nonscientific experiences—unless you really want the world so entirely intellectually oriented that in your most intimate personal moments you hear the "voice of reason" declaring "I'm so chemically receptive to your pheromones" instead of your heart whispering "I love you"!

Gloria J. Leitner, Boulder, Colo.

David Todd sent me Eden Phillpotts's lyric "Miniature," from *The Oxford Book of Short Poems*:

> The grey beards wag, the bald heads nod,
> And gather thick as bees,
> To talk electrons, gases, God,
> Old nebulae, new fleas.
> Each specialist, each dry-as-dust
> And professorial oaf,
> Holds up his little crumb of crust
> And cries, 'Behold the loaf!'

I came across Christopher Morley's "Autumn Colors" in his collection *Hide and Seek*:

> How tedious it seems, and strange,
> That poets should be raving still
> Of autumn tints: it's just the change
> From chlorophyll to xanthophyll.

I was somewhat appalled by angry letters from readers who seemed to think I had a chip on my shoulder against the arts. The point I wanted to make was simple. Most artists, poets, musicians, actors, and professors of liberal arts (especially in England) know less about science than scientists know about the arts.

It would not be easy to locate a physicist who didn't know the difference between a sonnet and an ode. How many Hollywood actors know the difference between a proton and a neutron? The percentage of actors who believe in astrology is huge compared to the percent of scientists. Indeed, among scientists the percentage is very close to zero. Remember, two astrology buffs, actors Ronald and Nancy Reagan, actually occupied the White House!

Obviously science and the arts complement each other, and no persons today should call themselves educated who are ignorant of either the yin or the yang. I fear that America is a long way from the time when education will give the two sides of our culture equal time.

6

THE PARADOX OF KNOWLEDGE
Lee Loevinger

(T) he greatest achievement of humankind in its long evolution from ancient hominoid ancestors to its present status is the acquisition and accumulation of a vast body of knowledge about itself, the world, and the universe. The products of this knowledge are all those things that, in the aggregate, we call "civilization," including language, science, literature, art, all the physical mechanisms, instruments, and structures we use, and the physical infrastructures on which society relies. Most of us assume that in modern society knowledge of all kinds is continually increasing and the aggregation of new information into the corpus of our social or collective knowledge is steadily reducing the area of ignorance about ourselves, the world, and the universe. But continuing reminders of the numerous areas of our present ignorance invite a critical analysis of this assumption.

In the popular view, intellectual evolution is similar to, although much more rapid than, somatic evolution. Biological evolution is often described by the statement that "ontogeny recapitulates phylogeny"—meaning that the individual embryo, in its development from a fertilized ovum into a human baby, passes through successive stages in which it resembles ancestral forms of the human species. The popular view is that humankind has progressed from a state of innocent ignorance, comparable to that of an infant, and gradually has acquired more and more knowledge, much as a child learns in passing through the several grades of the educational system. Implicit in this view is an assumption that phylogeny resembles ontogeny, so that there will ultimately be a stage in which the accumulation of knowledge is essentially complete, at least in specific fields, as if society had graduated with all the advanced degrees that signify mastery of important subjects.

Such views have, in fact, been expressed by some eminent scientists. In 1894

This article first appeared in *Skeptical Inquirer* 19, no. 5 (September/October 1995): 18–21. This is a revised version of an article originally published in *COSMOS 1994*. Copyright © 1995 by Lee Loevinger.

the great American physicist Albert Michelson said in a talk at the University of Chicago:

> While it is never safe to affirm that the future of Physical Science has no marvels in store even more astonishing than those of the past, it seems probable that most of the grand underlying principles have been firmly established and that further advances are to be sought chiefly in the rigorous application of these principles to all the phenomena which come under our notice. . . . The future truths of Physical Science are to be looked for in the sixth place of decimals.

In the century since Michelson's talk, scientists have discovered much more than the refinement of measurements in the sixth decimal place, and none is willing to make a similar statement today. However, many still cling to the notion that such a state of knowledge remains a possibility to be attained sooner or later. Stephen Hawking, the great English scientist, in his immensely popular book *A Brief History of Time* (1988), concludes with the speculation that we may "discover a complete theory" that "would be the ultimate triumph of human reason—for then we would know the mind of God." Paul Davies, an Australian physicist, echoes that view by suggesting that the human mind may be able to grasp some of the secrets encompassed by the title of his book *The Mind of God* (1992). Other contemporary scientists write of "theories of everything," meaning theories that explain all observable physical phenomena, and Nobel Laureate Steven Weinberg, one of the founders of the current standard model of physical theory, writes of his *Dreams of a Final Theory* (1992).

Despite the eminence and obvious yearning of these and many other contemporary scientists, there is nothing in the history of science to suggest that any addition of data or theories to the body of scientific knowledge will ever provide answers to all questions in any field. On the contrary, the history of science indicates that increasing knowledge brings awareness of new areas of ignorance and of new questions to be answered.

Astronomy is the most ancient of the sciences, and its development is a model of other fields of knowledge. People have been observing the stars and other celestial bodies since the dawn of recorded history. As early as 3000 B.C. the Babylonians recognized a number of the constellations. In the sixth century B.C., Pythagoras proposed the notion of a spherical Earth and of a universe with objects in it that moved in accordance with natural laws. Later Greek philosophers taught that the sky was a hollow globe surrounding the Earth, that it was supported on an axis running through the Earth, and that stars were inlaid on its inner surface, which rotated westward daily. In the second century A.D., Ptolemy propounded a theory of a geocentric (Earth-centered) universe in which the sun, planets, and stars moved in circular orbits of cycles and epicycles around the Earth, although the Earth was not at the precise center of these orbits. While somewhat awkward, the Ptolemaic system could produce reasonably reliable predictions of planetary positions, which were, however, good for only a few years and which developed substantial discrepancies from actual observations over a long period of time. Nevertheless, since there was no evidence then apparent to astronomers that the Earth itself moves, the Ptolemaic system remained unchallenged for more than 13 centuries.

In the sixteenth century Nicolaus Copernicus, who is said to have mastered all the knowledge of his day in mathematics, astronomy, medicine, and theology, became dissatisfied with the Ptolemaic system. He found that a heliocentric system was both mathematically possible and aesthetically more pleasing, and wrote a full exposition of his hypothesis, which was not published until 1543, shortly after his death. Early in the seventeenth century, Johannes Kepler became imperial mathematician of the Holy Roman Empire upon the death of Tycho Brahe, and he acquired a collection of meticulous naked-eye observations of the positions of celestial bodies that had been made by Brahe. On the basis of these data, Kepler calculated that both Ptolemy and Copernicus were in error in assuming that planets traveled in circular orbits, and in 1609 he published a book demonstrating mathematically that the planets travel around the sun in elliptical orbits. Kepler's laws of planetary motion are still regarded as basically valid.

In the first decade of the seventeenth century Galileo Galilei learned of the invention of the telescope and began to build such instruments, becoming the first person to use a telescope for astronomical observations, and thus discovering craters on the moon, phases of Venus, and the satellites of Jupiter. His observations convinced him of the validity of the Copernican system and resulted in the well-known conflict between Galileo and church authorities. In January 1642 Galileo died, and in December of that year Isaac Newton was born. Modern science derives largely from the work of these two men.

Newton's contributions to science are numerous. He laid the foundations for modern physical optics, formulated the basic laws of motion and the law of universal gravitation, and devised the infinitesimal calculus. Newton's laws of motion and gravitation are still used for calculations of such matters as trajectories of spacecraft and satellites and orbits of planets. In 1846, relying on such calculations as a guide to observation, astronomers discovered the planet Neptune.

While calculations based on Newton's laws are accurate, they are dismayingly complex when three or more bodies are involved. In 1915, Einstein announced his theory of general relativity, which led to a set of differential equations for planetary orbits identical to those based on Newtonian calculations, except for those relating to the planet Mercury. The elliptical orbit of Mercury rotates through the years, but so slowly that the change of position is less than one minute of arc each century. The equations of general relativity precisely accounted for this precession; Newtonian equations did not.

Einstein's equations also explained the red shift in the light from distant stars and the deflection of starlight as it passed near the sun. However, Einstein assumed that the universe was static, and, in order to permit a meaningful solution to the equations of relativity, in 1917 he added another term, called a "cosmological constant," to the equations. Although the existence and significance of a cosmological constant is still being debated, Einstein later declared that this was a major mistake, as Edwin Hubble established in the 1920s that the universe is expanding and galaxies are receding from one another at a speed proportionate to their distance.

Another important development in astronomy grew out of Newton's experimentation in optics, beginning with his demonstration that sunlight could be broken up by a prism into a spectrum of different colors, which led to the science of spectroscopy. In the twentieth century, spectroscopy was applied to astronomy to gain information about the chemical and physical condition of celestial bodies that was not disclosed by visual observation. In the 1920s, precise photographic photometry was introduced to astronomy and quantitative spectrochemical analysis became common. Also during the 1920s, scientists like Heisenberg, de Broglie, Schrödinger, and Dirac developed quantum mechanics, a branch of physics dealing with subatomic particles of matter and quanta of energy. Astronomers began to recognize that the properties of celestial bodies, including planets, could be well understood only in terms of physics, and the field began to be referred to as "astrophysics."

These developments created an explosive expansion in our knowledge of astronomy. During the first five thousand years or more of observing the heavens, observation was confined to the narrow band of visible light. In the last half of this century astronomical observations have been made across the spectrum of electromagnetic radiation, including radio waves, infrared, ultraviolet, X-rays, and gamma rays, and from satellites beyond the atmosphere. It is no exaggeration to say that since the end of World War II more astronomical data have been gathered than during all of the thousands of years of preceding human history.

However, despite all improvements in instrumentation, increasing sophistication of analysis and calculation augmented by the massive power of computers, and the huge aggregation of data, or knowledge, we still cannot predict future movements of planets and other elements of even the solar system with a high degree of certainty. Ivars Peterson, a highly trained science writer and an editor of *Science News*, writes in his book *Newton's Clock* (1993) that a surprisingly subtle chaos pervades the solar system. He states:

> In one way or another the problem of the solar system's stability has fascinated and tormented astronomers and mathematicians for more than 200 years. Somewhat to the embarrassment of contemporary experts, it remains one of the most perplexing, unsolved issues in celestial mechanics. Each step toward resolving this and related questions has only exposed additional uncertainties and even deeper mysteries.

Similar problems pervade astronomy. The two major theories of cosmology, general relativity and quantum mechanics, cannot be stated in the same mathematical language, and thus are inconsistent with one another, as the Ptolemaic and Copernican theories were in the sixteenth century, although both contemporary theories continue to be used, but for different calculations. Oxford mathematician Roger Penrose, in *The Emperor's New Mind* (1989), contends that this inconsistency requires a change in quantum theory to provide a new theory he calls "correct quantum gravity."

Furthermore, the observations astronomers make with new technologies disclose a total mass in the universe that is less than about 10 percent of the total mass that mathematical calculations require the universe to contain on the basis of its observed rate of expansion. If the universe contains no more mass than we have been able to observe directly, then according to all current theories it should have expanded in the past, and be expanding now, much more rapidly than the rate actually observed. It is therefore believed that 90 percent or more of the mass in the universe is some sort of "dark matter" that has not yet been observed and the nature of which is unknown. Current theories favor either WIMPs (weakly interacting massive particles) or MACHOs (massive compact halo objects). Other similar mysteries abound and increase in number as our ability to observe improves.

The progress of biological and life sciences has been similar to that of the physical sciences, except that it has occurred several centuries later. The theory of biological evolution first came to the attention of scientists with the publication of Darwin's *Origin of Species* in 1859. But Darwin lacked any explanation of the causes of variation and inheritance of characteristics. These were provided by Gregor Mendel, who laid the mathematical foundation of genetics with the publication of papers in 1865 and 1866.

Medicine, according to Lewis Thomas, is the youngest science, having become truly scientific only in the 1930s. Recent and ongoing research has created uncertainty about even such basic concepts as when and how life begins and when death occurs, and we are spending billions in an attempt to learn how much it may be possible to know about human genetics. Modern medicine has demonstrably improved both our life expectancies and our health, and further improvements continue to be made as research progresses. But new questions arise even more rapidly than our research resources grow, as the host of problems related to the Human Genome Project illustrates.

From even such an abbreviated and incomplete survey of science as this, it appears that increasing knowledge does not result in a commensurate decrease in ignorance, but, on the contrary, exposes new lacunae in our comprehension and confronts us with unforeseen questions disclosing areas of ignorance of which we were not previously aware.

Thus the concept of science as an expanding body of knowledge that will eventually encompass or dispel all significant areas of ignorance is an illusion. Scientists and philosophers are now observing that it is naive to regard science as a process that begins with observations that are organized into theories and are then subsequently tested by experiments. The late Karl Popper, a leading philosopher of science, wrote in *The Growth of Scientific Knowledge* (1960) that science starts from problems, not from observations, and that every worthwhile new theory raises new problems. Thus there is no danger that science will come to an end because it has completed its task, thanks to the "infinity of our ignorance."

At least since Thomas Kuhn published *The Structure of Scientific Revolutions* (1962), it has been generally recognized that observations are the result of theories (called paradigms by Kuhn and other philosophers), for without theories of relevance and irrelevance there would be no basis for determining what observations to make. Since no one can know everything, to be fully informed on any subject (a claim sometimes made by those in authority) is simply to reach a judgment that additional data are not important enough to be worth the trouble of securing or considering.

To carry the analysis another step, it must be recognized that theories are the result of questions and questions are the product of perceived ignorance. Thus it is that ignorance gives rise to inquiry that produces knowledge, which, in turn, discloses new areas of ignorance. This is the paradox of knowledge: As knowledge increases so does ignorance, and ignorance may increase more than its related knowledge.

My own metaphor to illustrate the relationship of knowledge and ignorance is based on a line from Matthew Arnold: "For we are here as on a darkling plain...." The dark that surrounds us, that, indeed, envelops our world, is ignorance. Knowledge is the illumination shed by whatever candles (or more technologically advanced light sources) we can provide. As we light more and more figurative candles, the area of illumination enlarges; but the area beyond illumination increases geometrically. We know that there is much we don't know; but we cannot know how much there is that we don't know. Thus knowledge is finite, but ignorance is infinite, and the finite cannot ever encompass the infinite.

PART TWO

SCIENCE AND ANTISCIENCE

7

THE ANTISCIENCE PROBLEM
Paul Kurtz

(I)t is paradoxical that today, when the sciences are advancing by leaps and bounds and when the earth is being transformed by scientific discoveries and technological applications, a strong antiscience counterculture has emerged. This contrasts markedly with attitudes toward science that existed in the nineteenth and the first half of the twentieth centuries. Albert Einstein perhaps best typified the high point of the public appreciation of scientists that prevailed at that time. Paul De Kruif (1926), in his book, *The Microbe Hunters,* described the dramatic results that scientists could now achieve in ameliorating pain and suffering and improving the human condition. John Dewey, perhaps the most influential American philosopher in the first half of this century, pointed out the great pragmatic benefits to humankind from the application of scientific methods of thinking to all aspects of human life. But today the mood has radically changed.

A recent essay in *Time* magazine (Overbye 1993) begins with the following ominous note:

> Scientists, it seems, are becoming the new villains of Western society. . . . We read about them in newspapers faking and stealing data, and we see them in front of congressional committees defending billion-dollar research budgets. We hear them in sound bites trampling our sensibilities by comparing the Big Bang or some subatomic particle to God.

An editorial in *Science* magazine (Nicholson 1993), referring to the *Time* essay, comments:

> Does this reflect a growing antiscience attitude? If so, the new movie Jurassic Park is not going to help. According to both the writer and the producer, the

This article first appeared in *Skeptical Inquirer* 18, no. 3 (Spring 1994): 255–63.

movie intentionally has antiscience undertones. Press accounts say that producer Steven Spielberg believes science is "intrusive" and "dangerous."

It is not only outsiders who are being critical. In recent speeches and publications, George Brown, chairman of the House Space, Science and Technology Committee, has seemed to question the very value of science. Brown has observed that despite our lead in science and technology, we still have many societal ills such as environmental degradation and unaffordable health care. Science, he says, has "promised more than it can deliver." Freeman Dyson seems to share some of this view. In a recent Princeton speech, he states, "I will not be surprised if attacks against science become more bitter and more widespread in the next few years, so long as the economic inequities in our society remain sharp and science continues to be predominantly engaged in building toys for the rich."[1]

"Are these just isolated events, or is something more going on?" asks Richard S. Nicholson in the editorial quoted above.[2]

A further sign that science has lost considerable prestige is the recent rejection of the superconducting supercollider project by the U.S. Congress. Although the chief reason given was the need to cut the national deficit, one cannot help but feel that this decision reflects the diminishing level of public confidence in scientific research.

II

There have always been two cultures existing side by side, as Lord C. P. Snow (1959) has shown. There has been a historic debate between those who wish to advance scientific culture and those who claim that there are "two truths." According to the latter, there exists, along with cognitive scientific knowledge, a mystical and spiritual realm and/or aesthetic and subjective aspects of experience. The two cultures do not live side by side in peaceful coexistence any longer; in recent decades there have been overt radical attacks on science that threaten its position in society.

From within philosophy dissent has come from two influential areas. First, many philosophers of science, from Kuhn to Feyerabend, have argued that there is no such thing as scientific method, that scientific knowledge is relative to sociocultural institutions, that paradigm shifts occur for extrarational causes, and that therefore the earlier confidence that there are objective methods for testing scientific claims is mistaken.

This critique is obviously greatly exaggerated. It is true that science functions in relation to the social and cultural conditions in which it emerges, and it is true that we cannot make absolute statements in science. Nonetheless, there are reliable standards for testing claims and some criteria of objectivity, and these transcend specific social and cultural contexts. How does one explain the vast body of scientific knowledge we possess? A specific claim in science cannot be said to be the same as a poetic metaphor or a religious tenet, for it is tested by its experimental consequences in the real world.

The second philosophical attack comes from the disciples of Heidegger, especially the French postmodernists, such as Derrida, Foucault, Lacan, and Lyotard. They argue that science is only one mythic system or narrative among many others. They maintain that by deconstructing scientific language, we discover that there are no real standards of objectivity. Heidegger complained that science and technology were dehumanizing. Foucault pointed out that science is often dominated by power structures, bureaucracy, and the state, and that the political and economic uses of science have undermined the pretensions of scientific neutrality. Some of these criticisms are no doubt valid, but they are overstated. If the alternative to objectivity is subjectivity, and if there are no warranted claims to truth, then the views of the postmodernists cannot be said to be true either. Surely we can maintain that the principles of mechanics are reliable, that Mars is a planet that orbits the sun, that cardiovascular diseases can be explained causally and preventive measures taken to lower the risk, that the structure of DNA is not simply a social artifact, nor insulin a cultural creation.

The postmodern critics of "modernity" are objecting to the rationalist or foundationalist interpretations of science that emerged in the sixteenth and seventeenth centuries, and perhaps rightly so. For the continuous growth and revision of scientific theories demonstrates that any "quest for certainty" or "ultimate first principles" within science is mistaken. Nonetheless, they go too far in abandoning the entire modern scientific enterprise. The scientific approach to understanding nature and human life has been vindicated by its success; and its premises, I submit, are still valid. What are some of the characteristics of this modern scientific outlook as it has evolved today?

First, science presupposes that there are objective methods by which reliable knowledge can be tested. Second, this means that hypotheses and theories can be formulated and that they can be warranted (a) by reference to the evidence, (b) by criteria of rational coherence, and (c) by their predicted experimental consequences. Third, modern scientists find that mathematical quantification is a powerful tool in establishing theories. Fourth, they hold that there are causal regularities and relationships in our interactions with nature that can be discovered. Fifth, although knowledge may not be universal, it is general in the sense that it goes beyond mere subjective or cultural relativity and is rooted in an intersubjective and intercultural community of inquirers. Sixth, as the progressive and fallible character of science is understood, it is seen that it is difficult to reach absolute or final statements, that science is tentative and probabilistic, and that scientific inquiry needs to be open to alternative explanations. Previous theories are therefore amenable to challenge and revision, and selective and constructive skepticism is an essential element in the scientific outlook. Seventh is the appreciation of the fact that knowledge of the probable causes of phenomena as discovered by scientific research can be applied, that powerful technological inventions can be discovered, and that these can be of enormous benefit to human beings.

III

Yet the scientific approach, which has had such powerful effectiveness in extending the frontiers of knowledge, is now under heavy attack. Of special concern has been the dramatic growth of the occult, the paranormal, and pseudosciences, and particularly the promotion of the irrational and sensational in these areas by the mass media. We allegedly have been living in the New Age. Side by side with astronomy there has been a return to astrology, and concomitant with psychology there was the growth of psychical research and parapsychology. The paranormal imagination soars; science fiction has no bounds. This is the age of space travel, and it includes abductions by extraterrestrial beings and unidentified flying objects from other worlds. The emergence of a paranormal worldview competes with the scientific worldview. Instead of tested causal explanations, the pseudosciences provide alternative explanations that compete in the public mind with genuine science. The huge increase in paranormal beliefs is symptomatic of a profound antiscience attitude, which has not emerged in isolation but is part of a wider spectrum of attitudes and beliefs.

The readers of the *Skeptical Inquirer* are no doubt familiar with the singular role of CSICOP in evaluating claims made about the paranormal and by fringe sciences. Our basic aim is to contribute to the public appreciation for scientific inquiry, critical thinking, and science education. We need to be equally concerned, I submit, about the growth of antiscience in general.

The most vitriolic attacks on science in recent decades have questioned its benefits to society. To a significant extent these criticisms are based on ethical considerations, for they question the value of scientific research and the scientific outlook to humankind. Here are 10 categories of such objections. There are no doubt others.[3]

1. After World War II great anxiety arose about a possible nuclear holocaust. This fear is not without foundation; for there is some danger of fallout from nuclear accidents and testing in the atmosphere, and there is the threat that political or military leaders might embark, consciously or accidentally, upon a devastating nuclear war. Fortunately, for the moment the danger of a thermonuclear holocaust has abated, though it surely has not disappeared. However, such critiques generated the fear of scientific research, and even, in some quarters, the view that physicists were diabolical beings who, in tinkering with the secrets of nature, held within their grasp the power to destroy all forms of life on this planet. The fear of nuclear radiation also applies to nuclear power plants. The accident at Chernobyl magnified the apprehension of large sectors of the world's population that nuclear energy is dangerous and that nuclear power plants should be closed down. In countries like the United States, no nuclear power plants are being built, although France and many other countries continue to construct them. The nuclear age has thus provoked an antinuclear reaction, and the beneficent symbol of the scientist of the past, Albert Einstein, has to some been transmogrified into a Dr. Strangelove. Although some of the apprehensions about nuclear radiation are no doubt warranted, to abandon nuclear fuel entirely, while the burning of

fossil fuels pollutes the atmosphere, leaves few alternatives for satisfying the energy needs of the world. This does not deny the need to find renewable resources, such as solar and wind power, but will these be sufficient?

2. The fear of science can also be traced to some excesses of the environmental movement. Although the environmentalists' emphasis on ecological preservation is a valid concern, it has led at times to the fear that human technology has irreparably destroyed the ozone layer and that the greenhouse effect will lead to the degradation of the entire planet. Such fears often lead to hysteria about all technologies.

3. In large sectors of the population, there is a phobia about any kind of chemical additive. From the 1930s to the 1950s, it was widely held that "better things and better living can be achieved through chemistry" and that chemicals would improve the human condition. Today there is, on the contrary, a widespread toxic terror—of PCBs and DDT, plastics and fertilizers, indeed of *any* kind of additive—and there is a worldwide movement calling for a return to nature, to organic foods and natural methods. No doubt we need to be cautious about untested chemical additives that may poison the ecosystem, but we should not forget that the skilled use of fertilizers led to the green revolution and a dramatic increase in food production that reduced famine and poverty worldwide.

4. Suspicion of biogenetic engineering is another dimension of the growth of antiscience. From its very inception biogenetic research has met opposition. Many feared that scientists would unleash a new, virulent strain of *E. coli* bacteria into sewer pipes—and then throughout the ecosystem—that would kill large numbers of people. Jeremy Rifkin (Rifkin 1990, 1991; Howard and Rifkin 1977) and others have demanded that all forms of biogenetic engineering research be banned because of its "dehumanizing" effect. A good illustration of this can be seen in the film *Jurassic Park*, produced by Steven Spielberg. Here not only does a Dr. Frankenstein seek to bring back the dead, but we are warned that a new diabolical scientist, in cloning dinosaurs, will unleash ominous forces across the planet. Although there may be some dangers in biogenetic engineering, it offers tremendous potential benefit for humankind—for the cure of genetic diseases as well as the creation of new products. Witness, for example, the production of synthetic insulin.

5. Another illustration of the growth of antiscience is the widespread attack on orthodox medicine. Some of these criticisms have some merit. With the advances of the scientific revolution and the growth of medical technology, we have been able to extend human life, yet many people are kept alive against their will and suffer excruciating pain; and the right to die has emerged as a basic ethical concern. Medical ethicists have correctly pointed out that the rights of patients have often been ignored by the medical and legal professions. In the past physicians were considered authoritarian figures, whose wisdom and skills were unquestioned. But to many vociferous critics today, doctors are demons rather than saviors. The widespread revolt against animal research is symptomatic of the attack on science. Granted that animals should not be abused or made to suffer unnecessary pain, but some animal rights advocates would ban all medical research on animals.

6. Another illustration of antiscience is the growing opposition to psychiatry. Thomas Szasz (1984; 1977) has no doubt played a key role here. As a result of his works, large numbers of mental patients were deinstitutionalized. *One Flew Over the Cuckoo's Nest,* by Ken Kesey (1962), dramatizes the view that it is often the psychiatrist himself who is disturbed rather than the patient. Many, like Szasz, even deny that there are mental illnesses, though there seems to be considerable evidence that some patients do suffer behavioral disorders and exhibit symptoms that can be alleviated by anti-psychotic drugs.

7. Concomitant with the undermining of public confidence in the practice of medicine and psychiatry has been the phenomenal growth in "alternative health cures," from faith healing and Christian Science to the relaxation response, iridology, homeopathy, and herbal medicines. This is paradoxical, because medical science has made heroic progress in the conquering of disease and the development of antibiotics and the highly successful techniques of surgical intervention. These have all been a boon to human health. But now the very viability of medical science itself has been questioned.

8. Another area of concern is the impact of Asian mysticism, particularly since World War II, whereby Yoga meditation, Chinese Qigong, gurus, and spiritualists have come into the Western world arguing that these ancient forms of knowledge and therapy can lead to spiritual growth and health in a way that modern medicine does not. Unfortunately, there are very few reliable clinical tests of these so-called spiritual cures. What we have are largely anecdotal accounts, but they hardly serve as objective tests of alternative therapies.

9. Another form of antiscience is the revival of fundamentalist religion even within advanced scientific and educational societies. Fundamentalists question the very foundation of scientific culture. Indeed, in the modern world, it is religion, not science, that seems to have emerged as the hope of humankind. Far more money is being poured into religion than into scientific research and education. Especially symptomatic is the continued growth of "scientific creationism" and widespread political opposition to the teaching of evolution in the schools, particularly in the United States.

10. A final area of antiscience is the growth of multicultural and feminist critiques of science education, particularly in the universities and colleges. The multiculturist view is that science is not universal or transcultural, but relative to the culture in which it emerges. There are, we are told, non-Western and primitive cultures that are as "true" and "valid" as the scientific culture of the Western world. This movement supports the complete relativization of scientific knowledge. The radical feminist indictment of "masculine bias" in science maintains that science has been the expression of "dead, white Anglo-Saxon males"— from Newton to Faraday, from Laplace to Heisenberg. What we must do, the extremists of these movements advise, is liberate humanity from cultural, racist, and sexist expressions of knowledge, and this means scientific objectivity as well. The positive contribution of these movements, of course, is that they seek to open science to more women and minorities. The negative dimension is that multiculturalist demands on education tend to weaken an understanding of the rigorous intellec-

tual standards essential for effective scientific inquiry. Clearly we need to appreciate the scientific contribution of many cultures and the role of women in science throughout history; on the other hand, some multicultural critics undermine the very possibility of objective science.

IV

What I have presented is a kaleidoscope illustration of many current trends that are undermining and threaten the future growth of science. They raise many questions. Why has this occurred? How shall those who believe in the value of scientific methods and the scientific outlook respond?

This is a complex problem, and I can only suggest some possible solutions. But unless the scientific community and those connected with it are willing to take the challenge to science seriously, then I fear that the tide of antiscience may continue to rise. Scientific research surely will not be rejected where there are obvious technological uses to be derived from it, at least insofar as economic, political, and military institutions find these profitable. But the decline in the appreciation of the methods of science and in the scientific outlook can only have deleterious effects upon the long-term role of science in civilization.

One reason for the growth of antiscience is a basic failure to educate the public about the nature of science itself. Of crucial significance is the need for public education in the aims of science. We need to develop an appreciation of the general methods of scientific inquiry, its relationship to skepticism and critical thinking, and its demand for evidence and reason in testing claims to truth. The most difficult task we face is to develop an awareness that the methods of science should not only be used in the narrow domains of the specialized sciences but should also be generalized, as far as possible, to other fields of human interest.

We also need to develop an appreciation for the cosmic outlook of science. Using the techniques of scientific inquiry, scientists have developed theories and generalizations about the universe and the human species. These theories often conflict with theological viewpoints that for the most part go unchallenged. They also often run counter to mystical, romantic, and aesthetic attitudes. Thus it is time for more scientists and interpreters of science to come forward to explain what science tells us about the universe: for example, they should demonstrate the evidence for evolution and point out that creationism does not account for the fossil record; that the evidence points to a biological basis for the mind and that there is no evidence for reincarnation or immortality. Until the scientific community is willing to explicate openly and defend what science tells us about life and the universe, then I fear it will continue to be undermined by the vast ignorance of those who oppose it.

In this process of education, what is crucial is the development of scientific literacy in the schools and in the communications media. Recent polls have indicated that a very small percentage of the U.S. population has any understanding of scientific principles. The figures are similar for Britain, France, and Germany, where large sectors of the population are abysmally unaware of the nature of the scientific outlook. Thus we need to educate the public about how science works and what it tells us about the world, and we should make sure this understanding is applied to all fields of human knowledge.

The growth of specialization has made this task enormously difficult. Specialization has enabled people to focus on one field, to pour their creative talents into solving specific problems, whether in biology or physics, mathematics or economics. But we need to develop generalists as well as specialists. Much of the fear and opposition to science is due to a failure to understand the nature of scientific inquiry. This understanding should include an appreciation for what we know and do *not* know. This means not only an appreciation of the body of reliable knowledge we now possess, but also an appreciation of the skeptical outlook and attitude. The interpreters of science must go beyond specialization to the general explication of what science tells us about the universe and our place in it. This is unsettling to many within society. In one sense, science is the most radical force in the modern world, because scientists need to be prepared to question everything and to demand verification or validation of any claims.

The broader public welcomes scientific innovation. Every new gadget or product and every new application in technology, where it is positive, is appreciated for its economic and social value. What is not appreciated is the nature of the scientific enterprise itself and the need to extend the critical methods of science further, especially to ethics, politics, and religion. Until those in the scientific community have sufficient courage to extend the methods of science and reason as far as they can to these other fields, then I feel that the growth of antiscience will continue.

Now it is not simply the task of scientists who work in the laboratory, who have a social responsibility to the greater society; it is also the task of philosophers, journalists, and those within the corporate and the political world who appreciate the contribution of science to humankind. For what is at stake in a sense is modernism itself. Unless corporate executives and those who wield political power recognize the central role that science and technology have played in the past four centuries, and can continue to play in the future, and unless science is defended, then I fear that the irrational growth of antiscience may undermine the viability of scientific research and the contributions of science in the future. The key is education—education within the schools, but education also within the media. We need to raise the level of appreciation, not simply among students, from grammar school through the university, but among those who control the mass media. And here, alas, the scientific outlook is often overwhelmed by violence, lurid sex, the paranormal, and religious bias.

The world today is a battlefield of ideas. In this context the partisans of science need to defend courageously the authentic role that science has played and can continue to play in human civilization. The growth of antiscience must be countered by a concomitant growth in advocacy of the virtues of science. Scientists are surely not infallible; they make mistakes. But the invaluable contributions of science need to be reiterated. We need public reenchantment with the ideals expressed by the scientific outlook.

NOTES

1. Dyson's comments have now been published in his essay in the Fall 1993 *American Scholar.*

2. For a useful overview, see the new book by Gerald Holton, *Science and Anti-Science* (Cambridge, Mass.: Harvard University Press, 1993).

3. For an extreme case of antiscience paranoia, see David Ehrenfeld's *The Arrogance of Humanism* (New York: Oxford University Press, 1975), where the author virtually equates humanism with science and modernism.

REFERENCES

De Kruif, Paul. 1926. *The Microbe Hunters.* New York: Harcourt, Brace.

Howard, Ted, and Jeremy Rifkin. 1977. *Who Should Play God? The Artificial Creation of Life and What It Means for the Future of the Human Race.* New York: Delacorte Press.

Kesey, Ken. 1962. *One Flew Over the Cuckoo's Nest.* New York: New American Library.

Nicholson, Richard S. 1993. Postmodernism. *Science* 261 (5118), July 9.

Overbye, D. *Time,* April 26.

Rifkin, Jeremy, ed. 1990. *The Green Lifestyle Handbook.* New York: Henry Holt.

———. 1959. *Biosphere Politics: A New Consciousness for a New Century.* New York: Crown.

Snow C. P. 1991. *The Two Cultures and the Scientific Revolution.* New York: Cambridge University Press.

Szasz, Thomas. 1977. *The Theology of Medicine.* New York: Harper and Row.

———. 1984. *The Myth of Mental Illness,* rev. ed. New York: Harper and Row.

8

KNOCKING SCIENCE FOR FUN AND PROFIT

Paul R. Gross and Norman Levitt

$\left(\mathbf{A}\right)$ new and fashionable cottage industry has appeared among the intelligentsia, especially among academics. Its principal activity is to issue quantities of arrogant and hostile criticism of science. Science, that is to say, as an institution and a way of thought, as an attitude toward knowledge. The specific content of the examined science rarely comes into it, for the simple reason that the trendier critics don't bother to study the science seriously. In our recently published book, *Higher Superstition: The Academic Left and Its Quarrels with Science*, we have examined this disturbing trend in some detail.

This style of criticism must be distinguished from the informed judgments that have by tradition been a part, explicitly or otherwise, of serious philosophy and history of science, and that have, at their best, encompassed the logic, the evidence, and the methodological constraints of the science under study. At the same time, the new fashion in science studies goes beyond the Romantic discontent and the edenic nostalgia that have for three hundred years (if not for more than two thousand) been antagonistic to science and to rationality in general.

Science criticism based solidly on knowledge of science is valuable. It has had positive effects on scientific progress and has influenced the way good scientists think about their work. Romantic antagonism has had little or no effect, despite great swings from approval to disapproval, and back again, in the general culture of the Western world. What concerns us, however, is the new brand of criticism, a hybrid discipline ("discipline" may be too generous a term) whose external semblances are variously sociological, historical, even philosophical, but whose underlying motive is none of these. Rather, it is ideology that is in the driver's seat—specifically, the ideology of what has been called the "academic" or "cultural" left.

This "academic left," we are careful to note, is by no means the same thing as the set of all academics with left-wing political views. It is one particular faction

This article first appeared in *Skeptical Inquirer* 19, no. 2 (March/April 1995): 38–42.

of that larger set, and, indeed, has come in for almost as much criticism from liberals as from conservatives. Nevertheless, it has been remarkably successful in winning recruits among partisans of various political causes. Its view of science is today highly influential within the academy (except among working scientists) and increasingly so outside the campus gates. In part, this is due to a tactical exploit of the new science-critics: they have avoided the professional scrutiny of scientists and others devoted to reason as a means of understanding the world. We believe this phenomenon to be dangerous, and not only to the vanity of scientists! In what follows—material taken with very slight modification from the closing chapter of our book—we discuss some of the reasons for our concern. Most important is the debasement of the public discourse of science and technology, a discourse already inadequate to the complexity of the global issues at whose heart lie scientific questions.

If, as seems obvious, scientific and technical issues will become increasingly and urgently relevant to public policy in the decades ahead, how well will such matters be debated in this country? Obviously, we cannot hold high hopes. The historical record of American education in making the general public conversant with basic science has always been poor, except for a brief flurry of serious effort in the post-Sputnik era. Superstition, whether about astrology, ancient astronauts, or alien abductions, has always had easy and profitable going. Fringe medicine and outright quackery, long endemic in American culture, have taken on new and ominous vigor, thanks in part to the rising costs and increasing impersonality of ordinary health care. The contrast between the incomparable virtuosity of professional American science and the general disregard of scientific substance, whether from complacency or hostility, on the part of the larger public grows ever more pronounced. It is one of the great social paradoxes of history.

Those on the left of the political spectrum are concerned, and rightly so, about the abridgment of democratic procedure and debate inherent in a system that delegates all responsibility for important policy matters to a technocratic elite. These misgivings are manifestations of a significant dilemma. How are such decisions to be made in a manner that takes serious and accurate account of technical and scientific matters without abrogating popular rule or reducing it to a mere symbol? Dozens of pressing issues, from AIDS to alternative energy sources, are complicated by this question. How do we permit a wide public to have a serious voice in such deliberations without inviting in gullibility, ignorance, and mere faddishness? The easy answer, of course, is to educate the great mass of citizens in such a way that thinking accurately about science is possible, if not quite second nature. The countervailing obstacle, however, is that widespread ignorance of science in and of itself prevents the development of an educational system that could dispel it.

It is clear that many or the left-wing thinkers whose ideas we find so unsatisfactory are, at bottom, obsessed by the same essential concern as the one we are now trying to address.

How do we democratize scientific and technological decision-making? How do we give the heretofore powerless some measure of control over the decisions, technological and otherwise, that so profoundly shape their lives? The unfortunate trajectory of academic radicalism has carried it to a position where this question is not so much answered as dispelled by a fog of philosophical conceits. Since science seems so difficult, so inaccessible and intimidating, when viewed on its own terms, the radical critics take the daring step of insisting that science can't be what it claims to be, no matter how well it backs up its claims with experiment and applied technology. Outwardly or covertly, they insist on supplanting standard science with other "ways of knowing" that, by their very nature, will be inclusive and welcoming. These ambitions are backed up by a "multiculturalism" insisting that science must be faulty because it has been done, for the most part, by white males; and by a kind of populism committed to setting New Age mysticism on a par with science because the former is so earnestly believed by so many well-meaning people.

The generosity of the democratic impulse when conjoined to this mode of thinking is instantly perverted to a kind of inverse intellectual snobbery, a form of coarse populism that is willing to exile the most stringent kinds of analytical thought and jettison the reliable devices of empiricism in the name of opening the doors of knowledge and driving the haughty priests of science from the temple. The theorizing done on behalf of this project is thus a species of incantation, a ritual rather than an argument. It does not conceive the need to examine science closely on the terms set by the logic of science because, in itself, that kind of examination would concede too much to the temperament and mind-set of the scientist. It would require precisely the kind of education, or self-education, that the critics, in the name of some kind of popular participation and empowerment, want desperately to prove superfluous.

Back-door utopianism, so characteristic of academic-leftist critics of science, is a sad and, ultimately, a woefully impatient business. Behind it stands the Romantic discontent that echoes perilously certain sentiments that were once recognized as reactionary. How much it will add, in the end, to the burden of outright superstition and ignorance that has always plagued the American democratic experiment, is difficult to say. It is plain, however, that the underlying disaffection is hostile to enlightenment as such, and not just—as the mandarins of this movement readily admit—to the Enlightenment. What is chiefly discouraging about its new ascendancy in academic life is the evidence it provides of a tradition of egalitarianism falling under the sway of obscurantism and muddle. We do not need to convict the paladins of the postmodern left of any particular superstitious foolishness, in the ordinary sense, to notice that they have an appalling tendency to condone such foolishness with a relativist nod and a deconstructionist wink.

Above all, the net effect of all this is to debase still further the already-corrupt coinage of public debate. The damage wrought by denatured language is all too apparent. Public health officials struggle to gain the credibility that should rightfully be theirs, and have to fight continually to be heard over a hubbub of voices stridently denouncing the arrogance of Western scientific and materialist paradigms, and offering to replace them with "alternative modes of healing" that promise to make us better faster and cheaper than stodgy old M.D.s. At the root of it all is an ancient amalgam of quackery and self-delusion, but now the fashionable shibboleths of postmodern academic discourse are available to array the old foolishness in up-to-date scholarly language. Discussion of environmental questions is now, at least to some degree, hostage to a rhetorical style and a technique of public relations in which unrelenting ecobabble plays an increasingly peremptory role. The locutions "environment-friendly" and "environmentally sane" cover a broad range of styles, practices, and products of commerce that are neither of those. This, too, is a language fostered in large measure by the peculiar intellectual gamesmanship of the academic left, and it is as often as not employed for purely political purposes.

As we have seen, practical measures for making discussion of scientific issues effectively more democratic by what should be the straightforward process of extending scientific literacy are continually subverted by the intrusion of "identity politics" into the pedagogy of science. In the case of "Afrocentric" science education, the phenomenon is nothing less than garish, although it remains strangely immune to criticism. It is clear that black youngsters who aspire to scientific careers will be in deep trouble if their early education is dominated by the Afrocentrism espoused by a number of now well-known writers toward whom the universities of the country have proved unaccountably friendly.

The feminist critique of science is subtler and, superficially, less provocative in style; but it may, ultimately, have even more widely exclusionary results. Young women—or men, for that matter—who try to embark on scientific vocations with the explicit aim of reconstituting science along the lines advocated by some of the best-known feminist science-critics are on a course leading to frustration and disillusion. We are not imagining such young people: we encounter them regularly in our classes and note their encouragement by colleagues in certain departments.

Science does not work the way the critics say it works, and the program of reforms mooted by the critics will turn it into something other, and less than, science. Enthusiastic recruits to the cause of "feminist" science will have to face this contradiction sooner or later—most likely sooner. They may come to take the view, shared by most women scientists of our acquaintance, that feminism, whatever its strengths as a moral stance and a social program, is not a methodology for doing science: it does *not* offer any privileged insights into scientific questions. That will be their victory. But if they attempt to hold fast to the most emphatic tenets of feminist dogma—for instance, the stylish assertion that "women" can't be "scientists" under the present order, because society constructs these as mutually exclusive categories, and therefore that scientific practice must be reconstituted along radical-feminist lines before women can participate—they will quickly find themselves effectively excluded from serious scientific work.

On the other hand, many women with scientific talent may not get even that far. They will be discouraged from the outset by the litany of the most prominent critics to the effect that science, as it is practiced, is innately antagonistic to women, that it reflects and embodies a system of "patriarchal" domination and "violent" subordination of nature. Thus, to the degree that this sort of thing actually happens, feminism will find itself in the position of frustrating an original, legitimate, and honorable feminist goal: that of augmenting the proportion, as well as the absolute number, of women in science.

We must also note that the left itself—not only the peculiar ideological tribe we have dubbed the "academic left," but the far broader and deeper tradition of egalitarian social criticism that properly deserves such a designation—is, potentially, one of the ironic victims of the doctrinaire science-criticism that has emerged, just as it has long been the victim of the worst kinds of Marxist, Leninist, and Stalinist cant. It is legitimate, for instance, to assert that socialist views have a place in the important debates about environmental questions. Without here endorsing—or rejecting—a socialist point of view, we appreciate that it exists and that it is distinguishable from alternative political visions. The underlying argument is that free-market capitalism, with its enthronement of profit and its tendency to insulate crucial economic decisions from democratic oversight, is, in itself, an obstacle to changes we should make in our uses of technology if we are to develop sound environmental practices. In itself, this is a view that can be argued and that cannot simply be dismissed without specific, and historically informed, criticism. If, however, "eco-socialists" are forced to carry the ideological baggage of the academic left—the relativism of the social constructivists, the sophomoric skepticism of the postmodernists, the incipient Lysenkoism of feminist critics, the millennialism of the radical environmentalists, the racial chauvinism of the Afrocentrists—then they will, in effect, greatly accommodate their opponents and facilitate the rapid dismissal of even their soundest points, for those will be embedded in a tissue of unscientific and antiscientific nonsense. Scientists, and the scientifically well-informed, will simply not accept any form of "socialism" whose agenda includes the subversion of science.

We are not calling for a purge of institutions of higher learning in this country. We don't advocate supplanting one regime of "political correctness" by another, even more odiously high-handed one. Having made that disclaimer, however, we can in good conscience urge that certain forms of vigilance are appropriate, troublesome as they may be to preoccupied teachers and scholars. First and most important is the necessity of seeing to it that whatever is labeled "science education" in our colleges and universities deserves that designation. Science courses must teach science. It's as simple as that. They should have substantive scientific content, validated by perfectly well known and legitimate modes of scientific inference. As educators, scarred in battle and wearing a few tarnished medals, we have experience of the attempts to label shaky theorizing and tendentious quibbling as "science" for the sake of introducing them into the curriculum.

"Creation Science" is an example with which most of us are familiar, although institutions of higher education (outside of sectarian colleges of dubious legitimacy) were rarely the targets of that campaign. We have also the example of various New Age confections peddled at some community colleges and in extension programs, under the pretext that they represent "science" of some kind. But the influence of the academic left is of a different kind, since it is seconded by the support, often enthusiastic, of many established senior members of the academic community. To date, it has merely nibbled at the fringes of the "hard" sciences, although, as we have seen, such heretofore honorable fields as anthropology and psychology have been gravely contaminated. We cannot help but feel, however, that there will be many more calls for "feminist" courses in biology and "Afrocentric" courses in mathematics. How much force such campaigns will have is hard to predict. In any event, scientists and science educators must, on their professional honor, be prepared to resist the insertion into the science curriculum of courses whose content is tailored to the demands of any ideological faction.

Beyond this, there is the matter of courses, seminars, symposia, and the like, that claim to address scientific matters while falling outside the official boundaries of science departments. We urge scientific professionals to scrutinize these offerings, whether or not invited to do so; to participate in them if possible, and with appropriate skepticism. We urge them not to fear making judgments, not to hesitate, for the sake of someone else's imagined good social intentions, to make their misgivings public. One can't assume in these matters that possession of an advanced degree or a professorship equates to intellectual legitimacy.

We realize, of course, that the greatest disincentive to participation in such controversies is the time and effort it takes, costs that will add greatly to the burden of sustaining a serious research program and fulfilling generously one's teaching responsibilities. Intellectually, these quarrels tend to be tiresome. Nature is the scientist's worthy adversary: we use the figure in defiance of the fact that science critics will sniff it out as evidence that we are slaves to the Western patriarchal paradigm of dominance and control. Academic leftists, on the other hand, tend to be unfocused bores, and a certain deliberate, cheerful, simple-mindedness is needed to hear them out sufficiently to catch the drift of the arguments and to formulate an apposite response. It is an unlikable chore, but one that a good many of us ought to be doing, out of loyalty to our own disciplines and to—forgive the thought—civilization.

9

MULTICULTURAL PSEUDOSCIENCE: SPREADING SCIENTIFIC ILLITERACY

Bernard Ortiz de Montellano

$\left(T \right)$ here is general agreement that minorities are underrepresented in science and engineering. There is also agreement that it would be useful to give young people in minority groups examples of the role minority researchers play and have played in science. Unfortunately, one widely distributed attempt to do this will increase scientific illiteracy and impede the recruitment of African-American children into scientific careers.

In 1987, the Portland, Oregon, school district published the *African-American Baseline Essays*, a set of six essays to be read by all teachers and whose contents are to be infused into the teaching of various subjects in all grades. The purpose of the essays is to provide resource materials and references for teachers so that they can use the knowledge and contributions of Africans and African Americans in their classes. The *Science Baseline Essay*, titled "African and African-American Contributions to Science and Technology" (Adams 1990), was written by Hunter Haviland Adams, who claims to be a research scientist at Argonne National Laboratory. Actually, Adams is an industrial-hygiene technician who "does no research on any topic at Argonne," and his highest degree is a high school diploma (Baurac 1991).

The *Science Baseline Essay* follows a pattern familiar to students of pseudoscience. It is a farrago of extraordinary claims with little or no evidence; it argues for the existence of the paranormal and advocates the use of religion as a part of the scientific paradigm. No distinction is made between information drawn from popular magazines, vanity press books, and the scientific literature, and quotations are often not attributed or are not accurate. There are a number of references to the existence and scientific validity of the paranormal in the context of its use by the ancient Egyptians. In this work, Egyptians are considered to be black and their culture is claimed as ancestral to African Americans.

Adams mentions the use of the zodiac and of "astropsychological treatises"

This article first appeared in *Skeptical Inquirer* 16, no. 1 (Fall 1991): 46–50.

by Egyptians. He clearly implies that it is science. Elsewhere he has stated that astrology is based in science and that "at birth every living thing has a celestial serial number or frequency power spectrum" (Adams 1987). The *Science Baseline Essay* also states that the ancient Egyptians were "famous as masters of psi, precognition, psychokinesis, remote viewing and other undeveloped human capabilities." It argues that there is a distinction between magic, which is not scientific, and "psychoenergetics," which is, but gives no basis to distinguish one from the other. It defines psychoenergetics as the "multidisciplinary study of the interface and interaction of human consciousness with energy and matter" and states that it is a true scientific discipline. The essay says that Egyptian professional psi engineers, *hekau,* were able to use these forces efficaciously and that psi has been researched and demonstrated in controlled laboratory and field experiments today. Apparently this is why Aaron T. Curtis, who is identified as an electrical engineer and psychoenergeticist, is included in a list of African-American contributors to science at the same level as Benjamin Banneker, George Washington Carver, and Ernest E. Just.

The essay is aimed primarily at grade-school teachers, many of whom suffer from science illiteracy. Thus this essay, endorsed by a school district, prepared by someone *identified as* a research scientist at Argonne, and written using scientific-sounding jargon, is certain to influence some teachers to accept psi as a scientifically valid concept. The effect will be that many minority—and majority—children will be taught in their science classes that psi is valid, in addition to being subjected to the usual barrage of "New Age" material.

Another fundamental problem is the claim in the *Baseline Essay* that Egyptian religion was supposed to be a key organizing principle of Egyptian society. This included beliefs such as: (1) Acknowledgment of a Supreme Consciousness or Creative Force, (2) Existence via Divine Self-Organization, (3) A Living Universe, (4) Material and Transmaterial Cause and Effect, (5) Consciousness Surviving the Dissolution of the Body, and (6) Emphasis on Inner Experiences for Acquiring Knowledge. According to the *Baseline Essay, Maat* represented the first set of scientific paradigms and was the basis from which "ancient Egyptians did all types of scientific investigations." Adams admits that *Maat*'s paradigms are antithetical to those of Western science. But an unsophisticated audience will see the long list of claimed early Egyptian discoveries and successes in science presented in the *Baseline Essay* as evidence that *Maat* is equivalent to or better than the standard scientific method. This approach, just like that of the "scientific" creationists, violates the First Amendment's clause on separation of church and state because it confuses the fundamental distinction between science, which can only use natural laws to explain observed phenomena, and religion. This distinction was crucial in the ruling of Judge Overton in *McLean* v. *Arkansas* that teaching "scientific" creationism violated the First Amendment.

The key question is whether children in public schools are going to be taught that religion (under the guise of "Egyptian science") equals science. Apart from the questionable constitutionality of teaching religion (be it Christian or Egyptian) in the public schools, it will be a great disservice to minority children to teach them such a distorted view of what constitutes science. Minorities are already greatly underrepresented in science and engineering. Teaching them pseudoscience will result in making it much more difficult for these young people to pursue scientific curricula because they will face ridicule at the point they encounter a true science class.

Egyptian religion and ethics can be taught in comparative religion courses or in social studies, but that is quite a distance from teaching that Egyptian religion is essential to Egyptian "science" and that it is superior or equal to "Western" science. Teaching morality and ethics is compatible with teaching science. Ethical principles like honesty, truth, and respect for others are involved in science. Science also involves others, such as justice, equality, and avoidance of harm to others, in evaluating the consequences of research. These factors, however, do not apply when explaining and understanding scientific phenomena, when only natural laws may be used. The Second Law of Thermodynamics does not have a supernatural or an ethical component. Its application in particular cases might have consequences that raise moral and ethical questions, and these might require discussion, but this is quite different from teaching that supernatural (or transmaterial) causes are acceptable explanations in science.

A very basic question is involved in the use of the *Baseline Essay*. What is its purpose in the curriculum? Do we want to teach science as it is usually conceived or do we wish to proselytize? How is a teacher to present this section to a class? To say that Egyptian science uses the supernatural, but that Western science does not, with the implication that Egyptian science is better or equal will only perpetuate scientific illiteracy. How will students learn to distinguish science from astrology, channeling, crystal healing, telekinesis, psychic surgery, and the myriad of other New Age pseudoscientific nonsense that is floating about?

We can also apply to the *Baseline Essay* the two basic principles to remember when confronting paranormal claims (cited by Gill 1991, 271): i.e., that the burden of proof is on the claimant and that extraordinary claims require extraordinary proofs. A number of outlandish claims about the accomplishments of Ancient Egypt are made with little or no evidence. For example, on the basis of a creation myth in which the word *evolved* is used, the *Baseline Essay* claims that Egyptians had a theory of species evolution "at least 2,000 years before Charles Darwin developed his theory...." On the basis of a $6'' \times 7''$ tailless, bird-shaped object found in the Cairo Museum, supposedly a scale model of a glider, Adams says that Egyptians had full-size gliders 4,000 years ago and "used their early planes for travel, expeditions, and recreation." The *Essay* credulously repeats assertions that certain dimensions of the Great Pyramid reveal and encode knowledge about the 26,000-year cycle of the equinoxes and the acceleration of gravity. He also claims that Egyptians electroplated gold and silver 4,000 years ago and had developed copper/iron batteries some 2,000 years ago.

One extended example illustrates the level of argument employed. Without citing a reference, the *Essay* states that the Egyptians had an effective pregnancy test: "A sample of a woman's urine was sprinkled on growing barley grains; if they failed to grow, the woman was considered not pregnant. Modern experiments show this method was effective in about 40 percent of tested cases. . . ." These purported "modern experiments" show that the Egyptian method is inferior to flipping a coin; the latter would have a 50-percent success rate. The Baseline Essay misquotes and distorts the true sense of the passage. A complete citation from the Berlin Papyrus (Manniche 1989) reads: "Barley [*Hordeum vulgare*] and emmer [*Triticum dicoccum*]. The woman must moisten it with urine every day like [she does the] dates and the sand, after it has been placed in two bags. If both grow, she will give birth. If the barley grows, it means a male child. If the emmer grows, it means a female child. If neither grows, she will not give birth." Since the word for barley in Egyptian is masculine and that for emmer is feminine, this is a classical example of magical procedure, which follows the Law of Similarities used in magical procedures everywhere.

The *Baseline Essay* is a classical example of pseudoscience, but because of the current pressure on school districts to incorporate multicultural material into the classroom and the dearth of this kind of material, it has been widely distributed. Hundreds of copies of the *Baseline Essays* have been sent to school districts across the country. Carolyn Leonard, Coordinator of Multicultural/Multiethnic Education for the Portland Public Schools, has given more than 50 presentations on the *Baseline Essays*. These have been adopted or are being seriously considered by school districts as diverse as Fort Lauderdale, Atlanta, Chicago, and Washington, D.C. The *Baseline Essay* has been used for several years in Portland, and has been adopted by the Detroit Public Schools. Hunter Adams, sponsored by D. C. Heath Publishers, was a featured speaker at a conference to stimulate and to train science teachers held by the Detroit Public Schools on April 27, 1991. A workshop on "African Contributions to Science and Technology" presented undiluted material from the *Baseline Essay*, including the use of gliders by Egyptians 4,000 years ago, without a murmur of dissent from an audience composed of grade-school science teachers.

Concerned scientists need to develop reliable and scientifically valid curricular material that deals with Africa and African Americans. There is much in Egypt that would be useful, shorn of its New Age accretions. For example, the building of the pyramids can be usefully developed into lessons about mechanics, there is interesting technology involved in irrigation, and we owe the division of the day and night into twelve hours each to the Egyptians. Scientists and concerned citizens need to question their schools to see if they have adopted or are considering the *Baseline Essay*. We should protest the inclusion of erroneous material that makes unsupported claims, introduces religion under the guise of science, and claims that the paranormal exists. The critical need to increase the supply of minority scientists requires that they be taught science at its best rather than a parody.

REFERENCES

Adams, H. H. 1990. African and African-American contributions to science and technology. In *African-American Baseline Essays*. Portland, Ore.: Multnomah School District (hereafter cited as *Baseline Essay*).

Adams, H. H. 1987. Lecture at First Melanin Conference, San Francisco broadcast by WDTR 90.9 FM, Detroit Public School Radio, on September 15,1990.

Baurac, Davis, Director of Public Information, Argonne National Laboratory. Letter to Christopher Trey, May 22, 1991.

Gill, S. 1991. Carrying the war into the never-never land of psi. *Skeptical Inquirer* 15: 269–73.

Manniche, L. 1989. *An Ancient Egyptian Herbal.* Museum Publication, pp. 107–108.

10

HOW FEMINISM IS NOW ALIENATING WOMEN FROM SCIENCE

Noretta Koertge

(T) wenty years ago the dominant mood of feminism could have been represented by the World War II poster of Rosie the Riveter. Activists were
rolling up their sleeves and demanding access to traditionally male jobs. Women
were no longer always willing to be nurses, or legal secretaries, or lab technicians.
They were demanding the opportunity to be electricians, engineers, forest
rangers, and astronauts—and gender stereotypes that implied that women
couldn't deal with machines or think analytically were anathema to them.

A feminist Rip Van Winkle who fell asleep during the seventies would be
amazed at the contrasting ethos prevalent within academic feminism today. The
tough-minded and strong-armed Rosie Riveters have been displaced by moralizing
Sensitive Susans, each desperately seeking to find a new ideological flaw in the so-
called hegemonic discourse of patriarchal, racist, colonial, Eurocentric culture.

All of this might be cheerfully relegated to the Ivory Tower's already over-
flowing dust bins if one were not concerned about its impact on idealistic young
women who are making curricular and career choices while struggling to construct
lifestyles that will be quite different from those of their mothers and grandmothers.

As Daphne Patai and I interviewed faculty, students, and staff from Women's
Studies programs for our book *Professing Feminism*, there emerged a complex picture of what we call "negative education"—a systematic undermining of the intellectual values of liberal education. And as Paul Gross and Norman Levitt have so
impressively documented in *Higher Superstition*, it is the natural sciences that are
under the heaviest fire.

Young women are being alienated from science in many ways. One strategy is
to try to redefine what counts as science. For example, instead of teaching about
the struggles—and triumphs—of great women scientists, such as Emmy Noether,
Marie and Irene Curie, and Kathleen Lonsdale, feminist accounts of the history of

This article first appeared in *Skeptical Inquirer* 19, no. 2 (March/April 1995): 42–43.

science now emphasize the contributions of midwives and the allegedly forgotten healing arts of herbalists and witches. More serious are the direct attempts to steer women away from the study of science. Thus, instead of exhorting young women to prepare themselves for a variety of technical subjects by studying science, logic, and mathematics, Women's Studies students are now being taught that logic is a tool of domination and that quantitative reasoning is incompatible with a humanistic appreciation of the qualitative aspects of the phenomenological world.

Feminists add a new twist to this old litany of repudiations of analytical reasoning by claiming that the standard norms and methods of scientific inquiry are sexist because they are incompatible with "women's ways of knowing." The authors of the prize-winning book with this title report that the majority of the women they interviewed fell into the category of "subjective knowers," characterized by a "passionate rejection of science and scientists." These "subjectivist" women see the methods of logic, analysis, and abstraction as "alien territory belonging to men" and "value intuition as a safer and more fruitful approach to truth" (*Women's Ways of Knowing*, by Mary Belenky et al., p. 71).

The authors, some of whom were trained as psychologists, admit that because of the high value Western technological societies place on objectivity and rationality, these women's ways of knowing have certain "maladaptive consequences," but they emphasize that even the most epistemologically mature women in their study continue to rely heavily on subjective experience. Even those they describe as "integrated knowers" have a need to "connect" in an empathetic way with the material they study. These women are uncomfortable with the idea of a detached, impartial observer; they dislike debates, and they find it impossible to separate a critique of ideas from a criticism of the people who hold them.

Given the traditional roles of nurturer and peacemaker ascribed to women, it is perhaps not too surprising that the researchers found these profiles, although one wonders how much such attitudes were reinforced by current feminism. But what is absolutely shocking to me are the conclusions that these feminists, who are presumably trying to better the lot of women, draw from their research. Instead of arguing that young women need special help in learning how to debate and deal with abstractions, instead of calling for studies of how best to alleviate math anxiety and science phobia, instead of deploring the forces that threaten to make many women innumerate and scientifically illiterate, the authors argue that society must simply place more value on "maternal thinking."

In addition to these generic repudiations of the methods of scientific inquiry, feminists have criticized the content of various sciences, taking a special dislike to biology. In our book we describe a series of recurring maneuvers that we call the game of *Biodenial*. Some feminists, for example, have claimed that the pain of childbirth is a social construction that would disappear in a more women-positive society and that the biological classification of human beings into two sexes is inspired by the political desire to clearly demarcate those who are to dominate from those who are to be oppressed. And of course there is the recurring talk about human parthenogenesis.

Once again it is difficult to imagine that even the perpetrators of these fantastical views take them seriously. Yet the effects are very real. Women who do decide to become scientists find themselves under attack from the self-proclaimed "echt" feminists, who call them "Athenas" and "Queen Bees." In many scientific disciplines, women are a tiny minority and find themselves in a climate where they could use a little feminist support as they seek to gain acceptance and equal treatment, but they may well not find it in today's feminist circles.

Even more troubling are the deleterious effects on the young women who buy into the feminist attacks on rationality and science. To give just one example: Traditional feminists often talked about the misogynist elements in Freud's theorizing and pointed out weaknesses in his methodology—the case of Dora was a favorite example of how Freud browbeat clients in his attempt to find the repressed memories he "knew" were there. What a painful irony that today's feminists have so uncritically endorsed the methods by which hypnotists and psychological counselors purport to unearth repressed memories of childhood sexual abuse and Satanic rituals of the most bizarre kinds. I cannot think of a better demonstration of how the credulous trust of subjective beliefs and the dismissal of the methods and content of science can turn out to be dangerous to all involved.

Feminism has a great past and there remains much to be done to ensure that women really do have an equal opportunity in all aspects of society. What would our great feminist foremothers say about what's happening today—Harriet Mill, Susan B. Anthony, Elizabeth Cady Stanton, all of the women who fought for the right to go to college and enter med school, the women who couldn't get into Caltech or the Chemical Society or the Royal Society until a relatively few years ago? I think they would be proud of the gains women have made in this century, but it might break their hearts to see the strange doctrines now being promulgated in the name of feminism.

EDITOR'S NOTE

Related reading: Barbara Walker, "Science: The Feminists' Scapegoat?" *Skeptical Inquirer* 18: 68-72, Fall 1993.

11

PHYSICIST ALAN SOKAL'S HILARIOUS HOAX

Martin Gardner

It is simply a logical fallacy to go from the observation that science is a social process to the conclusion that the final product, our scientific theories, is what it is because of the social and historical forces acting in this process. A party of mountain climbers may argue over the best path to the peak, and these arguments may be conditioned by the history and social structure of the expedition, but in the end either they find a good path to the peak or they do not, and when they get there they know it. (No one would give a book about mountain climbing the title *Constructing Everest.*)

—Steven Weinberg, *Dreams of a Final Theory,* chapter 7

In a Spring/Summer 1996 issue devoted to what they called "Science Wars," the editors of *Social Text,* a leading journal of cultural studies, revealed themselves to be unbelievably foolish. They published an article titled "Transgressing the Boundaries: Toward a Transformative Hermeneutics of Quantum Gravity." It was written by Alan Sokal, a physicist at New York University. His paper included thirteen pages of impressive endnotes and nine pages of references.

Why were the editors foolish? Because Sokal's paper was a deliberate hoax, so obvious in its gibberish that any undergraduate in physics would have at once recognized it as a hilarious spoof. Did the editors bother to check with another physicist? They did not. To their everlasting embarrassment, at the same time they published the hoax, *Lingua Franca,* in its May/June 1996 issue, ran an article by Sokal in which he revealed the joke and explained why he had concocted it.

Sokal opened his parody with a strong attack on the belief that there is "an external world whose properties are independent of any human individual and indeed of humanity as a whole." Science, he continued, cannot establish genuine knowledge, even tentative knowledge, by using a "so-called" scientific method.

"Physical reality . . . is at bottom a social and linguistic construct," Sokal main-

---

This article first appeared in *Skeptical Inquirer* 20, no. 6 (November/December 1996): 14–16.

tained in the next paragraph. In his *Lingua Franca* confessional he comments: "Not our *theories* of physical reality, mind you, but the reality itself. Fair enough: Anyone who believes the laws of physics are mere social conventions is invited to try transgressing those conventions from the windows of my apartment (I live on the twenty-first floor)."

Here are a few more absurdities defended in Sokal's magnificent spoof:

• Rupert Sheldrake's morphogenetic fields are at the "cutting edge" of quantum mechanics. (On Sheldrake's psychic fantasies see chapter 15 of my *The New Age*, published by Prometheus Books in 1991.)

• Jacques Lacan's Freudian speculations have been confirmed by quantum theory.

• The axiom that two sets are identical if they have the same elements is a product of "nineteenth-century liberalism."

• The theory of quantum gravity has enormous political implications.

• Jacques Derrida's deconstructionist doctrines are supported by general relativity, Lacan's views are boosted by topology, and the opinions of Ms. Luce Irigaray, France's philosopher of feminism, are closely related to quantum gravity.

The funniest part of Sokal's paper is its conclusion that science must emancipate itself from classical mathematics before it can become a "concrete tool of progressive political praxis." Mathematical constants are mere social constructs. Even pi is not a fixed number but a culturally determined variable!

I hope no reader tries to defend this by pointing out that pi has different numerals when expressed in a different notation. To say that a notation alters pi is like saying 3 has a different value in France because it is called *trois*.

Pi is precisely defined within the formal system of Euclidian geometry, and has the same value inside the sun or on a planet in Andromeda. The fact that space-time is non-Euclidian has not the slightest effect on pi. African tribesmen may think pi equals 3, but that's a matter not of pure math but of applied math. This confusion of the certainty of mathematics within a formal system and the uncertainty of its applications to the world is a common mistake often made by ignorant sociologists.

The media had a field day with Sokal's hoax. Edward Rothstein's article in the *New York Times* (May 26, 1996) was titled "When Wry Hits Your Pi from a Real Sneaky Guy." Janny Scott's piece "Postmodern Gravity Deconstructed, Slyly" ran on the front page of the *New York Times* (May 18). Roger Kimball wrote in the *Wall Street Journal* on "A Painful Sting within the Academic Hive." George Will, in his syndicated column, gloated over Sokal's flimflam. *Social Text*, he predicted, "will never again be called a 'learned journal.'"

The editors of *Social Text* were understandably furious. Stanley Aronowitz, cofounder of the journal, is a Marxist sociologist. He branded Sokal "ill-read and half-educated." Andrew Ross, another leftist and the editor responsible for putting together the special issue, said he and the other editors thought Sokal's piece "a little hokey" and "sophomoric." Why then did they publish it? Because they checked on Sokal and found he had good credentials as a scientist.

The strongest attack on the hoax came from Stanley Fish, an English professor at Duke University and executive director of the university's press, which publishes *Social Text.* Fish has long been under the spell of deconstructionism, an opaque and rapidly fading French movement that replaced existentialism as the latest French philosophical fad. In his "Professor Sokal's Bad Joke," on the *New York Times* Op-Ed page (May 21, 1996), Fish vigorously denied that sociologists of science think there is no external world independent of observations. Only a fool would think that, he said. The sociologists contend nothing more than what observers say about the real world is "relative to their capacities, education, training, etc. It is not the world or its properties but the vocabulary in whose terms we know them that are [sic] socially constructed."

In plain language, Fish is telling us that of course there is a structured world "out there," with objective properties, but the way scientists *talk* about those properties is cultural. Could anything be more trivial? The way scientists talk obviously is part of culture. Everything humans do and say is part of culture.

Having admitted that a huge universe not made by us is out there, independent of our little minds, Fish then proceeds to blur the distinction between scientific truth and language by likening science to baseball! He grants that baseball involves objective facts, such as the distance from the pitcher's mound to home plate. Then he asks: "Are there balls and strikes in nature (if by nature you understand physical reality independent of human actors)?" Fish answers: No. Are balls and strikes social constructs? Yes.

Let's examine this more closely. The sense in which balls and strikes are defined by a culture is obvious. Chimpanzees and (most) Englishmen don't play baseball. Like the rules of chess and bridge, the rules of baseball are not part of nature. Who could disagree? Nor would Fish deny that pitched baseballs are "out there" as they travel objective paths to be declared balls and strikes by an umpire. Even the umpire is not needed. A camera hooked to a computer can do the job just as well or better. The bases for such decisions are of course cultural rules, but the ball's trajectory, and whether it goes over the plate within certain boundaries, is as much part of nature as the path of a comet that "strikes" Jupiter.

The deeper question that lies behind the above banalities is whether the rules of baseball are similar to or radically different from the rules of science. Clearly they are radically different. Like the rules of chess and bridge, the rules of baseball are made by humans. But rules of science are not. They are discovered by observation, reason, and experiment. Newton didn't invent his laws of gravity except in the obvious sense that he thought of them and wrote them down. Biologists didn't "construct" the DNA helix; they observed it. The orbit of Mars is not a social construction. Einstein did not make up $E=mc^2$ the way game rules are made up. To see rules of science as similar to baseball rules, traffic rules, or fashions in dress is to make a false analogy that leads nowhere.

It goes without saying that sociologists are not such idiots as to deny an outside world, just as it goes without saying that physicists are not so foolish as to deny that culture influences science. To cite a familiar example, culture can determine to a large extent what sort of research should be funded. And there are indeed fashions in science. The latest fashion in physics is the superstring theory of particles. It could be decades before experiments, not now possible, decide whether superstring theory is fruitful or a dead end. But that science moves inexorably closer to finding objective truth can only be denied by peculiar philosophers, naive literary critics, and misguided social scientists. The fantastic success of science in explaining and predicting, above all in making incredible advances in technology, is proof that scientists are steadily learning more and more about how the universe behaves.

The claims of science lie on a continuum between a probability of 1 (certainty) and a probability of 0 (certainly false), but thousands of its discoveries have been confirmed to a degree expressed by a decimal point followed by a string of nines. When theories become this strongly confirmed they turn into "facts," such as the fact that the earth is round and circles the sun, or that life evolved on a planet older than a million years.

The curious notion that "truth" does not mean "correspondence with reality," but nothing more than the successful passing of tests for truth, was dealt a death blow by Alfred Tarski's famous semantic definition of truth: "Snow is white" is true if and only if snow is white. The definition goes back to Aristotle. Most philosophers of the past, all scientists, and all ordinary people accept this definition of what they *mean* when they say something is true. It is denied only by a small minority of pragmatists who still buy John Dewey's obsolete epistemology.

Those who see science as mythology rather than an increasingly successful search for objective truth have been roughly grouped under the term "postmoderns." It includes the French deconstructionists, some old-fashioned Marxists, and a few angry feminists and Afrocentrists who think the history of science has been severely distorted by male and white chauvinism. Why did men study the dynamics of solids before they turned their attention to fluid dynamics? It is hard to believe, but one radical feminist claims it was because male sex organs become rigid, whereas fluids suggest menstrual blood and vaginal secretions!

A typical example of postmodern antirealism is Bruce Gregory's *Inventing Reality: Physics as Language*. The title tells it all. See my *Skeptical Inquirer* column "Relativism in Science" (Summer 1990), reprinted in *On the Wild Side* (Prometheus, 1992), for a review of this peculiar book. For a more resounding attack on such baloney, I highly recommend the recently published *Einstein, History, and Other Passions: The Rebellion Against Science at the End of the Twentieth Century* (Addison Wesley, 1996) by the distinguished Harvard physicist and science historian Gerald Holton.

The late Thomas Kuhn's famous book *The Structure of Scientific Revolutions* has been responsible for much postmodern mischief. Pragmatist Kuhn saw the history of science as a series of constantly shifting "paradigms." The final chapter of his book contains the following incredible statement: "We may, to be more precise, have to relinquish the notion, explicit or implicit, that changes of paradigm carry scientists and those who learn from them closer and closer to the truth." As if Copernicus did not get closer than Ptolemy, or Einstein closer than Newton, or quantum theory closer than earlier theories of matter! It takes only a glance at a working television set to see the absurdity of Kuhn's remark.

Fish and his friends are not that extreme in rejecting objective truth. Where

In his *Lingua Franca* article Sokal selected this as a typical passage from his hoax:

FROM "TRANSGRESSING THE BOUNDARIES":

Thus, general relativity forces upon us radically new and counterintuitive notions of space, time, and causality; so it is not surprising that it has had a profound impact not only on the natural sciences but also on philosophy, literary criticism, and the human sciences. For example, in a celebrated symposium three decades ago on *Les Langages Critiques et les Sciences de l'Homme*, Jean Hyppolite raised an incisive question about Jacques Derrida's theory of structure and sign in scientific discourse. . . . Derrida's perceptive reply went to the heart of classical general relativity:

> The Einsteinian constant is not a constant, is not a center. It is the very concept of variability—it is, finally, the concept of the game. In other words, it is not the concept of some*thing*—of a center starting from which an observer could master the field—but the very concept of the game.

In mathematical terms, Derrida's observation relates to the invariance of the Einstein field equation $G_{mn} = 8pGT_{mn}$ under nonlinear space-time diffeomorphisms (self-mappings of the space-time manifold which are infinitely differentiable but not necessarily analytic). The key point is that this invariance group "acts transitively": this means that any space-time point, if it exists at all, can be transformed into any other. In this way the infinite-dimensional invariance group erodes the distinction between observer and observed; the p of Euclid and the G of Newton, formerly thought to be constant and universal, are now perceived in their ineluctable historicity; and the putative observer becomes fatally de-centered, disconnected from any epistemic link to a space-time point that can no longer be defined by geometry alone.

they go wrong is in their overemphasis on how *heavily* culture influences science, and above all, in their obfuscatory style of writing. Examining interactions between cultures and the history of science is a worthwhile undertaking that may even come up someday with valuable new insights. So far it has had little to say that wasn't said earlier by Karl Mannheim and other sociologists of knowledge. Meanwhile, it would be good if postmoderns learned to speak clearly. Scientists and ordinary people talk in a language that takes for granted an external world with structures and laws not made by us. The language of science distinguishes sharply between language and science. The language of the sociologists of science blurs this commonsense distinction.

   It is almost as if Fish were to astound everyone by declaring that fish are not part of nature but only cultural constructs. Pressed for clarification of such a bizarre view he would then clear the air by explaining that he wasn't referring to "real" fish out there in real water, but only to the *word* "fish." In a fundamental sense scientists and sociologists of science may not disagree. It's just that the sociologists and postmoderns talk funny. So funny that when Sokal talked even funnier in one of their journals they were unable to realize they had been had.

After this column was written, *Lingua Franca,* in its July/August 1996 issue, published an article by Bruce Robbins and Andrew Ross, coeditors of *Social Text,* in which they do their best to justify accepting Sokal's brilliant prank. Their reasons fail to mention the real one—their total ignorance of physics.

In an amusing rejoinder, Sokal writes: "[M]y goal isn't to defend science from the barbarian hordes of lit crit (we'll survive just fine, thank you), but to defend the Left from a trendy segment of itself." His reply is followed by a raft of letters from scholars, some praising Sokal, some condemning him. They add little substance to the debate.

POSTSCRIPT:

For a splendid lengthy discussion of Sokal's prank and its implications, see "Sokal's Hoax," by the noted physicist Steven Weinberg, in the *New York Review of Books,* August 8, 1996, and the letters that followed (October 8, 1996). See also "Science as a Cultural Context," an excellent article by Kurt Gottfried and Kenneth Wilson, in *Nature* (Vol. 386: 545, April 10, 1997).

PART THREE

SCIENCE, PSEUDOSCIENCE, AND PATHOLOGICAL SCIENCE

12

PATHOLOGICAL SCIENCE: AN UPDATE

Alan Cromer

(T) he Nobel chemist Irving Langmuir (1881–1957) used to give a cautionary talk on pathological science, and photocopies of a transcription of his December 13, 1953, colloquium circulated for years before being published in 1989 (Langmuir 1989). In this memorable talk, Langmuir told a number of stories of pathological science and listed the features they have in common. Here I would like to retell two of these stories and add one of my own. These case studies clearly show features common to all pathological science and how such pathology can arise even among competent scientists. With this background, the recent case of cold fusion is seen as a textbook example of pathological science.

One of Langmuir's stories goes back to 1903. Wilhelm Roentgen's discovery of X-rays in 1895 was a major event in science and initiated a burst of new research. While most scientists were content to learn as much as possible about this mysterious new emanation, others wanted the glory of discovering emanations of their own. So perhaps it wasn't surprising that in 1903, Prosper René Blondlot, a distinguished member of the French Academy of Sciences, announced the discovery of N-rays which he had produced by heating a wire inside an iron tube.

These rays didn't pass through the iron, but did pass through an aluminum window in the iron. They were detected by looking at a very faintly illuminated screen in an otherwise dark room. If the N-rays were there, the screen became more visible; of course a great deal of skill was needed for this because the screen was just on the edge of visibility. Under these conditions, he discovered that many different things give off N-rays, including people. Then he discovered negative N-rays, which decreased the visibility of the screen. He published many papers on the subject, and so did many others, confirming a multitude of unusual properties for these rays.

Among them was the fact that they could be broken up into a spectrum by passing

This article first appeared in *Skeptical Inquirer* 17, no. 4 (Summer 1993): 400–407.

Nobel laureate Irving Langmuir, who often warned of pathological science (photo courtesy of GE Research and Development Center).

them through a large aluminum prism. In 1904, the American physicist R. W. Wood visited Blondlot and found him measuring, to a tenth of a millimeter, the position of the N-rays as they came through an aluminum prism. "How is that possible," Wood asked, "when the original beam is coming from a slit two millimeters wide?"

"That's one of the fascinating things about the N-rays," Blondlot replied. "They don't follow the ordinary laws of science." So Wood, the room being very dark, removed the aluminum prism that was bending the N-rays onto Blondlot's screen and put it in his pocket. Blondlot, unaware of this, continued getting the same results. Wood published a report of this incident, in *Nature*, which put an end to N-rays.

In 1934, Langmuir himself visited the parapsychologist J. B. Rhine at Duke University and pointed out that Rhine's work had all the characteristic symptoms of pathological science. Rhine thought it would be great if Langmuir published this. "I'd have more graduates," he told Langmuir. "We ought to have more graduate students. This thing is so important that we should have more people realize its importance. This should be one of the biggest departments in the university."

Rhine had begun his studies in extrasensory perception at Duke University in 1930. Most of these were done with cards showing one of the five ESP symbols: a circle, a cross, wavy lines, a rectangle, and a star. Usually a deck was used that had five cards of each kind, 25 in all. The deck would be shuffled and cut, and the subject would call the cards in the order they were picked from the top of the deck. Since there are five different cards, there is a one-in-five chance of a correct call. Results are usually reported as the number of correct calls out of 25. If there is no extrasensory perception, the average of many scores would be 5, although individual variations of plus or minus 3 aren't unlikely.

Langmuir spent the whole day with Rhine, who was in a philosophical mood. "People don't like these experiments," he said. "They don't like me. Sometimes, to spite me, they made their scores purposely low [less than 5]....I took [these low results] and sealed them in envelopes and I put a code number on the outside, and I didn't trust anybody to know that code. Nobody."

Langmuir thought this interesting. "You said that you had published a summary of all your data and that the average was 7. Now you are saying you have additional data that, if added to your published data, would bring the average to 5. Will you do this?"

"Of course not," he said. "That would be dishonest. The low scores are just as significant as the high ones, aren't they? They proved that there is something there just as much, and therefore it wouldn't be fair [to combine negative and positive data]."

Rhine felt justified in withholding low scores from his average because he believed the low scorers were deliberately (or paranormally) producing their low scores. Such self-deception is a common human failing not restricted to occultists. Mainstream scientists sometimes delude themselves as well, as the following case shows.

The A2 is an elementary particle created in experiments in which pi mesons from a high-energy accelerator collide with the proton nuclei of ordinary hydrogen. The direction and energy of the recoil proton after each collision is measured by a complex array of electronic detectors. The data from the detectors are stored on magnetic tape for later computer analysis. If a particle is created in some of the collisions, it appears as a bump in a plot of the analyzed data. All such bumps, technically called resonances, have similar shapes, which follow from fundamental theory. But in 1967, a group at the European Center for Nuclear Research (CERN) in Geneva found that the resonance of the A2 had an anomalous dip in the center (Figure 1); the resonance was split (Chikovani et al. 1967).

This was a startling discovery, contrary to all experience. As seen in Figure 1, the CERN dip has only a few points in it, each with a probable error that is one-quarter the size of the dip. Still, the probability of this being a statistical fluctuation is less than 0.1 percent. Immediately, theorists started churning out papers to explain the greatest anomaly since parity violation.

The position in energy of the A2 dip had been so well determined by the CERN equipment that the experimental group from the Northeastern University Physics Department planned to use it to check an experiment they were doing at the Brookhaven National Laboratory on Long Island, New York. To their great disappointment, they didn't see the dip. A dip is something that could easily be missed if there were a problem with the experiment, but something that was unlikely to be created by a problem. Thus the Northeastern group first thought the problem was theirs, not CERN's. But after repeated checks of their equipment revealed no problems, and repeated experiments continued to show no dip, they announced their results at a stormy meeting of the American Physical Society in 1971 (Bowen et al. 1971).

A spokesman for the CERN experiment vigorously defended its result, claiming that the *fact* that CERN saw the dip *proved* that the CERN experiment had better resolution. To which Bernard Gottschalk, speaking for Northeastern, replied, "Seeing spots before your eyes doesn't mean you have better vision." And so, amid cheers and catcalls, the physicists argued their cases. Within a few months, Northeastern's results were confirmed by other groups, and the dip was never seen or heard of again.

This leaves the question of how a group of distinguished scientists, using the best equipment in the world, could see something that wasn't there. The CERN group, it seems, did exactly what Rhine had done—discarded data that didn't show what it wanted to find. As one CERN scientist so ingenuously explained to Gottschalk: "We broke the data into batches. Whenever we found a batch with no

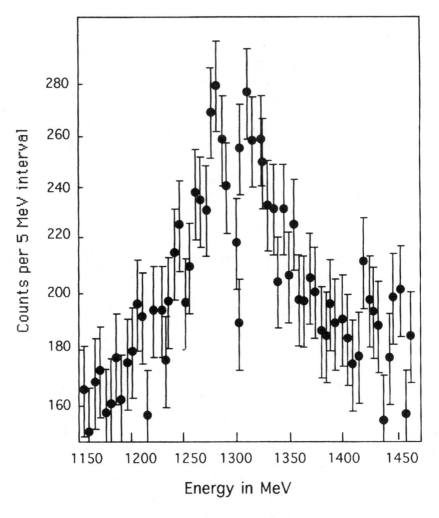

Figure 1. The A_2 resonance as reported by the experimental group from CERN (Chikovani et al. 1967). The anomalous (and erroneous) dip in the middle of the resonance was caused by a bias in the way the data were selected.

dip, we looked very carefully for something wrong and we always found something." Since they didn't look so carefully for trouble when a batch showed a dip, and since there is always something wrong in every run of a highly complex experiment, they managed to boost an initially insignificant glitch into pathological science.

This story is worth telling because we are dealing here, not with a few benighted occultists, but with a large team of highly trained scientists working with a mountain of electronic equipment and computers. But in the end, both Rhine and the CERN group were fishing in the noise for inexplicable phenomena.

Langmuir found that the cases of pathological science he had studied shared certain characteristics. They are generally claims, based on a weak or marginal effect, of fantastic phenomena contrary to all experience. There are conflicting reports from independent investigators. Reasonable explanations of the data, based on known science, are rejected. Interest rises rapidly for a time, then gradually fades away.

Nothing follows this description more closely than the case of cold fusion. [For a previous discussion, see Milton Rothman's "Cold Fusion: A Case History in 'Wishful Science'?" *Skeptical Inquirer,* Winter 1990, pp. 161–70.] It was at a press conference on March 23,1989, that Stanley Pons, a chemist at the University of Utah, and Martin Fleischmann, a chemist from the University of Southampton, England, first announced that they had obtained a controlled fusion reaction in a small electrolytic cell. This claim is certainly contrary to all experience or understanding of both nuclear and solid-state physics. Furthermore, Fleischmann and Pons didn't detect the lethal dose of neutrons that should have accompanied the amount of fusion they reported. Their claim was doubly fantastic. A miracle squared.

Immediately after the press conference, a group of physicists at Brigham Young University, headed by Steven Jones, made a similar announcement. The two groups had been working independently, but became aware of each other's work some months earlier. At a meeting that involved the presidents of these two Utah universities, an agreement was reached that on March 23, 1989, each group would submit a paper on its work to *Nature.* Pons and Fleischmann also announced their results at a press conference on the same day, which wasn't part of the agreement.

Nature accepted the Jones paper (Jones 1989), but not the Fleischmann and Pons paper. This didn't delay publication, however, because Fleischmann and Pons had already had their cold-fusion paper accepted by the *Journal of Electroanalytical Chemistry* (Fleischmann, Pons, and Hawkins 1989).

How inconsistent were the Utah results with previous experience? Fusion reactions—nuclear reactions in which two light nuclei combine to form a heavier one—have been intensely studied for more than 50 years. They are the basis of the hydrogen bomb, of the energy production of the sun, and of research efforts to produce controlled fusion on earth. In the Utah experiments, the reaction involved the fusing of two deuterium (hydrogen-2) nuclei, which generally produces either a tritium (hydrogen-3) nucleus and a proton or a helium-3 nucleus and a neutron. The fusion of two deuterium nuclei into a helium-4 nucleus and a gamma ray occurs less than one percent of the time.

A deuterium nucleus consists of one proton and one neutron; it is an isotope of the hydrogen nucleus, which is just a single proton. Normally the deuterium nucleus has an electron encircling it, in which case it is an atom of deuterium (heavy hydrogen). Deuterium, like hydrogen, is a gas at room temperature; in this gas the deuterium atoms are combined in pairs, forming a molecule.

The difficulty of achieving fusion comes from the fact that two deuterium nuclei repel each other electrically, because each is positively charged. They normally don't get close enough to interact. In a deuterium molecule, the electrons overcome this repulsion to a large extent, and the two nuclei in the molecule are 7-millionths of a millimeter apart. This is still too far apart for them to fuse, but Jones had previously achieved fusion by replacing one of the electrons in a deuterium molecule with a negatively charged muon (Rafelski and Jones 1987). The muon is 200 times heavier than an electron and brings the two deuterons 200 times closer together. At this separation, fusion occurs at a measurable rate, though not enough energy is released to pay for the cost of creating the muons.

It thus is clear that cold fusion is possible if two deuterium atoms can be squeezed together closer than they are in a deuterium molecule. It is also well known that many metals, including palladium, absorb hydrogen. Therefore, isn't it reasonable to suppose that, if deuterium were forcibly incorporated into palladium using an electrical current, deuterium atoms could be squeezed together close enough for their nuclei to fuse?

No, it isn't. The palladium atoms are themselves three times farther apart than are the two deuterium atoms in a deuterium molecule. The palladium is able to absorb deuterium molecules because the spacing between the palladium atoms is larger than the diameter of the deuterium molecule. No squeezing is involved. In fact, the deuterium molecule breaks apart inside the palladium, and its two deuterium atoms end up being farther apart in the palladium than they were in the free deuterium molecule.

Furthermore, Fleischmann and Pons claimed their fusion reaction generated a large quantity of heat. A simple calculation shows that if the heat they claimed were due to fusion, there would have been enough neutrons generated to have killed the experimenters. They took the absence of the neutrons as the discovery of a new type of nuclear reaction.

Scientists weren't immediately aware of all this when the announcement was made at a press conference. So when reporters asked scientists for their assessment of the Utah experiments, there were mixed responses. Philip Morrison said, "Based on the information I have, I feel it's a very good case." He said his confidence in the reality of the reaction was "high, but not conclusive." The Nobel physicist Sheldon Glashow said, "I don't believe a word of it" (Chandler 1989b). The amusing comment of Kim Molvig—"I am willing to be open-minded, but it's really inconceivable that there is anything there" (Pool 1989)— probably reflects the ambiguous use we often make of "open-minded." Most alarming were the comments from scientists who put extraordinary confidence in Fleischmann and Pons: "I'd be extremely surprised if they've done anything stupid. They have a very good track record in electrochemistry. I am pretty excited about this" (Chandler 1989a). In fact, stupidity, or human error, or self-delusion in science, is far less surprising than is the radical overthrow of well-established doctrine.

A number of confirming experiments were reported soon after cold fusion was announced, followed by a deluge of nonconfirming experiments. At one point, we had the peculiar situation that physicists largely rejected the cold-fusion claims, while many chemists accepted them. In July 1989, a U.S. Department of Energy panel stated that there wasn't sufficient evidence of cold fusion to warrant government funding. Still, Fleischmann and Pons—together with the remnants of their followers—carried on for another year, supported by the largesse of the State of Utah.

Jones and his group did detect a slight excess of neutrons above background radiation, but mostly in the first hour of the experiment. Some other groups also reported excess neutrons, but many groups didn't see any more neutrons than are usually present in background radiation. So this is clearly a marginal effect. The heat claimed by Pons and Fleischmann—though not in principle marginal—in fact resulted from heavily processing data obtained by incorrect procedures. They never actually generated the heat they talked about.

But for the layperson, unfamiliar with the science involved, the most characteristic sign of pathology was the language used. Fleischmann and Pons's insistence that "it is inconceivable that this [heat] could be due to anything but nuclear processes" isn't the language of science. It is the language of minds fixed on their own strongly held beliefs and unwilling to listen to the justified skepticism of others.

There are many lessons from this episode. First, scientists themselves are often poor judges of the scientific process. Many took Fleischmann and Pons's incredible conclusions about their own work at face value, before even reading their papers.

Second, scientific research is very difficult. Anything that can go wrong will go wrong. Fleischmann and Pons forgot to stir their cell while measuring its temperature, totally invalidating their measurements. Working in secrecy and isolation, even experienced scientists will be hindered by the lack of guidance and criticism of others.

Third, science isn't dependent on the honesty or wisdom of scientists. It is a collective enterprise that seeks to obtain the broadest possible consensus among its practitioners (Ziman 1968). It will survive Fleischmann and Pons, but only after the wasteful expenditure of hundreds of man-years of work and at least one death (Dye 1992).

Real discoveries of phenomena contrary to all previous scientific experience are very rare, while fraud, fakery, foolishness, and error resulting from overenthusiasm and delusion are all too common. Thus, Glashow's closed-minded "I don't believe a word of it" is going to be correct far more often than not. As Langmuir said about earlier nonexistent phenomena:

> These are cases where there is no dishonesty involved, but where people are tricked into false results by a lack of understanding about what human beings can do to themselves in the way of being led astray by subjective effects, wishful thinking, or threshold interactions. These are examples of pathological science. These are things that attracted a great deal of attention. Usually hundreds of papers have been published upon them....
>
> [But] the critics can't reproduce the effects. Only the supporters could do that. In the end, nothing was salvaged. Why should there be? There isn't anything there. There never was. (Langmuir 1989)

REFERENCES

Bowen, D., et al. 1971. Measurement of the A_{2-} and A_{2+} mass spectra. *Physical Review Letters* 26: 1663.

Chandler, David L. 1989a. Optimism grows over fusion report. *Boston Globe,* April 1, p. 1.
————. 1989b. Reports set off chain reaction. *Boston Globe,* April 3, p. 27.
Chikovani, G., et al. 1967. Evidence for a two-peak structure in the A_2 meson. *Physics Letters* 25B, 44.
Dye, Lee. 1992. Scientist killed, 3 hurt in explosion at research facility. *Los Angeles Times,* January 3, p. 3.
Fleischmann, Martin, Stanley Pons, and M. Hawkins. 1989. Electrochemically induced nuclear fusion of deuterium. *Journal of Electroanalytical Chemistry* 261: 301–308; erratum 263; 187.
Jones, S., et al. 1989. Observations of cold fusion in condensed matter. *Nature* 338: 737–40.
Langmuir, Irving. 1989. Pathological science. *Physics Today* 42: 36–48, October. Transcribed and edited by R. N. Hall from a microgrove disk found among Langmuir's papers in the Library of Congress of a colloquium given at the Knolls Research Laboratory, December 13, 1953.
Pool, Robert. 1989. Fusion breakthrough? *Science* 243: 1661.
Rafelski, J., and S. E. Jones. 1987. Cold nuclear fusion: The electronlike particles called muons can catalyze nuclear fusion reactions. *Scientific American* 257: 84, July.
Ziman, John M. 1968. *Public Knowledge.* Cambridge University Press.

13

SEVERE FLAWS IN SCIENTIFIC STUDY CRITICIZING EVOLUTION

Dan Larhammar

$\left(\begin{array}{c}\mathbf{E}\end{array}\right)$ volution is the process that has led to the life forms that inhabit Earth today and provides the fundamental basis for our understanding of all biology. The progress in molecular genetics over the past several years has provided overwhelming independent support for evolution and has helped clarify evolutionary processes. No genetic evidence has been found that argues against evolution as a phenomenon. A few creationist books have tried to interpret molecular data differently, but have only revealed their own fundamental misunderstandings, for instance Michael Denton's *Evolution: A Theory in Crisis.*

I am aware of only a single report in an established scientific journal that has claimed molecular data against evolution in modern times. The article was published in the *International Journal of Neuroscience* in 1989 (49: 43–59) by Dmitrii A. Kuznetsov (Moscow, Russia) and apparently went unnoticed for a long time since this type of work is not within the regular scope of that journal. I was recently made aware of Kuznetsov's article by a Swedish creationist. Because I found severe flaws in Kuznetsov's methodology and discovered that many of the references cited in his article appear to be nonexistent, I contacted the editor of the journal, who offered to publish my critique (*International Journal of Neuroscience,* 77: 199– 201, 1994). I summarize here Kuznetsov's study and describe the main flaws in his work.

Kuznetsov's article was published under the complex title "In vitro studies of interactions between frequent and unique mRNAs and cytoplasmic factors from brain tissue of several species of wild timber voles of Northern Eurasia, *Clethrionomys glareolus, Clethrionomys frater* and *Clethrionomys gapperi.* A new criticism to a modern molecular-genetic concept of biological evolution." Briefly, Kuznetsov studied the go-between that converts the information of the genes into functional proteins, the so-called messenger RNA (mRNA) molecules. He isolated such mRNA from three species of voles and used this to produce protein in test tubes. Surprisingly, each vole species was found

This article first appeared in *Skeptical Inquirer* 19, no. 2 (March/April 1995): 30–31.

to have a substance that blocked the production of protein from the other two species' mRNA (but it did not block its own mRNA). Remarkably, this inhibiting substance did not prevent protein synthesis from mRNAs of two distantly related species, namely, rabbit and human. Kuznetsov called his substance an "antievolutionary factor" that would serve to maintain constancy of species. He interpreted his results as "a new criticism to a modern molecular-genetic concept of biological evolution" that could be used to support "the general creationist concept on the problems of the origin of boundless multitudes of different and harmonically functioning forms of life."

The experimental approach used by Kuznetsov is extraordinary and obscure (for technical details, see my critique published in the *International Journal of Neuroscience*). None of his experiments were documented qualitatively; the report contains only tables and numbers, and the numerical results indicate experimental precision that is beyond normal accuracy for such assays. This could indicate that some of the results were fabricated. Kuznetsov disregarded that mRNA sequence variability between species displays a continuous spectrum from some highly similar molecules to some widely divergent molecules. In fact, a single mRNA molecule may be highly conserved in one part but dramatically divergent in other parts. Furthermore, Kuznetsov ignored the fact that one feature is shared by almost all mRNA molecules, the so-called poly(A) tail. Finally, the complex vole mRNA samples (many different molecules) were compared with individual mRNA molecules from the distantly related species rabbit and human. Thus, it is exceedingly unlikely that an "antievolutionary factor" would be able to recognize and block mRNA molecules from a closely related species (but not from distantly related species).

One of the procedures that Kuznetsov used was cited to be published by other researchers in "Uppsala University Research Reports" in 1974 (this is the university where I work), but no such journal has been heard of at this university and no persons with the names cited could be traced. Other important aspects of Kuznetsov's methodology were referenced to articles in four other scientific journals. None of these journals could be found, not even in the internationally used library indexes called Medline and CASSI (Chemical Abstracts Service Source Index). Two of the methodological references were purportedly published by Kuznetsov himself, but these journals were also impossible to find. Remarkably, these two references were absent from Kuznetsov's own list of publications distributed by a Swedish creationist before his recent visit to Sweden. Finally, an article ascribed to Holger Hydén (a member of the editorial board of the *International Journal of Neuroscience*) is unknown to Hydén himself. The Scandinavian journal where the article was said by Kuznetsov to be published does not exist. This purported article as well as several others in the list of references had illogical titles with grammatical errors. Taken together this strongly suggests that many of the references could have been fabricated by Kuznetsov.

In conclusion, Kuznetsov's obscure experimental approach, qualitatively undocumented results, and incomplete evaluation undermine all his conclusions. All key methodological references seem to be nonexistent. Thus, Kuznetsov's critique of evolution has no scientific basis. That his anti-evolutionary article was simply a bad joke is unlikely since it is indeed included in his list of scientific publications.

14

LYING ABOUT POLYGRAPH TESTS
Elie A. Shneour

In every culture, lying is proscribed, and great efforts are made in attempts to overcome it. This is not so much because people dislike being misled, but because lies pose a threat to the stability of society. Lying is tolerated only when life or property is in imminent danger. The ultimate deception, however, comes from claims for the effectiveness of lie detection using modern technology. This is nothing short of the assertion that it is now possible to read other persons' minds through the expert use of an instrument called the polygraph.

In its most recent incarnation, the polygraph is a small suitcase-sized machine that can measure, and continuously record on a strip chart, the following physiological variables: pulse rate; blood pressure; rate and depth of respiration; and galvanic skin response (GSR), also called electrodermal response (EDR), which is a measure of sweat production through electrical conductance.

The scientific basis for making these presumed lie-detecting measurements rests on the following premises: (1) that the transducers used in a polygraph are able to make, when properly calibrated and responsibly operated by a skilled operator, fairly accurate measurements of these variables, (2) that the physiological variables measured by a polygraph are related to physiological arousal, and (3) that psychological stimuli can be associated with these physiological responses.

No one seriously questions the first of these premises. The controversy swirling around polygraph tests focuses on premises (2) and (3) and the contention that it is possible to interpret them in ways that can detect deception. On this subject there is an extensive and, to a large extent, a confusing literature: It ranges all the way from anecdotal to peer-reviewed scientific reports, performed with protocols as complex as those used in the clinical trials of new drugs. In general, it can be concluded from these reports that the anecdotal data tend to support the polygraph tests as an effective method of deception detection, while the doubts

This article first appeared in *Skeptical Inquirer* 14, no. 3 (Spring 1990): 292–97.

get increasingly more significant as one moves ever closer to carefully controlled clinical experiments.[1]

Although there are important differences in how polygraph tests are actually administered, with conflicting claims made about which procedures and methods are optimal in the detection of deception, they all generally include three phases: (1) the pretest interview, (2) the questioning procedure, and (3) the post-test interview.

The *pretest interview* is intended to generate the psychological climate essential to optimize the effectiveness of the examination to follow. It consists, in the main, of a usually successful effort by the examiner to convince the subject that the examination is conducted by an expert and that any attempted deception by the subject will become immediately obvious to the examiner. The pretest interview also includes questions about whether the subject is under the influence of licit or illicit medication susceptible of affecting the results of the examination.

It is a remarkable fact that the medication assessment is almost never made the subject of independent verification by the collection and testing of blood and urine samples, thus resting solely on the word of a person whose credibility is the justification for the polygraph examination. The second part of a polygraph examination is the actual *questioning procedure*, which has been exhaustively reviewed by the leaders in the field, notably by Barland and Raskin and by Reid and Inbau.[2,3] This part of the examination begins with the subject being cuffed and strapped to the device. The considerable resulting discomfort is eased every 15 minutes or so while the examiner changes charts.

These interludes provide the examiner with opportunities to ask the subject about his reaction to the questions posed and allow refinement of the questions to be asked next. The examiner also performs stimulation tests to further convince the subject of the accuracy of the polygraph examination. There are many variants of the "stim" tests, but the most common ones involve the use of playing cards. The examiner unerringly identifies cards secretly chosen by the subject, on the basis of questions and answers recorded by the polygraph. These stim tests sometimes involve deception by the examiner: the use of marked cards to ensure a perfect score, something he or she knows the unaided polygraph can never produce.

The examination strategy consists of asking questions intended to reveal deception by the subject. They usually include *relevant* questions, such as "Did you steal the $1,000?" or, in a security investigation, "Did you ever have a contact with any foreign intelligence agent?" Since there is no known physiological response unique to lying, it is necessary to continuously reestablish a baseline evaluation of responses against *neutral* or *irrelevant* questions, such as "Is your name Jones?" or "Is today Monday?"

There are also *control* questions intended to elicit guilty responses to questions about lying by even honest people, such as "Have you ever lied about your age?" and *concealed information* questions aimed at detecting whether the subject is familiar with information that only a guilty person would know. For example, if a stolen car was blue, several successive questions are asked, going through a list of colors. It is presumed that a distinctive polygraph response will be obtained when the examiner asks whether the car stolen was blue.

Actually, this phase of the examination is even more complex than that. There are several arcane strategies of questioning being endlessly debated by the polygraph community. We need not dwell on them here. They include zone of comparison (ZOC) tests, peak of tension (POT) tests, guilty knowledge tests (GKT), and modified general questions tests (MGQT). There is no consensus about which of these strategies might be optimal, because no credible database has ever been developed to evaluate this or any other significant issue involving polygraph testing.

The final part of the polygraph examination is the *post-test interview*. Releasing the straps and changing the charts, mentioned earlier, are opportunities, albeit strained, for verbal exchanges between examiner and subject. This helps the examiner not only to formulate questions but, more significantly, to form an opinion of the subject's truthfulness. At the conclusion of the examination, the examiner usually makes an on-the-spot assessment of the subject as deceptive or truthful, an opinion that the examiner shares with the subject. The exception to this rule is the federal government's evaluation of national security cases, in which an official review of the results must be made prior to disclosure to the subject.

If the subject is judged to have been deceptive, the examiner will attempt to elicit a confession. This is usually done indirectly, to facilitate the opportunity for the subject to clarify, explain, or confess the meaning of the responses elicited by the examiner during the polygraph examination. Although few examiners will admit it, a good judge of human behavior will override the polygraph charts and generate a report that is more heavily weighed by the examiner's own perception of the subject. It can be argued in this context that nothing can substitute for an expert cross-examination, without the mumbo-jumbo associated with the use of the polygraph. This is a recognition of the fact that the best lie detector in existence since the dawn of human history has been, and remains, the perceptive human being. Alas, few polygraph examiners appear to fit that description.

Since the examiner is the key to the effective use of the polygraph, it is interesting to observe that very little attention has been paid to document a question that lies at the core of polygraph testing legitimacy: Who is the polygraph examiner and how, by whom, and where is he trained and accredited?

There are about six thousand people in the United States who call themselves polygraphers, and there are no formal licensing procedures for them. Anyone can put up a shingle, buy something simulating a polygraph, and conduct polygraph examinations on which people's livelihood, reputation, and freedom may depend. One national organization, the American Polygraph Association, has been trying, with largely indifferent results, to set standards for polygraphers. Only the U.S. government runs an international school for polygraphers, most of whose alumni are subsequently employed as polygraphers by the United States and its allies.

This school, claimed with some justification to be the best in the world, opened in 1951 and greatly expanded after 1981, is located in building 3165 of Fort McClellan in Alabama. Some three dozen students at a time, drawn mostly from the ranks of the FBI, the Secret Service, the National Security Agency, and the several military investigative branches of the U. S. and its allies, spend 14 weeks in training at the school. The first four weeks of lectures are on the subjects of law, semantics, ethics, physiology, pharmacology, psychology, and the operation, testing, and maintenance of the polygraph. This is followed by 10 weeks of actual practice, after which the successful candidate can conduct polygraph examinations for the government. By comparison, in most states at least a year's training is necessary to become a licensed barber.

Although polygraph evidence cannot be used in most courts of law, except by subterfuge, and polygraph testing as a condition of civilian employment has now been outlawed, the U.S. government still heavily depends on these examinations. The General Accounting Office (GAO) reported in 1987 that 2.2 million security clearances were held within 41 government agencies, exclusive of the NSA and the CIA. A significant proportion of these clearances were subject to polygraph examination. In the Department of Defense alone, for example, the number of polygraph tests administered more than doubled between 1981 (6,556) and 1985 (13,786). They exceeded 21,000 by 1987—in spite of the fact that there were 750,000 fewer federal contractors and workers with security clearances in 1985 than there were in 1984. In 1985, the U.S. government had 160 polygraph operators and had ordered 153 additional machines.

The heavy reliance on polygraph examinations in national security procedures has been reviewed, documented, and sharply criticized in several reports. A notable example is the one published by the U.S. House Select Committee on Intelligence of the 100th Congress. It concludes that the rapidly increasing use of, and excessive reliance on, polygraph examinations since 1981 creates a false sense of security and represents a dangerous trend that may increase rather than decrease the risks to our national security.[4]

The central premise of polygraph testing, the psychological assumption that guilt can *always* be inferred from emotional disturbance, is considered to be implausible by the majority of knowledgeable psychologists in the field. The American Polygraph Association claims that studies of polygraph examinations yield accuracy rates of from 87.2 to 96.2 percent. Although these undocumented figures are dubious at best, if for the sake of argument we accept a 90-percent figure, this means that out of the total of 21,000 examinations reported by the federal government (not including the CIA and NSA) in 1986, 2,100 got away with their guilt or were innocent when they "failed" the test, and some of those who got away with false negatives (i.e., labeled as truthful and reliable) may, at this very moment, still hold sensitive government positions of trust.

But the ultimate irony lies in the well-established observation that polygraph examinations tend to err on generating substantially more false positive than false negatives. This means that truthful persons incriminated as liars by the polygraph will outnumber actual liars. Good advice would be that if you are innocent, never take a lie-detector test. But if you are guilty, by all means take one: you may be exonerated. Acceptance of polygraph testing is a peculiarly unique U.S. phenomenon, where the procedure is rarely questioned and is accepted at face value by most people. It is almost unknown in the rest of the civilized world.

This underscores the issue of polygraph tests' jeopardy to the basic constitutional premise that a person is innocent until proved guilty, that it is better to let a guilty person evade the net of justice than to punish a single innocent person. Nearly all parties to polygraph testing, including U.S. government authorities responsible for such examinations and the quasi totality of responsible academic investigators in this field, have taken the position that polygraph testing should not be permitted as a condition for gaining and retaining employment. The U.S. Congress, under the leadership of Senator Orrin Hatch (R-Utah), passed legislation now in effect making the use of polygraph testing for most civilian employment illegal.

Polygraph testing uses the jargon and attributes of science for legitimacy, but it properly belongs to pseudoscience. Its main justification for existence is that it can be effective in getting at the truth through intimidation. It is not the technical data that provides such a determination, but the interpretation given to the data by the polygraph examiner. Objective criteria to make such determinations simply do not exist and there is as yet no known reliable method to get at the truth by the application of scientific principles.

The debate about polygraph examinations has raged for more than six decades, and still no consensus has emerged on their effectiveness or their justification in detecting deception.[5] Until modern technology develops credible methods, if this is at all possible, polygraph tests will remain the subject of continuing controversy.

NOTES

1. *Scientific Validity of Polygraph Testing: A Research Review and Evaluation* (Washington, D.C.: Office of Technology Assessment, November 1983).

2. G. H. Barland and D. C. Raskin, *Validity and Reliability of Polygraph Examinations of Criminal Suspects,* Report No. 76-1, Contract No. N1-99-0001 (Washington, D.C.: National Institutes of Justice, Department of Justice, 1976).

3. J. E. Reid and F. E. Inbau, *Truth and Deception: The Polygraph Technique,* 3d ed. (Baltimore: William Wilkins, 1977).

4. *Report by the Permanent Select Committee on Intelligence, House of Representatives, 100th Congress,* Report No. 100-3 (Washington, D.C.: U.S. Government Printing Office, 1987).

5. W. G. Iacono and C. J. Patrick, What psychologists should know about lie detection, chapter 17 of *Handbook of Forensic Psychology,* ed. A. Hess and I. Weiner (New York: Wiley, 1986).

15

DO HONESTY TESTS REALLY MEASURE HONESTY?

Scott O. Lilienfeld

$\left(\mathbf{L} \right)$ aypersons and scientists alike have long been intrigued by the prospect of developing a foolproof technique for assessing deception (Lykken 1981). Perhaps because lying is so prevalent in our daily experience (Saxe 1991) and because humans are such fallible lie detectors (even the most perspicacious among us are capable of distinguishing truths from falsehoods no more than about 70 to 85 percent of the time [Ekman 1985; Lykken 1981]), the public has maintained a fascination with methods purported to detect dishonesty with essentially perfect accuracy. Nevertheless, for decades the validity of such methods has been an ongoing source of controversy among researchers (e.g., Lykken 1979; Raskin and Podlesny 1979). In this article, I focus upon what has recently become the most widespread and lucrative technology in the detection-of-deception industry: the use of self-report "honesty" questionnaires for preemployment screening.

OVERVIEW OF "HONESTY" TECHNIQUES

Most of the early efforts to assess dishonesty were predicated upon the existence of a "Pinocchio response"—a supposed observable reaction that always occurs while telling lies but never at other times. Undoubtedly the greatest amount of public and scientific attention has focused upon the polygraph, or "lie detector," test, which has often been used by companies to identify individuals responsible for theft and other crimes in the workplace. In addition, the polygraph test has frequently been utilized as a pre-employment screening device to detect potential employees with a history of criminality (Lykken 1981). The polygraph test was invented by William Marston, whose other major creation was the comicbook character "Wonder Woman," who induced criminals to tell the truth by encircling

This article first appeared in *Skeptical Inquirer* 18, no. 1 (Fall 1993): 32–41.

their waists with a magical lasso. The standard polygraph test compares subjects' autonomic responses, such as their heart rate, respiration, and palmar sweating, to "relevant" questions ("Did you rob the bank?") with their autonomic responses to "irrelevant" questions ("Were you born in the U.S.?"). Many variants of the polygraph test also include "control" questions designed to gauge the subject's autonomic responsitivity to known lies (Lykken 1981). Subjects who consistently exhibit more pronounced autonomic reactions to relevant questions compared with other questions are presumed to have given dishonest responses to the former.

Nevertheless, as there is no scientific evidence for a specific autonomic lie response (Lykken 1981; but see Bashore and Rapp 1993 for a discussion of the potential use of brain electrical activity in lie detection), the polygraph test rests upon an untenable assumption. Because autonomic arousal can stem from a variety of sources, including surprise, anger, fear, disgust, guilt, and even amusement, individuals can exhibit more elevated autonomic responses to relevant questions than to other questions for a multitude of reasons other than lying. Thus the polygraph test suffers from what Ekman (1985) has termed the "Othello error," after the tragic Shakespearean character who mistook his wife's distress at his accusations as proof of her infidelity— namely, the error of interpreting arousal as evidence of deceit (also see Shneour 1990). For example, a subset of innocent subjects, sometimes referred to as "guilt-grabbers," appear to exhibit elevated autonomic responses to relevant questions as a consequence of an overly harsh conscience. Such people often show marked physiological arousal following questions like "Did you strangle your boss?" not because they are guilty of the act, but rather because they feel guilty for having entertained thoughts of doing so. Thus the "lie detector," paradoxically, may detect excessively guilt-prone individuals, who are probably among the least likely of all people to prevaricate.

Because autonomic arousal is not a specific indicator of lying, the polygraph test typically yields a high rate of *false positives* (Lykken 1981); in less technical terms, the test mistakenly classifies a large number of innocent individuals as guilty. In addition, there is suggestive evidence that the polygraph test yields a nontrivial rate of *false negatives*—guilty individuals mistakenly classified by the test as innocent. For example, psychopathic personalities, who lack a well-developed conscience and experience low levels of fear, may often "beat" the lie detector because of their low autonomic reactivity (Lykken 1978; cf. Patrick and Iacono 1989).

A variety of other methods have similarly been heralded as essentially infallible indicators of either truths or falsehoods. One technique that has long captured the public's imagination is the administration of so-called "truth serum," or sodium amobarbital (amytal), a barbiturate sometimes given during psychiatric interviews to uncover memories that the individual is unwilling or unable to reveal (Schatzberg and Cole 1986). There is no evidence that this so-called truth serum increases the probability of honest responding, and it seems likely that amytal simply lowers the threshold for reporting virtually all memories, both accurate and inaccurate. Popular misconceptions notwithstanding, hypnosis appears no better at recapturing veridical memories (Hilgard 1981). Voice stress analysis (Ekman 1985; Klass 1980), measures of pupillary diameter (Lykken 1981), graphology (Furnham 1988), "fidgetometers" (i.e., chairs measuring the extent to which subjects squirm during interrogation [Lykken 1981]), and numerous other methods for distinguishing honest from dishonest responses, suffer from the same dubious assumption that is the Achilles heel of the polygraph—the existence of a Pinocchio response. Not surprisingly, they have all been found to be poor indices of deception.

THE NEW FRONTIER: SELF-REPORT HONESTY TESTS

The growing recognition of the deficiencies of the polygraph test and of similar measures has led to intensified public and scientific pressure to place tight restrictions upon their use. Several state laws limit polygraph use (Sackett, Burris, and Callahan 1989), and the 1988 Federal Polygraph Protection Act banned most uses of the polygraph in the private sector. In response, an increasing number of organizations seeking an alternative to the polygraph have turned to questionnaire measures of honesty. Unlike the polygraph, which is usually administered to ascertain whether an individual is dissembling on a particular occasion, honesty tests are intended to assess whether individuals possess high levels of the trait of honesty (but see Hartshorne and May 1928 and Mischel 1968, for criticisms of the concept of an honesty trait). By a trait, psychologists typically mean an enduring disposition that is relatively stable across many situations. Honesty tests are typically used by companies to detect prospective employees at high risk for dishonest or so-called counterproductive behaviors (stealing, falsifying time-cards, goldbricking, taking long lunch breaks, making unauthorized telephone calls). More rarely, these tests are also administered to current employees (Camera and Lane 1992).

The desire of companies to detect would-be employees who may be prone to thievery and related behaviors in the workplace is certainly understandable. Estimates of the amount of money lost to employee theft and other defalcations in the United States range from $5 billion to $200 billion annually, with most estimates in the $40 billion range (Murphy 1993). In one recent survey, 62 percent of workers in the fast-food industry confessed to pilfering money or merchandise during their previous six months on the job (Slora 1991); comparable rates have been reported for a number of other businesses. Moreover, it has been estimated that approximately 30 percent of new companies fail as a result of employee theft (Morgenstern 1977).

The increased state and federal prohibitions on polygraph use, coupled with the need to control employee dishonesty on the job, have led to a marked upsurge in the use of self-report honesty measures. At least 5,000 companies administer honesty tests to approximately 5 million people every year (Sackett and Harris 1984). This makes them among the most frequently administered psychological tests in the United States. Moreover, in many organizations honesty tests have essentially become a substitute for the polygraph test. Indeed, many of these tests, such as the Reid Report (e.g., Ash 1975), have been developed by polygraph firms (Sackett et al. 1989). With few exceptions, honesty tests have been constructed by nonpsychologists who have no formal training in psychological assessment, test development, measurement, or statistics. For example, the Phase II profile, which has been administered to well over one million people, was developed by a former police officer with no background in psychological testing (Paajanen 1988).

Although a large number of honesty tests are currently on the market (see Sackett and Harris 1984, and Sackett et al. 1989), most are quite similar. Most honesty tests, for example, contain items assessing admissions of prior theft (e.g., "Have you ever stolen merchandise from your place of work?"). The rationale is that people who admit to having stolen in the past are at increased risk of stealing in the future. Another set of questions assesses attitudes and thoughts concerning theft and related behaviors (e.g., "Have you ever been tempted to steal a piece of jewelry from a store?"). The assumption here is that people who admit to frequent temptations are at heightened risk of succumbing to these temptations in the future.

Other items measure one's perceptions regarding the dishonesty of others (e.g., "Do you think that most people steal money from their workplace every now and then?"); a "true" response is scored in the direction of dishonesty. These questions are based upon the assumption that dishonest individuals tend to believe that most other people are similar to themselves (Goldberg et al. 1991).

Still other items assess what is sometimes referred to as "punitiveness"—the extent to which a person endorses strict punishments for others' misbehaviors (e.g., "A person has been a loyal and honest employee at a firm for 20 years. One day, after realizing that she has neglected to bring lunch money, she takes $10 from her workplace, but returns it the next day. Should she be fired?"). A response of "No" to such items is considered an indicator of dishonesty; the premise here is that dishonest individuals are less willing to favor punishments for persons like themselves (Lykken 1992).

Nevertheless, the assumptions behind many honesty-test items appear to be questionable. For example, items assessing admissions of past wrongdoings and items assessing temptations to misbehavior may reward respondents who consciously attempt to create a favorable impression, and punish those who are honest, open about their faults, and introspective (Lykken 1992). Thus, like the polygraph, many honesty tests, paradoxically, may penalize particularly moral individuals, many of whom may be the "guilt-grabbers" erroneously detected by the polygraph test. Although relatively little research has been conducted to evaluate this possibility, one of my graduate assistants and I (Andrews and Lilienfeld 1993) have found that college students with high ("more honest") scores on several commonly used honesty measures obtained slightly *lower* scores on a well-known measure of moral reasoning, Rest's Defining Issues Test (Rest 1979). Items assessing individuals' perceptions concerning the dishonesty of others may reward respondents who hold a naive and Polyannaish view of the world (Lykken 1992) and punish those who are "street smart," are perceptive regarding others' shortcomings, or live in areas where crime is rampant. Finally, the "punitiveness" items may reward respondents who are authoritarian and rule-bound and punish those who are forgiving and flexible.

Preliminary support for this latter conjecture derives from analyses we have recently conducted (Andrews and Lilienfeld 1992). Specifically, we found that a group of 41 monks and nuns obtained significantly *lower* (i.e., "more dishonest") scores on the Reid Report punitiveness items than a group of 226 college students, and even scored lower (although not significantly) than a group of incarcerated criminals. These findings raise the possibility that punitiveness items are biased against those who have been taught to preach and practice charity (also see Guastello and Rieke 1991 and Lykken 1981).

The claims made by many honesty-test publishers range from the unintentionally humorous to the plainly deceptive. The publishers of the Reid Report, for example, distributed an advertisement headlined "After years of catching thieves with the lie-detector, we've perfected a way to catch them with paper and pencil" (Lykken 1981). Other honesty-test publishers proclaim that their tests have been demonstrated to identify the majority of individuals who subsequently engage in theft, but they neglect to report the proportion of innocent individuals who are mistakenly classified as likely to steal (i.e., false positives). Detecting in advance all individuals who will engage in theft is no great accomplishment; a simple reductio ad absurdum shows why. By predicting that every employee will steal, one could successfully identify 100 percent of future thieves. In another case, the institutional affiliation of a researcher employed by a test publisher was listed as the well-known university at which he obtained his doctorate 30 years earlier, rather than his present affiliation (Sackett et al. 1989). The blatantly overblown and sometimes deceitful assertions of a number of honesty-test publishers might lead an impartial observer to wonder whether such publishers would succeed in passing their own tests.

RESEARCH ON THE VALIDITY OF HONESTY TESTS

Despite the increasingly widespread use of honesty tests, there is little adequate research on their *validity*—the extent to which they assess what they are purported to assess. Moreover, little consensus exists on the constructs, or underlying psychological attributes, assessed by honesty tests. Among the constructs honesty-test publishers claim their tests measure are honesty, integrity, counterproductivity, employee deviance, wayward impulses, organizational delinquency, absenteeism/tardiness, dependability/conscientiousness, theft-proneness, emotional stability, job performance, predictiveness, service orientation, and stress tolerance (Camera and Lane 1992). This proliferation of terms reflects pervasive confusion regarding what honesty tests measure.

The claims of test publishers aside, do honesty tests predict counterproductive behaviors in the workplace? A major difficulty with conducting research on the validity of honesty tests is that the scoring keys for most of these tests are proprietary; they are not in the public domain and can be obtained only with the permission of the test publisher. Independent investigators who wish to conduct research on proprietary tests must have their research proposal "approved" by the publisher. There is of course no guarantee that this approval process is impartial; in our experience, we have found that some publishers reject proposals aimed at subjecting their tests to close scrutiny. Consequently, most of the research on the validity of honesty tests has been performed by researchers directly affiliated with test publishers, who possess a vested interest in obtaining and reporting positive findings on their own measures.

Moreover, the proprietary nature of most honesty tests makes objective evaluation of the research on their validity difficult or impossible. Research on honesty testing is especially susceptible to the "file-drawer problem" (Rosenthal 1979)—the tendency of negative findings to remain unpublished. The file-drawer problem may contribute to a favorable but misleading evaluation of a test's validity. For example, Ones, Viswesvaran, and Schmidt (1991) recently used meta-analysis (a statistical technique for summarizing the results of studies) to conduct a comprehensive review of validity research on honesty tests. They reported that honesty tests tend to have moderate levels of validity for a number of criteria, such as theft on the job. Nevertheless, it is difficult to interpret the results of Ones et al.'s review because the majority of studies included in their analyses were conducted "in-house" (i.e., by test publishers), and the authors made no attempt to correct for the possibility that negative findings on honesty tests were selectively withheld.

Setting aside for a moment the file-drawer problem, what do the published data on the validity of honesty tests indicate? One piece of evidence often cited in support of their validity is that convicted felons tend to receive lower scores on such tests (note: low scores on honesty tests are in the direction of greater "dishonesty") than do individuals in the general population (Goldberg et al. 1991). Nevertheless, it should come as no great surprise that felons, many of whom are incarcerated for theft, report more instances of theft and more frequent thoughts about theft compared with the average person. Thus such data do not provide particularly compelling support for the validity of honesty tests.

A number of researchers have found that scores on honesty tests correlate negatively with admissions obtained from polygraph examinations, as well as with the results of these examinations. Nevertheless, these findings are flawed in two major respects. First, admissions derived from polygraph tests are highly problematic, because a substantial number of guilty subjects may not confess to wrongdoing. Consequently, the use of polygraph-elicited admissions may detect only those individuals honest enough to confess to minor misbehaviors, who in turn may be the same individuals who confess to such misbehaviors on honesty tests. The second problem with the polygraph studies is similar: because many individuals who fail the polygraph test are innocent, a correlation between honesty tests and the results of polygraph examinations may indicate only that the same innocent subjects ("guilt-grabbers"?) are misclassified by both tests.

Other studies reveal that honesty-test results are negatively correlated with admissions of theft and related behaviors obtained from self-report measures. These studies are unconvincing, however, for the same reason that renders the studies of admissions elicited during polygraph tests suspect: honest individuals may be especially likely to confess to minor wrongdoings, resulting in a spurious correlation between honesty-test results and admissions of wrongdoing. In addition, because many honesty tests themselves contain questions eliciting admissions of theft, this correlation may simply be a result of content overlap between the two measures.

Other researchers have examined the relation between the introduction of honesty tests in the workplace and institutional "shrinkage"—the loss of money ostensibly stemming from employee theft. The results of several studies indicate that once an organization begins to screen potential employees with honesty tests, shrinkage decreases (Sackett et al. 1989). Nevertheless, such studies are fatally flawed in one major respect—they lack a control group that did not receive the honesty test. This makes it impossible to attribute definitively the cause of the shrinkage reduction to the introduction of the test. Even with a control group, reductions in shrinkage may result from a number of factors unrelated to honesty tests themselves, such as perceived changes in an organization's attitudes and policies toward theft.

Superficially, the most persuasive support for the validity of honesty tests comes from studies examining their ability to predict counterproductive behaviors in the workplace. In general, honesty tests have been found to forecast theft and related misbehaviors in the workplace at better than chance levels (Goldberg et al. 1991). But the magnitude of these predictive correlations tends to be quite low; most are in the 0.1 to 0.3 range, indicating that honesty tests account for between roughly 1 to 9 percent of the differences among individuals in job-related dishonesty (one obtains these percentages by squaring the correlations). The remaining 90 percent or more of these differences typically remain unpredicted by honesty tests. Moreover, honesty tests yield a very high false positive rate. Using the cut-offs recommended by test publishers, between 73 and 97 percent of individuals who receive failing scores on such tests are not found to later commit theft on the job (Office of Technology Assessment 1990, cited in Guastello and Rieke 1991).

Interestingly, most honesty tests have been found to be moderately *positively* correlated with questionnaire measures of lying (for which high scores indicate a greater probability of lying); that is, individuals with "more honest" scores on honesty tests paradoxically tend to receive "more dishonest" scores on measures of lying (Guastello and Rieke 1991).

These latter measures contain items assessing denial of common or everyday frailties ("I occasionally make mistakes in my work," "I sometimes get more upset about things than I should"); subjects who deny a large number of such human frailties can generally be assumed to be dissimulating. The positive correlation between honesty tests and measures of lying is problematic, because individuals with higher levels of honesty would presumably be expected to admit to more trivial faults shared by virtually all of humanity. Instead, this finding again raises the possibility that honesty tests penalize individuals who are frank and introspective.

A related issue that has received surprisingly little attention is the extent to which honesty tests are susceptible to attempts at "faking good." Many of the items on these measures are highly "face-valid," meaning that most respondents can easily discern what answers will make them appear honest. Nevertheless, there has been virtually no research on the issue of fakability, leaving open the possibility that a number of high scorers on honesty tests are "seeing through" the test and consciously creating a favorable impression. Although a prominent proponent of honesty testing has asserted that "in fact, ['faking good'] does not seem to happen" (Ash 1975: 141), he provided no evidence to substantiate his claim. In an attempt to deal with the potential problem of fakability, some organizations have developed "disguised-purpose" honesty tests (Sackett et al. 1989; Murphy 1993). These tests consist primarily of items ("I would do almost anything for a thrill," "I usually enjoy watching a good brawl") that assess personality traits putatively related to dishonesty (e.g., impulsivity, aggressiveness). The extent to which such tests can circumvent the shortcomings of more traditional "clear-purpose" tests (Murphy 1993) remains to be seen. Similarly, no research has been conducted on the extent to which people can be trained to beat honesty tests; if such tests were highly susceptible to training, they would presumably be at risk of becoming obsolete soon after the formula for passing them became widely known.

REPRISE: DO HONESTY TESTS REALLY MEASURE HONESTY?

What can we say in response to the question constituting the title of this article? It appears that honesty tests assess a variety of characteristics largely or entirely unrelated to honesty, and are thus misnamed. Low scores on these tests probably reflect dishonesty in some cases, but they may just as often reflect other traits, such as street-wiseness, charity toward others, and extreme honesty (!) and openness concerning one's temptations and history of peccadilloes. The Reid Report correlates positively with a number of characteristics not traditionally considered important to the construct of honesty, such as freedom from stress, desire for interpersonal intimacy, fearfulness, and traditionalism (Lilienfeld, Andrews, Stone, and Stone, in press). Such findings raise the possibility that honesty tests may systematically exclude certain persons from some occupations because of their standing on personality traits that are irrelevant to honesty.

Moreover, honesty tests seem to suffer from many of the same liabilities that have bedeviled the polygraph test and other standard methods of lie detection.

Honesty tests, like the polygraph, yield a high rate of false positives. Ironically, many of the innocent individuals mistakenly identified as dishonest by both of these procedures may be especially moral people whose strong consciences lead them to take these tests more seriously and conscientiously than does the average person. In addition, both honesty tests and polygraph tests may yield a significant number of false negatives. Equally ironically, many of the guilty individuals missed by both procedures may be people whose absence of a strong conscience allows them to beat these tests with equanimity. Thus honesty tests, which in large measure are the progeny of polygraph tests, appear to have inherited many of their ancestors' deficiencies. Unless these shortcomings can be remedied, honesty tests seem destined to go the way of the polygraph and the other pseudoscientific dinosaurs of the lie-detection industry.

NOTE

I thank George Alliger, Brian Andrews, Lori Marino, and Vincent Marino for their helpful comments on an earlier draft of this article.

AUTHOR'S POSTSCRIPT

Since this article appeared in 1993, there have been several important developments in the field of honesty testing. Partly in recognition of findings suggesting that honesty tests are not pure indicators of honesty, these measures have increasingly come to be known as "integrity" tests.

The quantitative literature review by Ones, Viswesvaran, and Schmidt referred to in the article has since appeared in print (1993, *Journal of Applied Psychology* 78: 679–703). This article, which represents the most comprehensive analysis of the validity of integrity tests to date, presented data from more than 180 studies examining the relation between these tests and work-related counterproductive behaviors. Ones and colleagues found that integrity tests possess moderate validity for predicting overall job performance. The average correlation between integrity tests and future job-related theft, however, was only .09 (see Sackett and Wanek, 1996, *Personnel Psychology* 49: 787–829), indicating that these tests account for less than 1 percent of the differences among individuals in theft-proneness (as noted in the article, one obtains this 1 percent figure by squaring the correlation). The average correlation between integrity tests and future antisocial behaviors other than theft (e.g., unexcused absenteeism) ranged from .20 to .27 depending on the type of test used, indicating that these tests account for between 4 and 8 percent of the differences among individuals in their propensity for such behaviors. These percentages, although too large to be attributable to chance, indicate that integrity tests, like the polygraph, yield extremely high error rates when used to predict occupational misbehavior. Moreover, the interpretation of Ones's findings is complicated by the fact that 90 percent of their data points were derived from unpublished reports, most of which were supplied by integrity test publishers. Because Ones and colleagues did not distinguish between independent research and research sponsored by test companies, it is impossible to ascertain the extent to which their reported correlations were affected by the "file-drawer problem" (see article).

In addition, there has recently been increased attention to the fakability and trainability of integrity tests. Alliger, Lilienfeld, and Mitchell (1996, *Psychological Science* 7: 32–39) found that college students asked to "fake good" on a prototypical integrity test obtained higher (more honest) scores than students who were asked to be candid in their responses, although the scores of the former students were well below the maximum possible (most honest) score. Students who were given a brief (less than one-half hour) tutorial on the content of integrity tests, however, obtained extremely high scores, many of which approached the maximum possible score. Fifty-eight percent of students who received the tutorial achieved scores that were 90 percent of the maximum score, compared with 19 percent of students asked to fake good and 0 percent of students asked to be candid in their responses. These results suggest that standard integrity tests, like the polygraph, may be highly susceptible to straightforward and easily taught countermeasures.

REFERENCES

Alliger, G. M., S. O. Lilienfeld, and K. E. Mitchell. 1996. The susceptibility of overt and covert integrity tests to coaching and faking. *Psychological Science* 7: 32–39.

Andrews, B. P., and S. O. Lilienfeld. 1992. Are self-report honesty measures biased against religious individuals? Manuscript.

———. 1993. Self-report honesty tests, script in preparation.

Ash, P. 1975. Predicting dishonesty with the Reid Report. *Polygraph* 5: 139–53.

Bashore, T. R., and P. E. Rapp. 1993. Are there alternatives to traditional polygraph procedures? *Psychological Bulletin* 113: 3–22.

Camara, W. G., and D. Lane. 1992. What we know, and still do not know, about integrity testing. Manuscript.

Ekman P. 1985. *Telling Lies: Clues to Deceit in the Marketplace, Politics, and Marriage*. New York: W. W. Norton.

Furnham, A. 1988. The validity of graphological analysis. *Skeptical Inquirer* 13(1): 64–69, Fall.

Goldberg, L. R., J. R. Grenier, R. M. Guion, L. B. Sechrest, and H. Wing. 1991. *Questionnaires Used in the Prediction of Trustworthiness in Pre-employment Selection Decisions: An APA Task Force Report*. Washington, D.C.: American Psychological Association.

Guastello, S. J., and M. L. Rieke. 1991. A review and critique of honesty test research. *Behavioral Sciences and the Law* 9: 501–23.

Hartshorne, H., and M. A. May. *Studies in the Nature of Character*. Vol. 1, *Studies in Deceit*. New York: Macmillan.

Hilgard, E. R. 1981. Hypnosis gives rise to fantasy and is not a truth serum. *Skeptical Inquirer* 5(3): 25, Spring.

Klass, P. J. 1980. Beware of the "truth evaluator." *Skeptical Inquirer* 4(4): 44–51, Summer.

Lilienfeld, S. O., B. P. Andrews, E. F. Stone, and D. Stone. 1994. The relations between a self-report honesty test and personality measures in prison and college samples. *Journal of Research in Personality* 28: 154–69.

Lykken, D. T. 1978. The psychopath and the lie detector. *Psychophysiology* 15: 137–42.

———. 1979. The detection of deception. *Psychological Bulletin* 86: 47–53.

———. 1981. *A Tremor in the Blood: Uses and Abuses of the Lie Detector.* New York: McGraw Hill.

———. 1992. Honesty testing: An environmental impact assessment. Manuscript.

Mischel, W. 1968. *Personality and Assessment.* New York: Wiley.

Morgenstern, D. 1977. *Blue Collar Theft in Business and Industry.* Springfield, Va.: National Technical Information Service.

Murphy, K. R. 1993. *Honesty in the Workplace.* Pacific Grove, Calif.: Brooks/Cole.

Office of Technology Assessment. 1990. *The Use of Integrity Tests for Pre-employment Screening.* Washington, D.C.: U.S. Government Printing Office.

Ones, D., C. Viswesvaran, and F. L. Schmidt. 1991. Moderators of the validity of integrity tests: A meta-analysis. Paper presented to the Sixth Annual Conference of the Society of Industrial and Organizational Psychologists, St. Louis, April.

———. 1993. Comprehensive meta-analysis of integrity test validities: Findings and implications for personnel selection and theories of job performance. *Journal of Applied Psychology* 78: 679–703.

Paajanen, G. 1988. The prediction of counterproductive behavior by individual and organizational variables. Doctoral dissertation, University of Minnesota, Minneapolis.

Patrick, C. J., and W. G. Iacono. 1989. Psychopathy, threat, and polygraph test accuracy. *Journal of Applied Psychology* 74: 347–55.

Raskin, D. C., and J. A. Podlesny. 1979. Truth and deception: A reply to Lykken. *Psychological Bulletin* 86: 54–59.

Rest, J. 1979. *Revised Manual for the Defining Issues Test: An Objective Test of Moral Judgment.* Minneapolis: Minnesota Moral Research Project.

Rosenthal, R. 1979. The "file drawer problem" and tolerance for null results. *Psychological Bulletin* 86: 638–41.

Sackett, P. R., L. R. Burris, and C. Callahan. 1989. Integrity testing for personnel selection: An update. *Personnel Psychology* 42: 491–529.

Sackett, P. R., and M. E. Harris. 1984. Honesty testing for personnel selection. A review and critique. *Personnel Psychology* 37: 221–46.

Sackett, P. R., and J. E. Wanek. 1996. New developments in the use of measures of honesty, integrity, conscientiousness, dependability, trustworthiness, and reliability for personnel selection. *Personnel Psychology* 49: 787–829.Saxe, L. 1991. Lying: Thoughts of an applied social psychologist. *American Psychologist* 46: 409–415.

Schatzberg, A., and J. O. Cole. 1986. *Manual of Clinical Psychopharmacology.* Washington, D.C.: American Psychiatric Press.

Shneour, E. A. 1990. Lying about polygraph tests. *Skeptical Inquirer* 14: 292–97. [Chapter 14 in this book.]

Slora, K. B. 1991. "An Empirical Approach to Determining Employee Deviance Base Rates." In *Pre-employment Honesty Testing: Current Research and Future Directions,* edited by J. W. Jones, 21–38. New York: Quorum Books.

16

QUANTUM QUACKERY
Victor J. Stenger

(C) ertain interpretations of quantum mechanics, the revolutionary theory developed early in the century to account for the anomalous behavior of light and atoms, are being misconstrued so as to imply that only thoughts are real and that the physical universe is the product of a cosmic mind to which the human mind is linked throughout space and time. This interpretation has provided an ostensibly scientific basis for various mind-over-matter claims, from ESP to alternative medicine. "Quantum mysticism" also forms part of the intellectual backdrop for the postmodern assertion that science has no claim on *objective reality*.

The word "quantum" appears frequently in New Age and modern mystical literature. For example, physician Deepak Chopra (1989) has successfully promoted a notion he calls *quantum healing,* which suggests we can cure all our ills by the application of sufficient mental power.

According to Chopra, this profound conclusion can be drawn from quantum physics, which he says has demonstrated that "the physical world, including our bodies, is a response of the observer. We create our bodies as we create the experience of our world" (Chopra 1993, 5). Chopra also asserts that "beliefs, thoughts, and emotions create the chemical reactions that uphold life in every cell," and "the world you live in, including the experience of your body, is completely dictated by how you learn to perceive it" (Chopra 1993, 6). Thus illness and aging are an illusion and we can achieve what Chopra calls "ageless body, timeless mind" by the sheer force of consciousness.[1]

Amit Goswami, in *The Self-Aware Universe: How Consciousness Creates the Material World*, argues that the existence of paranormal phenomena is supported by quantum mechanics:

This article first appeared in *Skeptical Inquirer* 21, no. 1 (January/February 1997): 37–40.

... psychic phenomena, such as distant viewing and out-of-body experiences, are examples of the nonlocal operation of consciousness . . . . Quantum mechanics undergirds such a theory by providing crucial support for the case of nonlocality of consciousness. (Goswami 1993, 136)

Since no convincing, reproducible evidence for psychic phenomena has been found, despite 150 years of effort, this is a flimsy basis indeed for quantum consciousness.[2]

Although mysticism is said to exist in the writings of many of the early century's prominent physicists (Wilber 1984), the current fad of mystical physics began in earnest with the publication in 1975 of Fritjof Capra's *The Tao of Physics* (Capra 1975). There Capra asserted that quantum theory has confirmed the traditional teaching of Eastern mystics: that human consciousness and the universe form an interconnected, irreducible whole. An example:

To the enlightened man . . . whose consciousness embraces the universe, to him the universe becomes his "body," while the physical body becomes a manifestation of the Universal Mind, his inner vision an expression of the highest reality, and his speech an expression of eternal truth and mantric power.
Lama Anagarika Govinda, *Foundations of Tibetan Mysticism*[3]
(Capra 1975, 305)

Capra's book was an inspiration for the New Age, and "quantum" became a buzzword used to buttress the trendy, pseudoscientific spirituality that characterizes this movement.[4]

WAVE-PARTICLE DUALITY

Quantum mechanics is thought, even by many physicists, to be suffused with mysteries and paradoxes. Mystics seize upon these to support their views. The source of most of these claims can be traced to the so-called *wave-particle duality* of quantum physics: Physical objects, at the quantum level, seem to possess both local, reductionist particle and nonlocal, holistic wave properties that become manifest depending on whether the position or wavelength of the object is measured.

The two types of properties, wave and particle, are said to be incompatible. Measurement of one quantity will in general affect the value the other quantity will have in a future measurement. Furthermore, the value to be obtained in the future measurement is undetermined; that is, it is unpredictable—although the statistical distribution of an ensemble of similar measurements remains predictable. In this way, quantum mechanics obtains its indeterministic quality, usually expressed in terms of the *Heisenberg uncertainty principle*. In general, the mathematical formalism of quantum mechanics can only predict statistical distributions.[5]

Despite wave-particle duality, the particle picture is maintained in most quantum mechanical applications. Atoms, nuclei, electrons, and quarks are all regarded as particles at some level. At the same time, classical "waves" such as

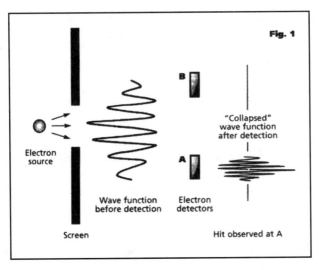

Figure 1. Wave function collapse in conventional quantum mechanics. An electron is localized by passing through an aperture. The probability that it will then be found at a particular position is determined by the wave function illustrated to the right of the aperture. When the electron is then detected at A, the wave function instantaneously collapses so that it is zero at B.

those of light and sound are replaced by localized *photons* and *phonons,* respectively, when quantum effects must be considered.

In conventional quantum mechanics, the wave properties of particles are formally represented by a mathematical quantity called the *wave function,* used to compute the probability that the particle will be found at a particular position. When a measurement is made, and its position is then known with greater accuracy, the wave function is said to "collapse," as illustrated in Figure 1.

Einstein never liked the notion of wave function collapse, calling it a "spooky action at a distance." In Figure 1, a signal would appear to propagate with infinite speed from A to B to tell the wave function to collapse to zero at B once the particle has been detected at A. Indeed, this signal must propagate at infinite speed throughout the universe since, prior to detection, the electron could in principle have been detected anywhere. This surely violates Einstein's assertion that no signals can move faster than the speed of light.

Although they are usually not so explicit, quantum mystics seem to interpret the wave function as some kind of vibration of a holistic ether that pervades the universe, as "real" as the vibration in air we call a sound wave. Wave function collapse, in their view, happens instantaneously throughout the universe by a willful act of cosmic consciousness.

In their book *The Conscious Universe,* Menas Kafatos and Robert Nadeau identify the wave function with "Being-In-Itself":

One could then conclude that Being, in its physical analogue at least, had been "revealed" in the wave function. . . . [A]ny sense we have of profound unity with the cosmos . . . could be presumed to correlate with the action of the deterministic wave function . . . . (Kafatos and Nadeau 1990, 124)

Thus they follow Capra in imagining that quantum mechanics unites mind with the universe. But our inner sense of "profound unity with the cosmos" is hardly scientific evidence.

The conventional interpretation of quantum mechanics, promulgated by Bohr and still held by most physicists, says nothing about consciousness. It concerns only what can be measured and what predictions can be made about the statistical distributions of ensembles of future measurements. As noted, the wave function is simply a mathematical object used to calculate probabilities. Mathematical constructs can be as magical as any other figment of the human imagination—like the Starship Enterprise or a Roadrunner cartoon. Nowhere does quantum mechanics imply that real matter or signals travel faster than light. In fact, superluminal signal propagation has been proven to be impossible in any theory consistent with conventional relativity and quantum mechanics (Eberhard and Ross 1989).

ROMANTIC INTERPRETATIONS

Not everyone has been happy with the conventional interpretation of quantum mechanics, which offers no real explanation for wave function collapse. The desire for consensus on an ontological interpretation of quantum mechanics has led to hundreds of proposals over the years, none gaining even a simple majority of support among physicists or philosophers.

Spurred on by Einstein's insistence that quantum mechanics is an incomplete theory, that "God does not play dice," subquantum theories involving "hidden variables" have been sought that provide for forces that lie below current levels of observation (Bohm and Hiley 1993). While such theories are possible, no evidence has yet been found for subquantum forces. Furthermore, experiments have made it almost certain that any such theory, if deterministic, must involve superluminal connections.[6]

Nevertheless, quantum mystics have greeted the possibility of nonlocal, holistic, hidden variables with the same enthusiasm they show for the conscious wave function. Likewise, they have embraced a third view: the *many worlds* interpretation of Hugh Everett (Everett 1957).

Everett usefully showed how it was formally possible to eliminate wave function collapse in a quantum theory of measurement. Everett proposed that all possible paths continue to exist in parallel universes which split off every time a measurement is made. This has left the door open for the quantum mystics to claim that the human mind acts as sort of a "channel selector" for the path that is followed in an individual universe while existing itself in all universes (Squires 1990). Needless to say, the idea of parallel universes has attracted its own circle of enthusiastic proponents, in all universes presumably.

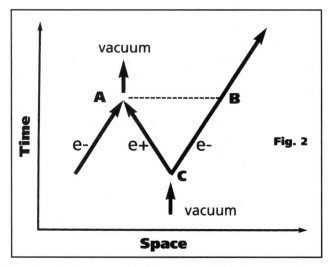

Figure 2. Effective nonlocality. How an apparent instantaneous "quantum leap" can be made between two points in space. An electron-positron pair is created at C by a quantum fluctuation of the vacuum. The positron annihilates an electron at A, undoing the original vacuum fluctuation so that there is zero net-energy change. The electron thus appears to make an instantaneous quantum leap from A to B. The distance AB is comparable to the wavelength associated with the particle, so "holistic" wave behavior results.

EFFECTIVE NONLOCALITY

Admittedly, the quantum world is different from the world of everyday experience that obeys the rules of classical Newtonian mechanics. Something beyond normal common sense and classical physics is necessary to describe the fundamental processes inside atoms and nuclei. In particular, an explanation must be given for the apparent nonlocality, the instantaneous "quantum leap," that typifies the non-commonsensical nature of quantum phenomena.

Despite the oft-heard statement that quantum particles do not follow well-defined paths in space-time, elementary-particle physicists have been utilizing just such a picture for fifty years. How is this reconciled with the quantum leap that seems to characterize atomic transitions and similar phenomena? We can see how, in the space-time diagram shown in Figure 2.

On the left, an electron (e^-) is moving along a well-defined path. An electron-positron pair (e^-e^+) is produced at point C by a quantum fluctuation of the vacuum, allowed by the uncertainty principle. The positron annihilates the original electron at point A while the electron from the pair continues past point B. Since all electrons are indistinguishable, it appears as if the original electron has jumped instantaneously from A to B.

In Figure 2, all the particles involved follow definite paths. None moves faster than the speed of light. Yet what is observed is operationally equivalent to an electron undergoing superluminal motion, disappearing at A and appearing simultaneously at a distant point B. No experiment can be performed in which the electron on the left can be distinguished from the one on the right. A simple calculation shows that the distance AB is of the order of the (de Broglie) wavelength of the particle. In this manner, the "holistic" wave nature of particles can be understood in a manner that requires no superluminal motion and certainly no intervention of human consciousness.

Furthermore, since the quantum jump is random, no signal or other causal effect is superluminally transmitted. On the other hand, a deterministic theory based on subquantum forces or hidden variables is necessarily superluminal.

Thus quantum mechanics, as conventionally practiced, describes quantum leaps without too drastic a quantum leap beyond common sense. Certainly no mystical assertions are justified by any observations concerning quantum processes.

CONCLUSION

Quantum mechanics, the centerpiece of modern physics, is misinterpreted as implying that the human mind controls reality and that the universe is one connected whole that cannot be understood by the usual reduction to parts. However, no compelling argument or evidence requires that quantum mechanics plays a central role in human consciousness or provides instantaneous, holistic connections across the universe. Modern physics, including quantum mechanics, remains completely materialistic and reductionistic while being consistent with all scientific observations.

The apparent holistic, nonlocal behavior of quantum phenomena, as exemplified by a particle's appearing to be in two places at once, can be understood without discarding the commonsense notion of particles following definite paths in space and time or requiring that signals travel faster than the speed of light.

No superluminal motion or signalling has ever been observed, in agreement with the limit set by the theory of relativity. Furthermore, interpretations of quantum effects need not so uproot classical physics, or common sense, as to render them inoperable on all scales—especially the macroscopic scale on which humans function. Newtonian physics, which successfully describes virtually all macroscopic phenomena, follows smoothly as the many-particle limit of quantum mechanics. And common sense continues to apply on the human scale.

NOTES

1. For a review of alternate medicine, including "quantum medicine," see Douglas Stalker and Clark Glymour, eds., *Examining Holistic Medicine* (Amherst, N.Y.: Prometheus Books, 1985).

2. For a fuller discussion and references, see Victor J. Stenger, *Physics and Psychics: The Search for a World Beyond the Senses* (Amherst, N.Y.: Prometheus Books, 1990).

3. L. A. Govinda, *Foundations of Tibetan Mysticism* (New York: Samuel Weiser, 1974), p. 225, as quoted in Capra 1975, p. 305.

4. See, for example, Marilyn Ferguson, *The Aquarian Conspiracy: Personal and Social Transformation in the 1980s* (Los Angeles: Tarcher, 1980).

5. Of course, in some cases those distributions may be highly peaked and thus an outcome can be predicted with high probability, that is, certainty for all practical purposes. In fact, this is precisely what happens in the case of systems of many particles, such as macroscopic objects. These systems then become describable by deterministic classical mechanics as the many-particle limit of quantum mechanics.

6. For a fuller discussion and references, see Victor J. Stenger, *The Unconscious Quantum: Metaphysics in Modern Physics and Cosmology* (Amherst, N.Y. : Prometheus Books, 1995).

REFERENCES

Bohm D., and B. J. Hiley. 1993. *The Undivided Universe: An Ontological Interpretation of Quantum Mechanics*. London: Routledge.

Capra, Fritjof. 1975. *The Tao of Physics*. Boulder, Colorado: Shambhala.

Chopra, Deepak. 1989. *Quantum Healing: Exploring the Frontiers of Mind/Body Medicine*. New York: Bantam.

———. 1993. *Ageless Body, Timeless Mind: The Quantum Alternative to Growing Old*. New York: Random House.

Eberhard, Phillippe H., and Ronald R. Ross. 1989. Quantum field theory cannot provide faster-than-light communication. *Foundations of Physics Letters* 2: 127–49.

Everett III, Hugh. 1957. "Relative state" formulation of quantum mechanics. *Reviews of Modern Physics* 29: 454–62.

Goswami, Amit. 1993. *The Self-Aware Universe: How Consciousness Creates the Material World*. New York: G. P. Putnam's Sons.

Kafatos, Menas, and Robert Nadeau. 1990. *The Conscious Universe: Part and Whole in Modern Physical Theory*. New York: Springer-Verlag.

Squires, Euan. 1990. *Conscious Mind in the Physical World*. New York: Adam Hilger.

Wilber, Ken, ed. 1984. *Quantum Questions: Mystical Writings of the World's Great Physicists*. Boulder, Colo.: Shambhala.

17

NOT–SO–SPONTANEOUS HUMAN COMBUSTION

Joe Nickell

(L)ike Count Dracula, the mythical specter of "spontaneous human combustion" (SHC) refuses to die. The latest book to fan the flames of belief, so to speak, is *Ablaze!* by Larry E. Arnold. The dust-jacket blurb states that the author "redirected a background in mechanical and electrical engineering to explore the Unconventional." Indeed, Arnold is a Pennsylvania school bus driver who has written a truly bizarre book—one that takes seriously such pseudoscientific nonsense as poltergeists and ley lines (Arnold 1995, 362–66), and that suggests that the Shroud of Turin's image was produced by "flash photolysis" from a body transformed by SHC "into a higher energy state" (463).

As if he were a trained physicist on a par with any Nobel laureate, Arnold blithely posits a subatomic "pyrotron" as the mechanism for SHC (99–106), and he casually opines that "extreme stress could be the trigger that sets a human being ablaze" (163). In the many cases in which the alleged SHC victim had been a careless cigarette smoker or in which the victim's body was found lying on a hearth, Arnold dodges the issue of SHC by invoking "preternatural combustibility" (84), an imagined state in which a body's cells reach a heightened susceptibility to ignition by an outside spark. To understand Arnold's approach we can look at a few of his major examples, those cases which are treated at chapter length.

Arnold leads off with the 1966 case of Dr. John Irving Bentley who was consumed by fire in the bathroom of his home in Coudersport, Pennsylvania. About all that was left of him—in recognizable form—was his lower leg that had burned off at the knee; it was lying at the edge of a hole about two and a half by four feet which had burned into the basement.

Spontaneous human combustion? Actually the infirm ninety-year-old physician had a habit of dropping matches and hot ashes from his pipe upon his robes which were spotted with burns from earlier occasions. He also kept wooden

This article first appeared in *Skeptical Inquirer* 20, no. 6 (November/December 1996): 17–20.

matches in both pockets of his day robe—a situation that could transform an ember into a fatal blaze. Apparently waking to find his clothing on fire, Dr. Bentley made his way into the bathroom with the aid of his aluminum walker—probably at an accelerated pace—where he vainly attempted to extinguish the flames. Broken remains of what was apparently a water pitcher were found in the toilet. Once the victim fell on the floor, his burning clothing could have ignited the flammable linoleum; beneath that was hardwood flooring and wooden beams—wood for a funeral pyre. Cool air drawn from the basement in what is known as the "chimney effect" could have kept the fire burning hotly (Arnold 1995, 1–12; Nickell and Fischer 1984).

In chapter 6, Arnold relates the fiery death of a widow, Mary Reeser, who perished in her efficiency apartment in St. Petersburg, Florida, in 1951. The case, a classic of SHC, has long been known as the "cinder woman" mystery. Except for a slippered foot, Mrs. Reeser's body was largely destroyed, along with the overstuffed chair in which she had been sitting and an adjacent end table and lamp (except for the latter's metal core). The rest of the apartment suffered little damage. "Nor," adds Arnold, "did the carpet beyond her incinerated chair show signs of fire damage!" (76).

In fact, the floors and walls of Mrs. Reeser's apartment were of concrete. When last seen by her physician son, Mrs. Reeser had been sitting in the big chair, wearing flammable nightclothes, and smoking a cigarette—after having taken two Seconal sleeping pills and stating her intention of taking two more. The official police report concluded, "Once the body became ignited, almost complete destruction occurred from the burning of its own fatty tissues." (Mrs. Reeser was a "plump" woman, and a quantity of "grease"—obviously fatty residue from her body—was left at the spot where the immolation occurred.) As the fat liquefied in the fire, it could have been absorbed into the chair stuffing to fuel still more fire to attack still more of the body (Arnold 1995, 73–91; Nickell and Fischer 1984). We will discuss the "candle effect" more fully later on.

In chapter 15, Arnold relates the case of one Jack Angel, who told him "an incredible incendiary tale." Angel stated that in mid-November 1974, while he was a self-employed traveling salesman, he awoke in his motorhome in Savannah, Georgia, to find that he had a severely burned hand, which later had to be amputated, plus a "hell of a hole" in his chest, and other burns—in the groin area, and on the legs and back "in spots!" Angel claimed one of his doctors said he had not been burned externally but rather internally, and he claimed to be a survivor of SHC. Interestingly, his clothing had not been burned, and there were no signs of burning in his motorhome.

Unfortunately, when Arnold and I appeared on a Canadian television show to debate SHC, Arnold was unaware of an earlier story about the injuries that Angel had told—in court. I revealed it on the show for the first time (courtesy of fellow investigator Phil Klass), thus publicly embarrassing Arnold, who has ever since been trying to rationalize away the evidence.

As it happens, a 1975 civil-action suit filed by Angel's attorney in Fulton County Superior Court tells how Angel (the plaintiff) was in his motorhome and "while

Plaintiff was in the process of taking a shower, the water suddenly stopped flowing from the shower plumbing." In attempting to learn why there was insufficient water pressure, Angel "exited said motorhome and attempted to inspect the hot water heater. In making said inspection, the pressure valve on the hot water heater released and as a result, scalding hot water under tremendous pressure was sprayed upon plaintiff." The complaint claimed that the defendant, the manufacturer of the motorhome, was negligent both in the design of the heater and valve and in failing to provide adequate warning of the damage. The suit was later transferred to federal court where it was eventually dismissed for costs paid by the defendant.

Arnold attempted to rebut this evidence, for example, by quoting some motorhome mechanics, but it does not seem that he gave the mechanics the full facts in soliciting their statements. For instance, forensic analyst John F. Fischer and I did not postulate "a bad valve" (as Arnold quoted the servicemen as stating we did in *Fate* magazine). Indeed, Arnold has repeatedly dodged—even outright omitted—powerful corroborative evidence, such as the water pump's drive belt being off, the water pump's drive pulley being loose, and the water heater's safety relief valve being in the open position! In our investigative report John Fischer and I listed more than a dozen additional corroborative factors, including the unburned clothes, which were especially consistent with scalding. We even included the opinions of two doctors whom Arnold cites as having diagnosed "electrical burns" as if their opinions—which were again apparently based on incomplete information—were more harmful to our position than his (Arnold 1995, 227–36; Nickell with Fischer 1992, 165–75).

Arnold's next major case is that of Helen Conway, who perished in 1964 in Delaware County, Pennsylvania. Except for her legs, her body was largely destroyed along with the upholstered chair in which she sat in her bedroom. The destruction took place in only twenty-one minutes (according to the fire marshal), although Arnold uses "commonsense deduction" (and an assumption or two) to whittle the time down to just six minutes (which becomes "a few seconds" in the caption to a photograph). Arnold asserts Mrs. Conway's body "exploded."

In fact Mrs. Conway was an infirm woman, who (according to the fire marshal) was also "reported to have been a heavy smoker with careless smoking habits." He added: "Cigarette burn marks were evident about the bedroom." (It is curious how people who are careless with fire are those who attract SHC.)

Apparently the fire took less time to destroy Mrs. Conway's torso than it did the body of Mary Reeser, but it may have begun at the base of the seated body and burned straight upward, fed by the fat in the torso, and may have thus been a much more intense fire—not unlike grease fires that all who cook are familiar with. Indeed, in searching through the dense smoke for the victim, an assistant chief sank his hand "into something greasy" that proved to be the woman's remains.

As to the bits of scattered debris that Arnold cites as evidence of "Spontaneous Human Explosion" (388), they could have been scattered by the chair's heavy right arm having fallen across the body at one point. Another possibility is revealed by the fact that the assistant fire marshal stated, "There wasn't debris scattered all over" (384), even though bits of debris are indeed shown in photos of the scene (illus. facing p. 212). In other words, the scattering may not have originally been present at the scene but could have been due to splashback from the firemen's high-pressure spray that was used to extinguish nearby flames. It is important to note that it is only Arnold—and not the fire officials, who actually blamed the fire on a dropped cigarette—who claimed the body exploded (378–92).

The fifth and last of Arnold's chapter-length cases is that of a fifty-eight-year-old retired fireman named George Mott. He died in 1986 in the bedroom of his home outside Crown Point, New York. His body was largely consumed along with the mattress of the bed on which he had lain. A leg, a shrunken skull (reported to have shrunk to an implausibly small size), and pieces of the rib cage were all that remained that were recognizably human. Arnold insists that there was no credible source for the ignition.

Whether or not we agree with Arnold's dismissal of the theories of two fire investigators—first, that an electric arc shot out of an outlet and ignited Mott's clothing, and second, that an "undetected" gas leak had been responsible—there are other possibilities. Mott was a man who formerly drank alcohol and smoked heavily. The day before he died he had been depressed over his illnesses which included respiratory problems and high blood pressure. What if, as could easily happen in such a state of mind, he became fatalistic and, shrugging off the consequences, opted for the enjoyment of a cigarette? This possibility gains credence from the fact that he was not wearing his oxygen mask although he was in bed and his oxygen-enricher unit was running. On top of the unit, next to the mask, was an otherwise puzzling canister of "barn burner" matches, yet there was no stove or other device in the room they would be used for. (At least Arnold does not mention a stove or other device being in the room. If there was, then we have another possible explanation for the fire, and there are additional potential explanations in any case—each more likely than SHC.) (Arnold 1995, 393–411)

Now Arnold cites the Mott case as a quintessential one of SHC, based on the process of elimination. He does not allow SHC to be eliminated, however, although there is no single instance that proves its existence and no known mechanism by which it could occur. And so he often dismisses what he feels is unlikely in favor of that which the best scientific evidence indicates is impossible. Such thinking has been called "straining at a gnat and swallowing a camel."

In fact Arnold's process-of-elimination approach here as elsewhere is based on a logical fallacy called "arguing from ignorance." As the great nineteenth-century scientist Justus von Liebig explained: "The opinion that a man can burn of himself is not founded on a knowledge of the circumstances of the death, but on the reverse of knowledge—on complete ignorance of all the causes or conditions which preceded the accident and caused it" (Liebig 1851).

In his relentless drive to foster any sort of mystery, in this and other cases, Arnold raises many attendant questions. For example, he wonders why extremities, such as a victim's leg, and nearby combustibles are not burned. The answer is that fire tends to burn upward; it burns laterally (sideways) with some difficulty. Anyone with camping experience has seen a log that was laid across a campfire reduced to ashes by the following morning while the butt ends of the log remained intact. Thus, outside the circle that burned through the carpet covering the concrete floor of Mary Reeser's apartment was found her slippered foot, because Mrs. Reeser had a stiff leg that she extended when she sat. Beyond the circle some newspapers did not ignite, while a lamp and table within it did burn. Similarly, Dr. Bentley's intact lower leg extended outside the edge of the hole that burned through his bathroom floor.

Beyond this matter of proximity, Arnold cites other examples of fire's "selectivity" that puzzle him. For example, in the Mott case, he wonders why matches near the burning bed did not ignite, while objects in other rooms suffered severe heat damage. The answer is one of elevation: Heat rises. In Mrs. Reeser's apartment, due to the accumulation of hot gases, soot had blackened the ceiling and walls above an almost level line some three and a half feet above the floor, there being negligible heat damage below the smoke line but significant damage above it: for example, plastic electrical switches had melted. Thus, in George Mott's house, reports Arnold, "On the counter directly beneath the melted towel holder sits an unopened roll of Bounty towels, upright. Ironically, it and its plastic wrapping were undamaged except for a glazed film on the top!" (Arnold 1995, 398).

Other factors relevant to heat-damage "selectivity" include the object's composition, density, confinement (for example, in a cupboard), placement on a surface that either radiates or retains heat, or its placement relative to convective currents, cinders carried aloft, and so forth.

While acknowledging that there is often a source for the ignition of the body, Arnold points to the sometimes extreme destruction—of the torso especially—as evidence, if not of SHC, then of preternatural combustibility, the imagined heightening of the body's flammability. In the nineteenth century, alcohol consumption was thought to cause increased flammability, but we now know that its only effect is in making people more careless with fire and less effective in responding to it (Nickell and Fischer 1984).

Arnold and other SHC advocates are quick to suggest that bodies are difficult to burn (which is true under certain circumstances). According to popular SHC writer Vincent Gaddis, "the notion that fluid-saturated fatty tissues, ignited by an outside flame, will burn and produce enough heat to destroy the rest of the body is nonsense" (Gaddis 1967).

Actually the reference to "fluid-saturated" tissues is correct but misleading in Gaddis's attempt to suggest that an external source of ignition could not cause such extreme destruction to a body because the great amount of water would retard burning. In fact the argument works more strongly *against* the concept of SHC than for it, there being no known means by which such fluid-saturated tissue could *self*-ignite. On the other hand, it is a fact that human fatty tissue will burn, the water it contains being boiled off ahead of the advancing fire.

Referring specifically to claims of SHC (and favorably citing research done by John F. Fischer and me), a standard forensic text, *Kirk's Fire Investigation*, states:

> Most significantly, there are almost always furnishings, bedding, or carpets involved. Such materials would not only provide a continuous source of fuel but

also promote a slow, smoldering fire and a layer of insulation around any fire once ignited. With this combination of features, the investigator can appreciate the basics—fuel, in the form of clothing or bedding as first ignition, and then furnishings as well as the body to feed later stages; an ignition source—smoking materials or heating appliances; and finally, the dynamics of heat, fuel, and ventilation to promote a slow, steady fire which may generate little open flame and insufficient radiant heat to encourage fire growth. In some circumstances the fat rendered from a burning body can act in the same manner as the fuel in an oil lamp or candle. If the body is positioned so that oils rendered from it can drip or drain onto an ignition source, it will continue to fuel the flames. This effect is enhanced if there are combustible fuels—carpet padding, bedding, upholstery stuffing—that can absorb the oils and act as a wick. (DeHaan 1991, 305)

Dr. Dougal Drysdale of Edinborough University agrees:

> The idea that the body can burn like a candle isn't so far fetched at all. In a way, a body is like a candle—inside out. With a candle the wick is on the inside, and the fat on the outside. As the wick burns the candle becomes molten and the liquid is drawn onto the wick and burns. With a body, which consists of a large amount of fat, the fat melts and is drawn onto the clothing which acts as a wick, and then continues to burn. (Drysdale 1989)

Experiments show that liquefied human fat burns at a temperature of about two hundred and fifty degrees Celsius; however, a cloth wick placed in such fat will burn even when the temperature falls as low as twenty-four degrees Celsius (Dee 1965). In an 1854 English case, a woman's body had been partially destroyed in the span of two hours; it was explained that "beneath the body there was a hempen mat, so combustible, owing to the melted human fat with which it was impregnated, that when ignited it burnt like a link [i.e., a pitch torch]" (Stevenson 1883, 718–27).

Even a lean body contains a significant amount of fat, which is present even in the bone marrow (Snyder 1967, 233, 242). Indeed, "once the body starts to burn, there is enough fat and inflammable substances to permit varying amounts of destruction to take place. Sometimes this destruction by burning will proceed to a degree which results in almost complete combustion of the body," as police officials reported in the Mary Reeser case (Blizin 1951). Moreover, in general, "women burn hotter and quicker than men, because proportionally, women carry more fat" (Bennett n.d.).

Arnold tries to compare favorably the partial destruction of bodies that occurs in his SHC cases (in which limbs, large segments of bone, and other matter may remain, although that which does is rarely quantified or described scientifically) with the more complete destruction typical of crematories. But this is an apples-versus-oranges comparison at best. As Drysdale (1989) explains:

> In a crematorium you need high temperatures—around 1,300 degrees C, or even higher—to reduce the body to ash in a relatively short period of time. But it's a

misconception to think you need those temperatures within a living room to reduce a body to ash in this way. You can produce local, high temperatures, by means of the wick effect and a combination of smouldering and flaming to reduce even bones to ash. At relatively low temperatures of 500 degrees C—and if given enough time—the bone will transform into something approaching a powder in composition.

It is interesting that the major proponents of SHC—Michael Harrison (*Fire from Heaven*, 1978), Jenny Randles and Peter Hough (*Spontaneous Human Combustion*, 1992), and Larry E. Arnold (*Ablaze!*, 1995)—are all popular writers who are credulous as to other paranormal claims. They stand in contrast to the physicists and chemists, the forensics specialists, and other scientists who question—on the evidence—the reality of spontaneous human combustion.

REFERENCES

Arnold, Larry E. *Ablaze! The Mysterious Fires of Spontaneous Human Combustion.* New York: M. Evans and Company.

Bennett, Valerie (crematorium superintendent). n.d. Quoted in Randles and Hough 1992, p. 50.

Blizin, Jerry. 1951. The Reeser case. *St. Petersburg Times* (Florida), August 9.

Dee, D. J. 1965. A case of "spontaneous combustion." *Medicine, Science and the Law* 5: 37–8.

DeHaan, John D. 1991. *Kirk's Fire Investigation,* 3d ed. Englewood Cliffs, N.J.: Brady.

Drysdale, Dougal. 1989. Quoted in Randles and Hough 1992, p. 43.

Gaddis, Vincent. 1967. *Mysterious Fires and Lights.* New York: David McKay.

Harrison, Michael. 1978. *Fire from Heaven.* New York: Methuen.

Liebig, Justus von. 1851. *Familiar Letters on Chemistry,* letter 22. London: Taylor, Walton and Maberly.

Nickell, Joe, and John F. Fischer. 1984. Spontaneous human combustion. *The Fire and Arson Investigator* 34 (March): 4–11; (June): 3–8. This was published in abridged form in Joe Nickell with John Fischer, *Secrets of the Supernatural* (Amherst, N.Y.: Prometheus Books, 1988, 149–57, 161–71).

Nickell, Joe, with John F. Fischer. 1992. *Mysterious Realms.* Amherst, N.Y.: Prometheus Books.

Randles, Jenny, and Peter Hough. 1992. *Spontaneous Human Combustion.* London: Robert Hale.

Snyder, Lemoyne. 1967. *Homicide investigation,* 2d ed. Springfield, Ill.: C.C. Thomas.

Stevenson, Thomas. 1883. *The Principles and Practice of Medical Jurisprudence,* 3d ed. Philadelphia: Lea.

PART FOUR

PSEUDOSCIENCE AND PATHOLOGICAL CULTS: HEAVEN'S GATE

18

UFO MYTHOLOGY: THE ESCAPE TO OBLIVION

Paul Kurtz

(**H**)eaven's Gate has stunned the world. Why would thirty-nine seemingly gentle and earnest people in Rancho Santa Fe, California, voluntarily commit collective suicide? They left us eerie messages on videotapes, conveying their motives: they wished to leave their "containers" (physical bodies) in order to ascend to a new plane of existence, a Level Above Human.

It was a celestial omen, Comet Hale-Bopp, that provoked their departure. For they thought that it carried with it a UFO spacecraft—an event already proclaimed on the nationally syndicated Art Bell radio show when Whitley Strieber and Courtney Brown maintained that a spaceship "extraterrestrial in origin" and under "intelligent control" was tracking the comet. According to astronomer Alan Hale, co-discoverer of the comet, what they probably saw was a star behind the comet. Interestingly, the twenty-one women and eighteen men, ranging in ages from twenty-one to seventy-two, seemed like a cross section of American citizens— though they demonstrated some degree of technical and engineering skills, and some even described themselves as "computer nerds." They sought to convey their bizarre UFO theology on the Internet. Were these people crazy, a fringe group, over-come by paranoia? Or were there other, deeper causes at work in their behavior?

Heaven's Gate was led by Marshall Herff Applewhite and Bonnie Lu Nettles (who died in 1985), who taught their followers how to enter the Kingdom of God. They believed that some 2,000 years ago beings from an Evolutionary Level Above Human sent Jesus to teach people how to reach the true Kingdom of God. But these efforts failed. According to documents left on the Heaven's Gate Web site, "In the early 1970s, two members of the Kingdom of Heaven (or what some might call two aliens from space) incarnated into two unsuspecting humans in Houston [Applewhite and Nettles]. . . ." Over the next twenty-five years Applewhite and Nettles transmitted their message to hundreds of followers. Those who killed

This article first appeared in *Skeptical Inquirer* 21, no. 4 (July/August 1997): 12–14.

Marshall Herff Applewhite, leader of Heaven's Gate, in his videotaped farewell message to the world.

themselves at Rancho Sante Fe (including Applewhite)—plus the two former members who subsequently attempted to take their lives on May 6, one of them succeeding—did so to achieve a higher level of existence.

Reading about the strange behavior of this cult of unreason, one is struck by the unquestioning obedience that Applewhite was able to elicit from his faithful flock. There was a rigid authoritarian code of behavior imposed upon everyone, a form of mind control. Strict rules and rituals governed all aspects of their monastic lives. They were to give up all their worldly possessions, their diets were regulated, and sex was strictly forbidden (seven members, including Applewhite, were castrated). The entire effort focused on squelching the personal self. Independent thinking was discouraged.

The followers of Heaven's Gate lived under a siege morality; they were super-secretive, attempting to hide their personal identities. They were like nomads wandering in the wilderness, seeking the truths of a Higher Revelation from extraterrestrial semi-divine beings. What has puzzled so many commentators is the depth of their conviction that space aliens were sending envoys to Earth and abducting humans. They kept vigils at night, peering for streaks of light that might be UFOs, waiting for spacecraft to arrive.

We read on their Web page: "We suspect that many of us arrived in staged spacecraft (UFO) crashes, and many of our discarded bodies (genderless, not belonging to the human species), were retrieved by human authorities (government and military)."

This form of irrational behavior should be no surprise to the readers of the *Skeptical Inquirer*. I submit that the mass media deserve a large share of the blame for this UFO mythology. Book publishers and TV and movie producers have fed the public a steady diet of science fiction fantasy packaged and sold as real. Alarmed by the steady stream of irresponsible programming spewing forth claims that were patently false, last year CSICOP (the Committee for the Scientific Investigation of Claims of the Paranormal), publisher of *Skeptical Inquirer,* established the Council for Media Integrity, calling for some balanced presentation of science. We said that, given massive media misinformation, it is difficult for large sectors of the public to distinguish between science and pseudoscience, particularly since there is a heavy dose of "quasi-documentary" films. Why worry about these programs? Because, I reply, the public, with few exceptions, does not have careful, critical knowledge of paranormal and pseudoscientific claims. So far, the Council for Media Integrity's warnings have gone largely unheeded. What drivel

NBC, Fox, and other networks have produced! (A notable exception to this is ABC, which we are glad to say has called upon CSICOP skeptics to present alternative views on "20/20," "PrimeTime Live," and other shows.) TV is a powerful medium; and when it enters the home with high drama and the stamp of authenticity, it is difficult for ordinary persons to distinguish purely imaginative fantasies from reality. Many people blame the Internet. I think the media conglomerates, who sell their ideas as products, are to blame, not the Internet. We are surely not calling for censorship, only that some measure of responsibility be exercised by editors and producers. Interestingly, the Heaven's Gaters were avid watchers of TV paranormal programs.

CSICOP and the *Skeptical Inquirer* have been dealing with UFO claims on a scientific basis for more than twenty years. We have attempted to provide, wherever we could, scientific evaluations of the claims. We have never denied that it is possible, indeed probable, that other forms of life, even intelligent life, exist in the universe. And we support any effort to verify such an exciting hypothesis. But this is different from the belief that we are now being visited by extraterrestrial beings in spacecraft, that they are abducting people, and that there is a vast government coverup of these alien invasions—a "Luciferian" conspiracy, according to Heaven's Gate.

In my view, what we are dealing with is "the transcendental temptation," the tendency of many human beings to leap beyond this world to other dimensions, impervious to the tests of evidence and the standards of logical coherence, the temptation to engage in magical thinking. UFO mythology is similar to the message of the classical religions where God sends his Angels as emissaries who offer salvation to those who accept the faith and obey his Prophets. Today, the chariots of the gods are UFOs. What we are witnessing in the past half century is the spawning of a New Age religion. (Nineteen ninety-seven marked the fiftieth anniversary of Kenneth Arnold's sighting of the first flying saucers over the State of Washington in 1947.)

There are many other signs that UFO mythology has become a space-age religion and that it is not based on scientific evidence so much as emotional commitment. Witness the revival of astrology today; or the growth of Scientology, which proposes space-age reincarnation to their Thetans and attracts famous movie stars such as Tom Cruise and John Travolta; or the Order of the Solar Temple, in which seventy-four people committed suicide in Switzerland, Quebec, and France, waiting to be transported to the star Sirius, nine light-years away. Perhaps one of the most graphic illustrations of this phenomenon is what occurred on April 21, 1997, when the cremated remains of twenty-four people, including Gene Roddenberry (father of *Star Trek*), Timothy Leary (former Harvard guru), and Gerard O'Neill (scientific promoter of space colonies), were catapulted into space from the Grand Canary island off of the Moroccan coast aboard an American Pegasus rocket. This celestial burial is symptomatic of the New Age religion, in which our sacred church is outer space. The religious temptation enters when romantic expectations outreach empirical capacities.

Science is based on factual observation and verification. It was perhaps best illustrated by the discovery of Comet Hale-Bopp. That the comet has been captured by the paranormal imagination and transformed into a religious symbol is unfortunate. Alan Hale deplored this extrapolation of his observations. Yet the transcendental temptation can at times be so powerful that it knows no bounds.

Incidentally, the paranormal—which means, literally, that which is alongside of or beside normal scientific explanation—was involved in other aspects of the Heaven's Gate theology. The members expressed beliefs in astrology, tarot cards, psychic channeling, telepathy, resurrection, and reincarnation. That is why it is often difficult to ferret out and examine these claims dispassionately, for New Agers are dealing with faith, credulity, and a deep desire to believe, rather than with falsifiable facts; and they are resistant to any attempt to apply critical thinking to such spiritual questions.

Quotations from the Heaven's Gate videotape are instructive. Those who committed suicide affirmed that: "We are looking forward to this. We are happy and excited." "I think everyone in this class wants something more than this human world has to offer." "I just can't wait to get up there." These testimonials sound like those of born-again fundamentalists who are waiting for the Rapture and whose beliefs are self-validating. These confirmations of faith are not necessarily true; they are accepted because they have a profound impact on the believers' lives. Heaven's Gate gave meaning and purpose to the lives of its followers. As such, it performed an existential, psychological function similar to that of other religious belief systems. Obedience to a charismatic leader offered a kind of sociological unity similar to that provided by traditional belief systems.

One might well ask, what is the difference between the myth of salvation of Heaven's Gate and many orthodox religious belief systems that likewise promise salvation to the countless millions who suppress their sexual passions, submit to ritual and dogma, and abandon their personal autonomy, all in quest of immortality? Their behavior is similar to the more than nine hundred Jewish Zealots who committed suicide at Masada in 73 C.E., or the early Christians who willingly died for the faith, or the young Muslim Palestinians today who strap explosives to their bodies and blow themselves to kingdom come in the hope of attaining heaven. I recently visited Cairo and the Great Pyramid of Gizeh, where a ship of the dead had been uncovered. The Pharaohs had equipped a vessel to take them to the underworld, hoping thereby to achieve immortality after death. This has been transformed into a UFO craft in modern-day lingo.

The bizarre apocalyptic theology of Heaven's Gate is interpreted by its critics as absurd and ridiculous; yet it was taken deadly serious by its devotees, and a significant part of the UFO scenario is now accepted by large sectors of the public.

In one sense the New Age paranormal religions are no more fanciful than the old-time religions. Considered cults in their own day, they were passed down from generation to generation, but perhaps they are no less queer than the new paranormal cults. No doubt many in our culture will not agree with my application of skepticism to traditional religion—CSICOP itself has avoided criticizing the classical systems of religious belief, since its focus is on empirical scientific inquiry, not faith.

I am struck by the fact that the Seventh-day Adventists, Jehovah's Witnesses,

Mormons, and Chassidic Jews were considered radical fringe groups when first proclaimed; today they are part of the conventional religious landscape, and growing by leaps and bounds. Perhaps the major difference between the established religions and the new cults of unreason is that the former religions have deeper roots in human history.

The Aum Shinri Kyo cult in Japan, which in 1995 released poison gas into a crowded subway station, killing twelve people, was made up of highly educated young people, many with advanced degrees. Unable to apply their critical thinking outside of their specialties, they accepted the concocted promises of their guru. Thus an unbridled cult of unreason can attract otherwise rational people.

The one thing I have discovered in more than two decades of studying paranormal claims is that a system of beliefs does not have to be true in order to be believed, and that the validation of such intensely held beliefs is in the eyes of the believer. There are profound psychological and sociological motives at work here. The desire to escape the trials and tribulations of this life and the desire to transcend death are common features of the salvation myths of many religious creeds. And they appear with special power and eloquence in the case of the misguided acolytes of Heaven's Gate, who, fed by an irresponsible media that dramatizes UFO mythology as true, found solace in a New Age religion of salvation, a religion whose path led them to oblivion.

19

HEAVEN'S GATE: THE UFO CULT OF BO AND BEEP

Martin Gardner

> For there are some eunuchs, which are so born from their mother's womb; and there are some eunuchs, which were made eunuchs of men; and there be eunuchs which have made themselves eunuchs for the kingdom of heaven's sake. He that is able to receive it, let him receive it.
>
> —Jesus, Matthew 19:12

(T)he nation's shocked reaction to the suicides in March 1997 of thirty-eight happy brainwashed innocents and their demented leader at Rancho Santa Fe, California, has been twofold. It has reawakened public awareness of the enormous power of charismatic gurus over the minds of cult followers, and it has focused attention on the extent to which the myth of alien spacecraft has become the dominant delusion of our times. A recent *Newsweek* poll revealed that almost half of Americans believe UFOs are for real and that our government knows it. As if a secret this monumental could be kept by our political leaders for more than a few hours!

Rumors about space aliens and their snatching of humans show no signs of abating. Harvard psychiatrist John Mack has published a book about the abductions of his patients. The rumors are magnified mightily by endless other books, lurid movies, and shameless radio and television shows. Ed Dames, who runs Psi-Tech, a psychic research center in Beverly Hills, California, was the first to proclaim that his "remote viewers" had spotted a massive spacecraft trailing Comet Hale-Bopp.

Dames's claim was "confirmed" by three psychics at The Farsight Institute in Atlanta, headed by Courtney Brown, Dames's former pupil. Brown teaches political science at Emory University. He is as embarrassing to Emory as Mack is to Harvard, but both have tenure and can't be fired. In the May/June 1997 *Skeptical Inquirer* I reviewed Brown's *Cosmic Voyage*, a crazy book telling how aliens are being shuttled to Earth to live under a mountain near Santa Fe, New Mexico.

This article first appeared in *Skeptical Inquirer* 21, no. 4 (July/August 1997): 15–17.

Who was most responsible for the Rancho Santa Fe horrors? They were two neurotic, self-deluded occultists: Marshall Herff Applewhite and his platonic companion Bonnie Lu Trousdale Nettles. Their story reads like bad science fiction.

Born in Spur, Texas, in 1931, the son of a Presbyterian minister, Applewhite graduated as a philosophy major at Austin (Texas) College in 1952. He had brief stays in seminary school and in the Army Signal Corps. But gifted with good looks and a fine baritone voice, his chosen career was in singing and music. He received a master's degree in music from the University of Colorado, starring in numerous operas produced in Houston and in Boulder while pursuing his degree. Throughout his musical career, he held a number of teaching positions and conducted numerous church choirs.

For several years in the sixties Applewhite taught music at St. Thomas University, a small Catholic school in Houston. The university fired him in 1970 over an affair with a male student. Struggling to control his homosexual impulses, depressed, and hearing voices, he checked into a psychiatric hospital in 1971. He told his sister he had suffered a heart attack and had had a near-death experience.

It was in this hospital that Applewhite's life took its fateful turn. His registered nurse, Bonnie Nettles[1] (she was forty-four, he forty), was a former Baptist, then deep into occultism, astrology, and reincarnation. Somehow she managed to convince Applewhite that they were aliens from a higher level of reality who had known each other in previous earthly incarnations. In the coming months and years, they would develop their bizarre religion, believing they had been sent to Earth to warn humanity that our civilization was about to collapse as foretold in Revelation, to be replaced by a new one after the battle of Armageddon and the destruction of Lucifer. They believed that Lucifer (a cut or two below Satan), aided by his "Luciferians," had long controlled our planet. It is, in fact, Lucifer's demons who have been piloting those spaceships that are abducting humans.

How can one escape the coming holocaust? Not by being "raptured," as Protestant fundamentalists teach, but by being beamed up to spacecraft operated by benign superbeings and taken to the gates of heaven. (Judging by the recent tragedy at Rancho Santa Fe, if you are male, the best way to make this journey is to cut off your testicles then kill yourself!)

Shortly after their meeting, and fired with the divine mandate to rescue as many humans as possible from the destruction of our world as we know it, Applewhite and Nettles embarked on their mission, Nettles abandoning a husband and four children in the process. (Applewhite, a father of two, was already divorced.) The pair quickly became insep-

"How a Member of the Kingdom of Heaven might appear" (from the Heaven's Gate Web site).

arable in a strange bond that psychiatrists call the "insanity of two." It develops when two neurotic persons live together and reinforce each other's delusions.

Indeed, Applewhite and Nettles began calling themselves The Two. They came to believe they were the "two witnesses" described in chapter 11 of Revelation. Verse 7 predicts that when the two witnesses "finish their testimony" they will be murdered. After three and a half days, God will resurrect them. A voice from heaven will say "Come up hither," and their enemies will see them taken to heaven by a "cloud."

"I'm not saying we are Jesus," Nettles wrote to her daughter. "It's nothing as beautiful but it is almost as big. . . . We have found out, baby, we have this mission before coming into this life. . . . It's in the Bible in Revelation."

The Two's first move was to open an occult bookstore in Houston. After it failed in 1973, they took to the road to gather converts. A group was started in Los Angeles called Guinea Pig. Applewhite was Guinea, Nettles was Pig. Soon they were calling their movement HIM (Human Individual Metamorphosis). Later it became TOA (Total Overcomers Anonymous). Because they considered themselves shepherds to a flock of sheep, Applewhite took the name of Bo, and Nettles became Peep. Over the years they liked to give themselves other whimsical names such as Him and Her, Winnie and Pooh, Tweedle and Dee, Chip and Dale, Nincom and Poop, Tiddly and Wink. Eventually they settled on the musical notes Do and Ti.

In a 1972 interview in the *Houston Post,* Nettles said her astrological work was assisted by Brother Francis, a nineteenth-century monk. "He stands beside me," she said, "when I interpret the charts." Both Do and Ti constantly channeled voices from superbeings who lived on the Evolutionary Level Above Human, or Next Level (the Kingdom of Heaven).

It all sounds so childish and insane, yet those who attended early cult meetings, mostly on college campuses, have testified to the pair's persuasive rhetoric. Early converts were mainly young hippies, drifters, and New Agers, disenchanted with other cults, eager to be told what to believe and do.

In 1975 about twenty followers were recruited in the seaside village of Waldport, Oregon, then taken to eastern Colorado where they expected to board a flying saucer and be carried to the Next Level. It was a vague region ruled by the great EGB (Energy God Being). When the spacecraft failed to appear, it was such a blow to Bo and Peep that they plunged underground for seventeen years.

There was a period when The Two preached that the "Demonstration" foretold in Revelation 11 would occur. As I said earlier, this would be their assassination, followed by their resurrection and journey to the Heavenly Kingdom in a spacecraft that the Bible called a cloud. "The chances it won't happen," Applewhite told a *New York Times* reporter in 1976, "are about as great as that a rain will wash all the red dirt out of Oklahoma." The interview got him fired as choir director of St. Mark's Episcopal Church in Houston.

Never sexual lovers, the peculiar pair surfaced in the mid-seventies as leaders of about fifty followers who wandered with them here and there. They camped out or lived in motels with funds donated by wealthy recruits or obtained from odd jobs and occasional begging. For several years they hunkered down in a camp near

Laramie, Wyoming. HIM was now a full-blown cult with members strictly regulated by what they called the Process. Recruits assumed new names. Sex, alcohol, tobacco, and pot were taboo. Ti, whom Do always considered his superior, died of cancer in 1985 after losing an eye to the disease. Until his suicide, Do said he was in constant communication with Ti, who had reached the Next Level.

Precise details about the cult's nomadic history remain obscure. Do convinced his sheep that they, too, were aliens from the Next Level, now incarnated in a body they called the soul's container, vehicle, instrument, or vessel. When the time was right, they would all be teleported to one of the spaceships operated by angels.

Time reported (August 27, 1979) that cult members were then wearing hoods and gloves, obeying "thousands" of rules, studying the Bible intensely, and undergoing periods in which they communicated with each other only by writing. It's not easy to believe, but the cult received so much media attention in the late seventies that a TV series called *The Mysterious Two* was planned. A pilot actually aired in 1982 featuring John Forsythe and Pricilla Pointer as The Two.

After the Internet became widely available to the public, the cult intensified its recruiting by way of a Web site called Heaven's Gate. A few followers had developed sufficient skills not only to go online, but also to run a service called Higher Source that designed Web sites for customers.

In 1996 the cult rented a sprawling, Spanish-style villa, with pool and tennis court, in Rancho Santa Fe, a few miles north of San Diego. The rent was $7,000 a month. Members began the day with prayers at 3 A.M., ate only two meals a day, had their hair cropped short, and wore baggy clothes to make themselves look genderless and unsexy. Their lives were more regulated than the lives of soldiers. Guns were stored just in case government forces attacked them as they had done to the Branch Davidians in Waco. Meticulous plans were drawn for a mass suicide as soon as the higher beings gave them a "marker" in the heavens. The marker, Do decided, was the giant spacecraft that psychics had convinced him (perhaps verified by the voice of Ti) was following Comet Hale-Bopp. A lunar eclipse on March 23, 1997, may have strengthened the sign.

Haunting videotapes were made on which the smiling and happy sheep said how joyfully they were looking forward to escaping from their vehicles and from a doomed planet. "We are happily prepared to leave 'this world' and go with Ti's crew," they posted on their Web site. Evidently they believed their beloved Ti was on the Hale-Bopp spacecraft!

As everyone now knows, eighteen men and twenty-one women put themselves to sleep with phenobarbital mixed into pudding or applesauce, and washed down with vodka. Plastic bags were tied over their heads so they would suffocate in their sleep. Faces and upper bodies of the "monks," as they called themselves, were neatly covered with a square of purple cloth. All thirty-nine were dressed alike—black shirts, black pants, and black Nike running shoes. The last two to die were women with bags on their heads but unshrouded by a purple covering.

The most perplexing aspect of these ritualized deaths were the neatly packed travel bags beside their bunks, and a five-dollar bill and some quarters in each of their pockets. Did they expect the superbeings to take the bags along with their souls? And what use did they suppose the money would have when they reached the spaceship?

The odor of rotting "vehicles" was so strong that the first police at the scene on March 26 suspected poison gas.

To me the saddest aspect of this insane event was the firm belief, expressed on the incredible videotapes, that cult members were killing themselves of their own free will. Nothing could have been more false. Although Do always told his robots they were free to go at any time—and hundreds had done just that—so powerful was his control over the minds of those who stayed that they believed anything he said, obeyed every order. Autopsies showed that Do and seven of his followers had been surgically castrated.

Do said he was dying of cancer. Yet his autopsy showed no sign of cancer or any other fatal illness. The wild-eyed expression on his face, reproduced on the covers of both *Time* and *Newsweek*, was not a look of illness. It was a look of madness.

Media reports have made fun of the belief that our bodies are mere containers and that in our next life we will be given glorious new bodies. This, of course, is exactly what Saint Paul taught, and what conservative Christians, Jews, Muslims, and most Eastern faiths believe. Similar mixtures of New Testament doctrine with New Age nonsense is what makes so many recent cults appealing to converts with Christian backgrounds. Members of Heaven's Gate firmly believed that Jesus was an extraterrestrial sent to Earth like Do and Ti to collect as many souls as possible and lead them upward to acquire new containers. When Jesus finished his work, he went back to heaven in a UFO.

The great adventist movements in America—Seventh-day Adventism, Jehovah's Witnesses, and Mormonism—are flourishing today as never before despite the long delay in Jesus' Second Coming. As the year 2000 approaches, we can be sure that lesser doomsday cults will be popping up, many of them recruiting followers on the fast-growing Internet. None of the major adventist faiths recommend suicide, but will there be more suicides by other weird little cults that surely will be capturing the minds of lonely, gullible souls in the remaining years before 2000?[2]

So pervasive is the worldwide belief in alien UFOs that a London company recently offered to insure anyone against abduction, impregnation, or attack by aliens. About four thousand people, mostly in England and the United States, bought policies. Last October Heaven's Gate paid a thousand dollars for a policy covering up to fifty members for $1 million each. After their mass suicides, the London firm decided to abandon its insurance against space aliens.

Some insight into the sort of people who were followers of Bo and Peep can be gained from a sad story distributed by the Associated Press in early April. Lorraine Webster, age seventy-eight, now living in Rolla, Missouri, abandoned her husband in 1978 to help found Heaven's Gate. She left the cult only because of a health problem. Her daughter was among those who died at Rancho Santa Fe.

Was Lorraine Webster disturbed by the suicides? Not in the least. Like all cult members, she doesn't like to call her cult a cult. It was a "movement." Do, she told the reporter, was a "kind and wonderful man." She misses her daughter but admires her for acting "like an angel." Ti frequently talks to Ms. Webster. She recently appeared at Webster's window in the form of a "chirping bird."

At the fifth annual Gulf Breeze (Florida) UFO Conference, March 21–23, Courtney Brown announced that his psychics' most recent remote viewing of Comet Hale-Bopp showed that the spacecraft was no longer there. It had moved, he said, to a spot behind the sun. Evidently this news failed to reach Do and his sheep. However, if Do was in touch with Ti on the ship, he probably would have taken her word over Brown's.

The *Village Voice*, covering the cult in its December 1, 1975, issue, included a prophetic passage: "The whole operation has lost that silvery crazy glitter. Now it seems black, dark, and a little ugly. It has the smell of ordinary death."

NOTES

1. Some sources say that Applewhite met Nettles when he was visiting a friend in the hospital. Nettles's daughter says they met at a drama school. (*New York Times*, April 28, 1997)

2. At the time I write there have been a total of five copycat suicide attempts, two of which were successful. Most notably, on Tuesday, May 6, 1997, two former Heaven's Gate cult members, Wayne Cooke and Charles Humphrey, were found in a hotel room not far from Rancho Santa Fe in an attempted double suicide. Cooke, whose wife was one of the original thirty-nine, was found dead; Humphrey was found unconscious and taken to the hospital.

PART FIVE

SCIENCE AND THE PSYCHOLOGY OF BELIEF

20

THE BELIEF ENGINE
James E. Alcock

(T) he following beliefs are strongly held by large numbers of people. Each of them has been hotly disputed by others:

- Through hypnosis, one can access past lives.
- Horoscopes provide useful information about the future.
- Spiritual healing sometimes succeeds where conventional medicine fails.
- A widespread, transgenerational Satanic conspiracy is afoot in society.
- Certain gifted people have been able to use their psychic powers to help police solve crimes.
- We can sometimes communicate with others via mental telepathy.
- Some people have been abducted by UFOs and then returned to earth.
- Elvis lives.
- Vitamin C can ward off or cure the common cold.
- Immigrants are stealing our jobs.
- Certain racial groups are intellectually inferior.
- Certain racial groups are athletically superior, at least in some specific sports.
- Crime and violence are linked to the breakdown of the traditional family.
- North Korea's developing nuclear capability poses a threat to world peace.

Despite high confidence on the part of both believers and disbelievers, in most instances, neither side has much—if any—objective evidence to back its position. Some of these beliefs, such as telepathy and astrology, stand in contradiction to the current scientific worldview and are therefore considered by many scientists to be "irrational." Others are not at all inconsistent with science, and whether or not they are based in fact, no one would consider them to be irrational.

This article first appeared in *Skeptical Inquirer* 19, no. 3 (May/June 1995): 14–18.

Nineteenth-century rationalists predicted that superstition and irrationality would be defeated by universal education. However, this has not happened. High literacy rates and universal education have done little to decrease such belief, and poll after poll indicates that a large majority of the public believe in the reality of "occult" or "paranormal" or "supernatural" phenomena. Why should this be so? Why is it that in this highly scientific and technological age superstition and irrationality abound?

It is because our brains and nervous systems constitute a belief-generating machine, an engine that produces beliefs without any particular respect for what is real or true and what is not. This belief engine selects information from the environment, shapes it, combines it with information from memory, and produces beliefs that are generally consistent with beliefs already held. This system is as capable of generating fallacious beliefs as it is of generating beliefs that are in line with truth. These beliefs guide future actions and, whether correct or erroneous, they may prove functional for the individual who holds them. Whether or not there is really a Heaven for worthy souls does nothing to detract from the usefulness of such a belief for people who are searching for meaning in life.

Nothing is fundamentally different about what we might think of as "irrational" beliefs—they are generated in the same manner as are other beliefs. We may not have an evidential basis for belief in irrational concepts, but neither do we have such a basis for most of our beliefs. For example, you probably believe that brushing your teeth is good for you, but it is unlikely that you have any evidence to back up this belief, unless you are a dentist. You have been taught this, it makes some sense, and you have never been led to question it.

If we were to conceptualize the brain and nervous system as a belief engine, it would need to comprise several components, each reflecting some basic aspect of belief generation. Among the components, the following units figure importantly:

1. The learning unit
2. The critical thinking unit
3. The yearning unit
4. The input unit
5. The emotional response unit
6. The memory unit
7. The environmental feedback unit.

THE LEARNING UNIT

The learning unit is the key to understanding the belief engine. It is tied to the physical architecture of the brain and nervous system; and by its very nature, we are condemned to a virtually automatic process of magical thinking. "Magical thinking" is the interpreting of two closely occurring events as though one caused the other, without any concern for the causal link. For example, if you believe that crossing your fingers brought you good fortune, you have associated the act of finger-crossing with the subsequent welcome event and imputed a causal link between the two.

Our brain and nervous system have evolved over millions of years. It is important to recognize that natural selection does not select directly on the basis of reason or truth; it selects for reproductive success. Nothing in our cerebral apparatus gives any particular status to truth. Consider a rabbit in the tall grass, and grant for a moment a modicum of conscious and logical intellect to it. It detects a rustling in the tall grass, and having in the past learned that this occasionally signals the presence of a hungry fox, the rabbit wonders if there really is a fox this time or if a gust of wind caused the grass to rustle. It awaits more conclusive evidence. Although motivated by a search for truth, that rabbit does not live long. Compare the late rabbit to the rabbit that responds to the rustle with a strong autonomic nervous-system reaction and runs away as fast as it can. It is more likely to live and reproduce. So, seeking truth does not always promote survival, and fleeing on the basis of erroneous belief is not always such a bad thing to do. However, while this avoidance strategy may succeed in the forest, it may be quite dangerous to pursue in the nuclear age.

The learning unit is set up in such a way as to learn very quickly from the association of two significant events—such as touching a hot stove and feeling pain. It is set up so that significant pairings produce a lasting effect, while nonpairings of the same two events are not nearly so influential. If a child were to touch a stove once and be burned, then if the child were to touch it again without being burned, the association between pain and stove would not automatically be unlearned. This basic asymmetry—pairing of two stimuli has an important effect, while presenting the stimuli unpaired (that is, individually) has a much lesser effect—is important for survival.

This asymmetry in learning also underlies much of the error that colors our thinking about events that occur together from time to time. Humans are very poor at accurately judging the relationship between events that only sometimes co-occur. For example, if we think of Uncle Harry, and then he telephones us a few minutes later, this might seem to demand some explanation in terms of telepathy or precognition. However, we can only properly evaluate the co-occurrence of these two events if we also consider the number of times that we thought of Harry and he did not call, or we did not think of him but he called anyway. These latter circumstances—these nonpairings—have little impact on our learning system. Because we are overly influenced by pairings of significant events, we can come to infer an association, and even a causal one, between two events even if there is none. Thus, dreams may correspond with subsequent events only every so often by chance, and yet this pairing may have a dramatic effect on belief. Or we feel a cold coming on, take vitamin C, and then when the cold does not get to be too bad we infer a causal link. The world around us abounds with coincidental occurrences, some of which are meaningful but the vast majority of which are not. This provides a fertile ground for the growth of fallacious beliefs. We readily learn that associations exist between events, even when they do not. We are often led by co-occurring events to infer that the one that occurred first somehow caused the one that succeeded it.

We are all even more prone to error when rare or emotionally laden events are involved. We are always looking for causal explanations, and we tend to infer causality even when none exists. You might be puzzled or even distressed if you heard a loud noise in your living room but could find no source for it.

THE CRITICAL-THINKING UNIT

The critical-thinking unit is the second component of the belief engine, and it is acquired—acquired through experience and explicit education. Because of the nervous-system architecture that I have described, we are born to magical thinking. The infant who smiles just before a breeze causes a mobile above her head to move will smile again and again, as though the smile had magically caused the desired motion of the mobile. We have to labor to overcome such magical pre-disposition, and we never do so entirely. It is through experience and direct teaching that we come to understand the limits of our immediate magical intuitive interpretations. We are taught common logic by parents and teachers, and since it often serves us well, we use it where it seems appropriate. Indeed, the cultural parallel of this developmental process is the development of the formal method of logic and scientific inquiry. We come to realize that we cannot trust our automatic inferences about co-occurrence and causality.

We learn to use simple tests of reason to evaluate events around us, but we also learn that certain classes of events are not to be subjected to reason but should be accepted on faith. Every society teaches about transcendental things—ghosts, gods, bogeymen, and so on; and here we are often explicitly taught to ignore logic and accept such things on faith or on the basis of other people's experiences. By the time we are adults, we can respond to an event either in a logical, critical mode or in an experiential, intuitive mode. The events themselves often determine which way we will respond. If I were to tell you that I went home last night and found a cow in my living room, you would be more likely to laugh than to believe me, even though there is certainly nothing impossible about such an event. If, on the other hand, I were to tell you that I went into my living room and was startled by an eerie glow over my late grandfather's armchair, and that the room went cold, you may be less likely to disbelieve and more likely to perk up your ears and listen to the details, possibly suspending the critical acumen that you would bring to the cow story. Sometimes strong emotion interferes with the application of critical thought. Other times we are cleverly gulled.

Rationality is often at a disadvantage to intuitive thought. The late psychologist Graham Reed spoke of the example of the gambler's fallacy: Suppose you are observing a roulette wheel. It has come up black ten times in a row, and a powerful intuitive feeling is growing in you that it must soon come up red. It cannot keep coming up black forever. Yet your rational mind tells you that the wheel has no memory, that each outcome is independent of those that preceded. In such a case, the struggle between intuition and rationality is not always won by rationality.

Note that we can switch this critical thinking unit on or off. As I noted earlier, we may switch it off entirely if dealing with religious or other transcendental matters. Sometimes, we deliberately switch it on: "Hold it a minute, let me think this out," we might say to ourselves when someone tries to extract money from us for an apparently worthy cause.

THE YEARNING UNIT

Learning does not occur in a vacuum. We are not passive receivers of information. We actively seek out information to satisfy our many needs. We may yearn to find meaning in life. We may yearn for a sense of identity. We may yearn for recovery from disease. We may yearn to be in touch with deceased loved ones.

In general we yearn to reduce anxiety. Beliefs, be they correct or false, can assuage these yearnings. Often beliefs that might be categorized as irrational by scientists are the most efficient at reducing these yearnings. Rationality and scientific truth have little to offer for most people as remedies for existential anxiety. However, belief in reincarnation, supernatural intervention, and everlasting life can overcome such anxiety to some extent.

When we are yearning most, when we are in the greatest need, we are even more vulnerable to fallacious beliefs that can serve to satisfy those yearnings.

THE INPUT UNIT

Information enters the belief engine sometimes in the form of raw sensory experience and other times in the form of organized, codified information presented through word of mouth, books, or films. We are wonderful pattern detectors, but not all the patterns we detect are meaningful ones. Our perceptual processes work in such a way as to make sense of the environment around us, but they do *make* sense—perception is not a passive gathering of information but, rather, an active construction of a representation of what is going on in our sensory world. Our perceptual apparatus selects and organizes information from the environment, and this process is subject to many well-known biases that can lead to distorted beliefs. Indeed, we are less likely to be influenced by incoming information if it does not already correspond to deeply held beliefs. Thus, the very spiritual Christian may be quite prepared to see the Virgin Mary; information or perceptual experience that suggests that she has appeared may be more easily accepted without critical scrutiny than it would be by someone who is an atheist. It is similar with regard to experiences that might be considered paranormal in nature.

THE EMOTIONAL RESPONSE UNIT

Experiences accompanied by strong emotion may leave an unshakable belief in whatever explanation appealed to the individual at the time. If one is overwhelmed by an apparent case of telepathy, or an ostensible UFO, then later thinking may well be dominated by the awareness that the emotional reaction was intense, leading to the conclusion that something unusual really did happen. And emotion in turn may directly influence both perception and learning. Something may be interpreted as bizarre or unusual because of the emotional responses triggered.

Evidence is accumulating that our emotional responses may be triggered by information from the outside world even before we are consciously aware that something has happened. Take this example, provided by LeDoux (1994) in his recent article in *Scientific American* (1994, 270, pp. 50–57):

> An individual is walking through the woods when she picks up information— either auditory, such as rustling leaves, or visual, such as the sight of a slender curved object on the ground—which triggers a fear response. This information, even before it reaches the cortex, is processed in the amygdala, which arouses the body to an alarm footing. Somewhat later, when the cortex has had enough time to decide whether or not the object really is a snake, this cognitive information processing will either augment the fear response and corresponding evasive behaviour, or will serve to bring that response to a halt.

This is relevant to our understanding of paranormal experience, for very often an emotional experience accompanies the putatively paranormal. A strong coincidence may produce an emotional "zing" that points us toward a paranormal explanation, because normal events would not be expected to produce such emotion.

Our brains are also capable of generating wonderful and fantastic perceptual experiences for which we are rarely prepared. Out-of-body experiences (OBEs), hallucinations, near-death experiences (NDEs), peak experiences—these are all likely to be based, not in some external transcendental reality, but rather in the brain itself. We are not always able to distinguish material originating in the brain from material from the outside world, and thus we can falsely attribute to the external world perceptions and experiences that are created within the brain. We have little training with regard to such experience. As children, we do learn to distrust, for the most part, dreams and nightmares. Our parents and our culture tell us that they are products of our own brains. We are not prepared for more arcane experiences, such as OBEs or hallucinations or NDEs or peak experiences, and may be so unprepared that we are overwhelmed by the emotion and come to see such experience as deeply significant and "real" whether or not it is.

Ray Hyman has always cautioned skeptics not to be surprised should they one day have a very strong emotional experience that seems to cry out for paranormal explanation. Given the ways our brains work, we would expect such experiences from time to time. Unprepared for them, they could become conversion experiences that lead to strong belief. When I was a graduate student, another graduate student who shared my office, and who was equally as skeptical as I was about the paranormal, came to school one day overwhelmed by the realism and clarity of a dream he had had the night before. In it, his uncle in Connecticut had died. It had been a very emotional dream, and was so striking that Jack told me that if his uncle died anytime soon, he would no longer be able to maintain his skepticism about precognition—the dream experience was that powerful. Ten years later, his uncle was still alive, and Jack's skepticism had survived intact.

THE MEMORY UNIT

Through our own experience, we come to believe in the reliability of our memories and in our ability to judge whether a given memory is reliable or not. However, memory is a constructive process rather than a literal rendering of past experience, and memories are subject to serious biases and distortions.

Not only does memory involve itself in the processing of incoming information and the shaping of beliefs; it is itself influenced strongly by current perceptions and beliefs. Yet it is very difficult for an individual to reject the products of his or her own memory process, for memory can seem to be so "real."

THE ENVIRONMENTAL FEEDBACK UNIT

Beliefs help us to function. They guide our actions and increase or reduce our anxieties. If we operate on the basis of a belief, and if it "works" for us, even though faulty, why would we be inclined to change it? Feedback from the external world reinforces or weakens our beliefs, but since the beliefs themselves influence how that feedback is perceived, beliefs can become very resistant to contrary information and experience. If you really believe that alien abductions occur, then any evidence against that belief can be rationalized away—in terms of conspiracy theories, other people's ignorance, or whatever.

As mentioned earlier, fallacious beliefs can often be even more functional than those based in truth. For example, Shelley Taylor, in her book *Positive Illusions*, reports research showing that mildly depressed people are often more realistic about the world than are happy people. Emotionally healthy people live to some extent by erecting false beliefs—illusions—that reduce anxiety and aid well-being, whereas depressed individuals to some degree see the world more accurately. Happy people may underestimate the likelihood of getting cancer or being killed, and may avoid thinking about the ultimate reality of death, while depressed people may be much more accurate with regard to such concerns.

An important way in which to run reality checks on our perceptions and beliefs is to compare them with those of others. If I am the only one who interpreted a strange glow as an apparition, I am more likely to reconsider this interpretation than if several others share the same view. We often seek out people who agree with us, or selectively choose literature supporting our belief. If the majority doubts us, then even if only part of a minority we can collectively work to dispel doubt and find certainty. We can invoke conspiracies and coverups to explain an absence of confirmatory evidence. We may work to inculcate our beliefs in others, especially children. Shared beliefs can promote social solidarity and even a sense of importance for the individual and group.

IN CONCLUSION

Beliefs are generated by the belief engine without any automatic concern for truth. Concern for truth is a higher order acquired cognitive orientation that reflects an underlying philosophy which presupposes an objective reality that is not always perceived by our senses.

The belief engine chugs away, strengthening old beliefs, spewing out new ones, rarely discarding any. We can sometimes see the error or foolishness in other people's beliefs. It is very difficult to see the same in our own. We believe in all sorts of things, abstract and concrete—in the existence of the solar system, atoms, pizza, and five-star restaurants in Paris. Such beliefs are no different in principle from beliefs in fairies at the end of the garden, in ghosts in some deserted abbey, in werewolves, in satanic conspiracies, in miraculous cures, and so on. Such beliefs are all similar in form, all products of the same process, even though they vary widely in content. They may, however, involve greater or lesser involvement of the critical-thinking and emotional-response units.

Critical thinking, logic, reason, science—these are all terms that apply in one way or another to the deliberate attempt to ferret out truth from the tangle of intuition, distorted perception, and fallible memory. The true critical thinker accepts what few people ever accept—that one cannot routinely trust perceptions and memories. Figments of our imagination and reflections of our emotional needs can often interfere with or supplant the perception of truth and reality. Through teaching and encouraging critical thought our society will move away from irrationality, but we will never succeed in completely abandoning irrational tendencies, again because of the basic nature of the belief engine.

Experience is often a poor guide to reality. Skepticism helps us to question our experience and to avoid being too readily led to believe what is not so. We should try to remember the words of the late P. J. Bailey (in *Festus: A Country Town*): "Where doubt, there truth is—'tis her shadow."

21

HOW TO SELL A PSEUDOSCIENCE
Anthony R. Pratkanis

(E) very time I read the reports of new pseudosciences in the *Skeptical Inquirer* or watch the latest "In Search Of"-style television show I have one cognitive response, "Holy cow, how can anyone believe that?" Some recent examples include: "Holy cow, why do people spend $3.95 a minute to talk on the telephone with a 'psychic' who has never foretold the future?" "Holy cow, why do people believe that an all-uncooked vegan diet is natural and therefore nutritious?" "Holy cow, why would two state troopers chase the planet Venus across state lines thinking it was an alien spacecraft?" "Holy cow, why do people spend millions of dollars each year on subliminal tapes that just don't work?"

There are, of course, many different answers to these "holy cow" questions. Conjurers can duplicate pseudoscientific feats and thus show us how sleights of hand and misdirections can mislead (e.g., Randi 1982a, 1982b, 1989). Sociologists can point to social conditions that increase the prevalence of pseudoscientific beliefs (e.g., Lett 1992; Padgett and Jorgenson 1982; Victor 1993). Natural scientists can describe the physical properties of objects to show that what may appear to be supernatural is natural (e.g., Culver and Ianna 1988; Nickell 1983, 1993). Cognitive psychologists have identified common mental biases that often lead us to misinterpret social reality and to conclude in favor of supernatural phenomena (e.g., Blackmore 1992; Gilovich 1991; Hines 1988). These perspectives are useful in addressing the "holy cow" question; all give us a piece of the puzzle in unraveling this mystery.

I will describe how a social psychologist answers the holy cow question. Social psychology is the study of social influence—how human beings and their institutions influence and affect each other (see Aronson 1992; Aronson and Pratkanis 1993). For the past seven decades, social psychologists have been developing theories of social influence and have been testing the effectiveness of various persuasion tac-

---

This article first appeared in *Skeptical Inquirer* 19, no. 4 (July/August 1995): 19–25.

tics in their labs (see Cialdini 1984; Pratkanis and Aronson, 1992). It is my thesis that many persuasion tactics discovered by social psychologists are used every day, perhaps not totally consciously, by the promoters of pseudoscience (see Feynman 1985 or Hines 1988 for a definition of pseudoscience).

To see how these tactics can be used to sell flimflam, let's pretend for a moment that we wish to have our very own pseudoscience. Here are nine common propaganda tactics that should result in success.

Gerald Fried

1. CREATE A PHANTOM

The first thing we need to do is to create a phantom—an unavailable goal that looks real and possible; it looks as if it might be obtained with just the right effort, just the right belief, or just the right amount of money, but in reality it can't be obtained. Most pseudosciences are based on belief in a distant or phantom goal. Some examples of pseudoscience phantoms: meeting a space alien, contacting a dead relative at a séance, receiving the wisdom of the universe from a channeled dolphin, and improving one's bowling game or overcoming the trauma of rape with a subliminal tape.

Phantoms can serve as effective propaganda devices (Pratkanis and Farquhar 1992). If I don't have a desired phantom, I feel deprived and somehow less of a person. A pseudoscientist can take advantage of these feelings of inferiority by appearing to offer a means to obtain that goal. In a rush to enhance self-esteem, we suspend better judgment and readily accept the offering of the pseudoscience.

The trick, of course, is to get the new seeker to believe that the phantom is possible. Often the mere mention of the delights of a phantom will be enough to dazzle the new pseudoscience recruit (see Lund's 1925 discussion of wishful thinking). After all, who wouldn't want a better sex life, better health, and peace of mind, all from a $14.95 subliminal tape? The fear of loss of a phantom also can motivate us to accept it as real. The thought that I will never speak again to a cherished but dead loved one or that next month I may die of cancer can be so painful as to cause me to suspend my better judgment and hold out hope against hope that the medium can contact the dead or that Laetrile works. But at times the sell is harder, and that calls for our next set of persuasion tactics.

2. SET A RATIONALIZATION TRAP

The rationalization trap is based on the premise: Get the person committed to the cause as soon as possible. Once a commitment is made, the nature of thought changes. The committed heart is not so much interested in a careful evaluation of the merits of a course of action but in proving that he or she is right.

To see how commitment to a pseudoscience can be established, let's look at a bizarre case—mass suicides at the direction of cult leader Jim Jones. This is the ultimate "holy cow" question: "Why kill yourself and your children on another's command?" From outside the cult, it appears strange, but from the inside it seems natural. Jones began by having his followers make easy commitments (a gift to the church, attending Wednesday night service) and then increased the level of commitment—more tithes, more time in service, loyalty oaths, public admission of sins and punishment, selling of homes, forced sex, moving to Guyana, and then the suicide. Each step was really a small one. Outsiders saw the strange end-product; insiders experienced an ever increasing spiral of escalating commitment. (See Pratkanis and Aronson 1992 for other tactics used by Jones.)

This is a dramatic example, but not all belief in pseudoscience is so extreme. For example, there are those who occasionally consult a psychic or listen to a subliminal tape. In such cases, commitment can be secured by what social psychologists call the foot-in-the-door technique (Freedman and Fraser 1966). It works this way: You start with a small request, such as accepting a free chiropractic spine exam (Barrett 1993a), taking a sample of vitamins, or completing a free personality inventory. Then a larger request follows—a $1,000 chiropractic realignment, a vitamin regime, or an expensive seminar series. The first small request sets the commitment: Why did you get that bone exam, take those vitamins, or complete that test if you weren't interested and didn't think there might be something to it? An all too common response, "Well gosh, I guess I am interested." The rationalization trap is sprung.

Now that we have secured the target's commitment to a phantom goal, we need some social support for the newfound pseudoscientific beliefs. The next tactics are designed to bolster those beliefs.

3. MANUFACTURE SOURCE CREDIBILITY AND SINCERITY

Our third tactic is to manufacture source credibility and sincerity. In other words, create a guru, leader, mystic, lord, or other generally likable and powerful authority, one who people would be just plain nuts if they didn't believe. For example, practitioners of alternative medicine often have "degrees" as chiropractors or in homeopathy. Subliminal tape sellers claim specialized knowledge and training in such arts as hypnosis. Advocates of UFO sightings often become directors of "research centers." "Psychic detectives" come with long résumés of police service. Prophets claim past successes. For example, most of us "know" that Jeane Dixon predicted the assassination of President Kennedy but probably don't know that she also predicted a Nixon win in 1960. As modern public relations has shown us, credibility is easier to manufacture than we might normally think (see Ailes 1988; Dilenschneider 1990).

Source credibility is an effective propaganda device for at least two reasons. First, we often process persuasive messages in a half-mindless state—either because we are not motivated to think, don't have the time to consider, or lack the abilities to understand the issues (Petty and Cacioppo 1986). In such cases, the presence of a credible source can lead one to quickly infer that the message has merit and should be accepted.

Second, source credibility can stop questioning (Kramer and Alstad 1993). After all, what gives you the right to question a guru, a prophet, the image of the Mother Mary, or a sincere seeker of life's hidden potentials? I'll clarify this point with an example. Suppose I told you that the following statement is a prediction of the development of the atomic bomb and the fighter aircraft (see Hines 1988):

> They will think they have seen the Sun at night
> When they will see the pig half-man:
> Noise, song, battle fighting in the sky perceived,
> And one will hear brute beasts talking.

You probably would respond: "Huh? I don't see how you get the atomic bomb from that. This could just as well be a prediction of an in-flight showing of the Dr. Doolittle movie or the advent of night baseball at Wrigley field." However, attribute the statement to Nostradamus and the dynamics change. Nostradamus was a man who supposedly cured plague victims, predicted who would be pope, foretold the future of kings and queens, and even found a poor dog lost by the king's page (Randi 1993). Such a great seer and prophet can't be wrong. The implied message: The problem is with you; instead of questioning, why don't you suspend your faulty, linear mind until you gain the needed insight?

4. ESTABLISH A GRANFALLOON

Where would a leader be without something to lead? Our next tactic supplies the answer: Establish what Kurt Vonnegut (1976) terms a "granfalloon," a proud and meaningless association of human beings. One of social psychology's most remarkable findings is the ease with which granfalloons can be created. For example, the social psychologist Henri Tajfel merely brought subjects into his lab, flipped a coin, and randomly assigned them to be labeled either Xs or Ws (Tajfel 1981; Turner 1987). At the end of the study, total strangers were acting as if those in their granfalloon were their close kin and those in the other group were their worst enemies.

Granfalloons are powerful propaganda devices because they are easy to create and, once established, the granfalloon defines social reality and maintains social identities. Information is dependent on the granfalloon. Since most granfalloons quickly develop out-groups, criticisms can be attributed to those "evil ones" outside the group, who are thus stifled. To maintain a desired social identity, such as that of a seeker or a New Age rebel, one must obey the dictates of the granfalloon and its leaders.

The classic séance can be viewed as an ad-hoc granfalloon. Note what happens as you sit in the dark and hear a thud. You are dependent on the group led by a medium for the interpretation of this sound. "What is it? A knee against the table or my long lost Uncle Ned? The group believes it is Uncle Ned. Rocking the boat would be impolite. Besides, I came here to be a seeker."

Essential to the success of the granfalloon tactic is the creation of a shared social identity. In creating this identity, here are some things you might want to include:

(a) *rituals and symbols* (e.g., a dowser's rod, secret symbols, and special ways of preparing food): these not only create an identity, but provide items for sale at a profit.

(b) *jargon and beliefs* that only the in-group understands and accepts (e.g., thetans are impeded by engrams, you are on a cusp with Jupiter rising): jargon is an effective means of social control since it can be used to frame the interpretation of events.

(c) *shared goals* (e.g., to end all war, to sell the faith and related products, or to realize one's human potential): such goals not only define the group, but motivate action as believers attempt to reach them.

(d) *shared feelings* (e.g., the excitement of a prophecy that might appear to be true or the collective rationalization of strange beliefs to others): shared feelings aid in the *we* feeling.

(e) *specialized information* (e.g., the U.S. government is in a conspiracy to cover up UFOs): this helps the target feel special because he or she is "in the know."

(f) *enemies* (e.g., alternative medicine opposing the AMA and the FDA, subliminal-tape companies spurning academic psychologists, and spiritualists condemning Randi and other investigators): enemies are very important because you as a pseudoscientist will need scapegoats to blame for your problems and failures.

5. USE SELF-GENERATED PERSUASION

Another tactic for promoting pseudoscience and one of the most powerful tactics identified by social psychologists is self-generated persuasion—the subtle design of the situation so that the targets persuade themselves. During World War II, Kurt Lewin (1947) was able to get Americans to eat more sweetbreads (veal and beef organ meats) by having them form groups to discuss how they could persuade others to eat sweetbreads.

Retailers selling so-called nutritional products have discovered this technique by turning customers into salespersons (Jarvis and Barrett 1993). To create a multilevel sales organization, the "nutrition" retailer recruits customers (who recruit still more customers) to serve as sales agents for the product. Customers are recruited as a test of their belief in the product or with the hope of making lots of money (often to buy more products). By trying to sell the product, the customer-turned-salesperson becomes more convinced of its worth. One multilevel leader tells his new sales agents to "answer all objections with testimonials. That's the secret to motivating people" (Jarvis and Barrett 1993), and it is also the secret to convincing yourself.

6. CONSTRUCT VIVID APPEALS

Joseph Stalin once remarked: "The death of a single Russian soldier is a tragedy. A million deaths is a statistic." (See Nisbett and Ross 1980.) In other words, a vividly presented case study or example can make a lasting impression. For example, the pseudosciences are replete with graphic stories of ships and planes caught in the Bermuda Triangle, space aliens examining the sexual parts of humans, weird goings-on in Borley Rectory or Amityville, New York, and psychic surgeons removing cancerous tumors.

A vivid presentation is likely to be very memorable and hard to refute. No matter how many logical arguments can be mustered to counter the pseudoscience claim, there remains that one graphic incident that comes quickly to mind to prompt the response: "Yeah, but what about that haunted house in New York? Hard to explain that." By the way, one of the best ways to counter a vivid appeal is with an equally vivid counter appeal. For example, to counter stories about psychic surgeons in the Philippines, Randi (1982a) tells an equally vivid story of a psychic surgeon palming chicken guts and then pretending to remove them from a sick and now less wealthy patient.

7. USE PRE-PERSUASION

Pre-persuasion is defining the situation or setting the stage so you win, and sometimes without raising so much as a valid argument. How does one do this? At least three steps are important.

First, establish the nature of the issue. For example, to avoid the wrath of the FDA, advocates of alternative medicine define the issue as health freedom (you should have the right to the health alternative of your choice) as opposed to consumer protection or quality care. If the issue is defined as freedom, the alternative medicine advocate will win because "Who is opposed to freedom?" Another example of this technique is to create a problem or disease, such as reactive hypoglycemia or yeast allergy, that then just happens to be "curable" with whatever quackery you have to sell (Jarvis and Barrett 1993).

Another way to define an issue is through differentiation. Subliminal-tape companies use product differentiation to respond to negative subliminal-tape studies. The claim: "Our tapes have a special technique that makes them superior to other tapes that have been used in studies that failed to show the therapeutic value of subliminal tapes." Thus, null results are used to make a given subliminal tape look superior. The psychic network has taken a similar approach—"Tired of those phony psychics? Ours are certified," says the advertisement.

Second, set expectations. Expectations can lead us to interpret ambiguous information in a way that supports an original hypothesis (Greenwald, Pratkanis, Leippe, and Baumgardner 1986). For example, a belief in the Bermuda Triangle may lead us to interpret a plane crash off the coast of New York City as evidence for the Triangle's sinister effects (Kusche 1986; Randi 1982a). We recently conducted a study that showed how an expectation can lead people to think that subliminal tapes work when in fact they do not (Greenwald, Spangenberg, Pratkanis, and Eskenazi 1991; Pratkanis, Eskenazi, and Greenwald 1994; for a summary see Pratkanis 1992). In our study, expectations were established by mislabeling half the tapes (e.g., some subjects thought they had a subliminal tape to improve memory but really had one designed to increase self-esteem). The results showed that about half the subjects thought they improved (though they did not) based on how the tape was labeled (and not the actual content). The label led them to interpret their behavior in support of expectations, or what we termed an "illusory placebo" effect.

A third way to pre-persuade is to specify the decision criteria. For example, psychic supporters have developed guidelines on what should be viewed as acceptable evidence for paranormal abilities—such as using personal experiences as data, placing the burden of proof on the critic and not the claimant (see Beloff 1985), and above all else keeping James Randi and other psi-inhibitors out of the testing room. Accept these criteria and one must conclude that psi is a reality. The collaboration of Hyman and Honorton is one positive attempt to establish a fair playing field (Hyman and Honorton 1986).

8. FREQUENTLY USE HEURISTICS AND COMMONPLACES

My next recommendation to the would-be pseudoscientist is to use heuristics and commonplaces. Heuristics are simple if-then rules or norms that are widely accepted; for example, if it costs more it must be more valuable. Commonplaces are widely accepted beliefs that can serve as the basis of an appeal; for example, government health-reform should be rejected because politicians are corrupt (assuming political corruption is a widely accepted belief). Heuristics and commonplaces gain their power because they are widely accepted and thus induce little thought about whether the rule or argument is appropriate.

To sell a pseudoscience, liberally sprinkle your appeal with heuristics and commonplaces. Here are some common examples.

(a) The *scarcity heuristic,* or if it is rare it is valuable. The Psychic Friends Network costs a pricey $3.95 a minute and therefore must be valuable. On the other hand, an average University of California professor goes for about 27 cents per minute and is thus of little value.[1]

(b) The *consensus* or *bandwagon* heuristic, or if everyone agrees it must be true.

Subliminal tapes, psychic phone ads, and quack medicine (Jarvis and Barrett 1993) feature testimonials of people who have found what they are looking for (see Hyman 1993 for a critique of this practice).

(c) The *message length* heuristic, or if the message is long it is strong. Subliminal-tape brochures often list hundreds of subliminal studies in support of their claims. Yet most of these studies do not deal with subliminal influence and thus are irrelevant. An uninformed observer would be impressed by the weight of the evidence.

(d) The *representative* heuristic or if an object resembles another (on some salient dimension) then they act similarly. For example, in folk medicines the cure often resembles the apparent cause of the disease. Homeopathy is based on the notion that small amounts of substances that can cause a disease's symptoms will cure the disease (Barrett 1993b). The Chinese Doctrine of Signatures claims that similarity of shape and form determine therapeutic value; thus rhinoceros horns, deer antlers, and ginseng root look phallic and supposedly improve vitality (Tyler 1993).

(e) The *natural* commonplace, or what is natural is good and what is made by humans is bad. Alternative medicines are promoted with the word "natural." Psychic abilities are portrayed as natural, but lost, abilities. Organic food is natural. Of course mistletoe berries are natural too, and I don't recommend a steady diet of these morsels.

(f) The *goddess-within* commonplace, or humans have a spiritual side that is neglected by modern materialistic science. This commonplace stems from the medieval notion of the soul, which was modernized by Mesmer as animal magnetism and then converted by psychoanalysis into the powerful, hidden unconscious (see Fuller 1982, 1986). Pseudoscience plays to this commonplace by offering ways to tap the unconscious, such as subliminal tapes, to prove this hidden power exists through extrasensory perception (ESP) and psi, or to talk with the remnants of this hidden spirituality through channeling and the séance.

(g) The *science* commonplaces. Pseudosciences use the word "science" in a contradictory manner. On the one hand, the word "science" is sprinkled liberally throughout most pseudosciences: subliminal tapes make use of the "latest scientific technology"; psychics are "scientifically tested"; health fads are "on the cutting edge of science." On the other hand, science is often portrayed as limited. For example, one article in *Self* magazine (Sharp 1993) reported our subliminal-tapes studies (Greenwald et al. 1992; Pratkanis et al. 1994) showing no evidence that the tapes worked and then stated: "Tape makers dispute the objectivity of the studies. They also point out that science can't always explain the results of mainstream medicine either" (p. 194). In each case a commonplace about science is used: (1) "Science is powerful" and (2) "Science is limited and can't replace the personal." The selective use of these commonplaces allows a pseudoscience to claim the power of science but have a convenient out should science fail to promote the pseudoscience.

9. ATTACK OPPONENTS THROUGH INNUENDO AND CHARACTER ASSASSINATION

Finally, you would like your pseudoscience to be safe from harm and external attack. Given that the best defense is a good offense, I offer the advice of Cicero: "If you don't have a good argument, attack the plaintiff."

Let me give a personal example of this tactic in action. After our research showing that subliminal tapes have no therapeutic value was reported, my co-authors, Tony Greenwald, Eric Spangenberg, Jay Eskenazi, and I were the target of many innuendoes. One subliminal newsletter edited by Eldon Taylor, Michael Urban, and others (see the *International Society of Peripheral Learning Specialists Newsletter*, August 1991) claimed that our research was a marketing study designed not to test the tapes but to "demonstrate the influence of marketing practices on consumer perceptions." The article points out that the entire body of data presented by Greenwald represents a marketing dissertation by Spangenberg and questions why Greenwald is even an author. The newsletter makes other attacks as well, claiming that our research design lacked a control group, that we really found significant effects of the tapes, that we violated American Psychological Association ethics with a hint that an investigation would follow, that we prematurely reported our findings in a manner similar to those who prematurely announced cold fusion, and that we were conducting a "Willie Horton"-style smear campaign against those who seek to help Americans achieve their personal goals.

Many skeptics can point to similar types of attacks. In the fourteenth century, Bishop Pierre d'Arcis, one of the first to contest the authenticity of the Shroud of Turin, was accused by shroud promoters as being motivated by jealousy and a desire to possess the shroud (Nickell 1983: 15). Today, James Randi is described by supporters of Uri Geller as "a powerful psychic trying to convince the world that such powers don't exist so he can take the lead role in the psychic world" (Hines 1988: 91).

Why is innuendo such a powerful propaganda device? Social psychologists point to three classes of answers. First, innuendoes change the agenda of discussion. Note the "new" discussion on subliminal tapes isn't about whether these tapes are worth your money or not. Instead, we are discussing whether I am ethical or not, whether I am a competent researcher, and whether I even did the research.

Second, innuendoes raise a glimmer of doubt about the character of the person under attack. That doubt can be especially powerful when there is little other information on which to base a judgment. For example, the average reader of the subliminal newsletter I quoted probably knows little about me—knows little about the research and little about the peer review process that evaluated it, and doesn't know that I make my living from teaching college and not from the sale of subliminal tapes. This average reader is left with the impression of an unethical and incompetent scientist who is out of control. Who in their right mind would accept what that person has to say?

Finally, innuendoes can have a chilling effect (Kurtz 1992). The recipient begins to wonder about his or her reputation and whether the fight is worth it. The frivolous lawsuit is an effective way to magnify this chilling effect.

CAN SCIENCE BE SOLD WITH PROPAGANDA?

I would be remiss if I didn't address one more issue: Can we sell science with the persuasion tactics of pseudoscience? Let's be honest; science sometimes uses these tactics. For example, I carry in my wallet a membership card to the Monterey Bay Aquarium with a picture of the cutest little otter you'll ever see. I am in the otter granfalloon. On some occasions skeptics have played a little loose with their arguments and their name-calling. As just one example, see George Price's (1955) *Science* article attacking Rhine's and Soal's work on ESP—an attack that went well beyond the then available data. (See Hyman's [1985] discussion.)

I can somewhat understand the use of such tactics. If a cute otter can inspire a young child to seek to understand nature, then so be it. But we should remember that such tactics can be ineffective in promoting science if they are not followed up by involvement in the process of science—the process of questioning and discovering. And we should be mindful that the use of propaganda techniques has its costs. If we base our claims on cheap propaganda tactics, then it is an easy task for the pseudoscientist to develop even more effective propaganda tactics and carry the day.

More fundamentally, propaganda works best when we are half mindless, simplistic thinkers trying to rationalize our behavior and beliefs to ourselves and others. Science works best when we are thoughtful and critical and scrutinize claims carefully. Our job should be to promote such thought and scrutiny. We should be careful to select our persuasion strategies to be consistent with that goal.

NOTES

I thank Craig Abbott, Elizabeth A. Turner, and Marlene E. Turner for helpful comments on an earlier draft of this article.

1. Based on 50 weeks a year at an average salary of $49,000 and a work week of 61 hours (as reported in recent surveys of the average UC faculty work load). Assuming a work week of 40 hours, the average faculty makes 41 cents a minute.

REFERENCES

Ailes, R. 1988. *You Are the Message.* New York: Doubleday.

Aronson, E. 1992. *The Social Animal,* 6th ed. New York: W. H. Freeman.

Aronson, E., and A. R. Pratkanis. 1993. "What Is Social Psychology?" In *Social Psychology,* vol. 1, edited by E. Aronson and A. R. Pratkanis, xiii–xx. Cheltenham, Gloucestershire: Edward Elgar Publishing.

Barrett, S. 1993a. "The Spine Salesmen." In *The Health Robbers,* edited by S. Barrett and W. T. Jarvis, 161–90. Amherst, N.Y.: Prometheus Books.

———. 1993b. "Homeopathy: Is It Medicine?" In *The Health Robbers,* edited by S. Barrett and W. T. Jarvis, 191–202. Amherst, N.Y.: Prometheus Books.

Beloff, J. 1985. "What Is Your Counter-explanation? A Plea to Skeptics to Think Again." In *A Skeptic's Handbook of Parapsychology,* edited by P. Kurtz, 359–77. Amherst, N.Y.: Prometheus Books.

Blackmore, S. 1992. Psychic experiences: Psychic illusions. *Skeptical Inquirer* 16: 367–76. [Chapter 25 in this book.]

Cialdini, R. B. 1984. *Influence*. New York: William Morrow.

Culver, R. B., and P. A. Ianna. 1988. *Astrology: True or False?* Amherst, N.Y.: Prometheus Books.

Dilenschneider, R. L. 1990. *Power and Influence*. New York: Prentice-Hall.

Feynman, R. P. 1985. *Surely You're Joking, Mr. Feynman!* New York: Bantam Books.

Freedman, J., and S. Fraser. 1966. Compliance without pressure: The foot-in-the-door technique. *Journal of Personality and Social Psychology* 4: 195–202.

Fuller, R. C. 1982. *Mesmerism and the American Cure of Souls*. Philadelphia: University of Pennsylvania Press.

———. 1986. *Americans and the Unconscious*. New York: Oxford University Press.

Gilovich, T. 1991. *How We Know What Isn't So*. New York: Free Press.

Greenwald, A. G., E. R. Spangenberg, A. R. Pratkanis, and J. Eskenazi. 1991. Double-blind tests of subliminal self-help audiotapes. *Psychological Science* 2: 119–22.

Greenwald, A. G., A. R. Pratkanis, M. R. Leippe, and M. H. Baumgardner. 1986. Under what conditions does theory obstruct research progress? *Psychological Review* 93: 216–29.

Hines, T. 1988. *Pseudoscience and the Paranormal*. Amherst, N.Y.: Prometheus Books.

Hyman, R. 1985. "A Critical Historical Overview of Parapsychology." In *A Skeptic's Handbook of Parapsychology*, edited by P. Kurtz, 3–96. Amherst, N.Y.: Prometheus Books.

———. 1993. "Occult Health Practices." In *The Health Robbers*, edited by S. Barrett and W. T. Jarvis, 55–66. Amherst, N.Y.: Prometheus Books.

Hyman, R., and C. Honorton. 1986. A joint communique: The Psi Ganzfeld controversy. *Journal of Parapsychology* 56: 351–64.

Jarvis, W. T., and S. Barrett. 1993. "How Quackery Sells." In *The Health Robbers*, edited by S. Barrett and W. T. Jarvis, 1–22. Amherst, N.Y.: Prometheus Books.

Kramer, J., and D. Alstad. 1993. *The Guru Papers: Masks of Authoritarian Power*. Berkeley, Calif.: North Atlantic Books.

Kurtz, P. 1992. On being sued: The chilling of freedom of expression. *Skeptical Inquirer* 16: 114–17.

Kusche, L. 1986. *The Bermuda Triangle Mystery Solved*. Amherst, N.Y.: Prometheus Books.

Lett, J. 1992. The persistent popularity of the paranormal. *Skeptical Inquirer* 16, 381–88. [Chapter 23 in this book.]

Lewin, K. 1947. "Group Decision and Social Change." In *Readings in Social Psychology*, edited by T. M. Newcomb and E. L. Hartley, 330–44. New York: Holt.

Lund, F. H. 1925. The psychology of belief. *Journal of Abnormal and Social Psychology* 20: 63–81, 174–96.

Nickell, J. 1983. *Inquest on the Shroud of Turin*. Amherst, N.Y.: Prometheus Books.

———. 1993. *Looking for a Miracle*. Amherst, N.Y.: Prometheus Books.

Nisbett, R., and L. Ross. 1980. *Human Inference: Strategies and Shortcomings of Social Judgment*. Englewood Cliffs, N.J.: Prentice-Hall.

Padgett, V. R., and D. O. Jorgenson. 1982. Superstition and economic threat: Germany 1918–1940. *Personality and Social Psychology Bulletin* 8: 736–41.

Petty, R. E., and J. T. Cacioppo. 1986. *Communication and Persuasion: Central and Peripheral Routes to Attitude Change*. New York: Springer-Verlag.

Pratkanis, A. R. 1992. The cargo-cult science of subliminal persuasion. *Skeptical Inquirer* 16: 260–72. [Chapter 30 in this book.]

Pratkanis, A. R., and E. Aronson. 1992. *Age of Propaganda: Everyday Use and Abuse of Persuasion*. New York: W. H. Freeman.

Pratkanis, A. R., J. Eskenazi, and A. G. Greenwald. 1994. What you expect is what you believe (but not necessarily what you get): A test of the effectiveness of subliminal self-help audiotapes. *Basic and Applied Social Psychology* 15: 251–76.

Pratkanis, A. R., and P. H. Farquhar. 1992. A brief history of research on phantom alternatives: Evidence for seven empirical generalizations about phantoms. *Basic and Applied Social Psychology* 13: 103–22.

Price, G. R. 1955. Science and the supernatural. *Science* 122: 359–67.

Randi, J. 1982a. *Flim-Flam!* Amherst, N.Y.: Prometheus Books.

———. 1982b. *The Truth About Uri Geller.* Amherst, N.Y.: Prometheus Books.

———. 1989. *The Faith Healers.* Amherst, N.Y.: Prometheus Books.

———. 1993. *The Mask of Nostradamus.* Amherst, N.Y.: Prometheus Books.

Sharp, K. 1993. The new hidden persuaders. *Self,* March: 174–75, 194.

Tajfel, H. 1981. *Human Groups and Social Categories.* Cambridge, U.K.: Cambridge University Press.

Turner, J. C. 1987. *Rediscovering the Social Group.* New York: Blackwell.

Tyler, V. E. 1993. "The Overselling of Herbs." In *The Health Robbers,* edited by S. Barrett and W. T. Jarvis, 213–24. Amherst, N.Y.: Prometheus Books.

Victor, J. S. 1993. *Satanic Panic: The Creation of a Contemporary Legend.* Chicago: Open Court.

Vonnegut, K. 1976. *Wampeters, Foma, and Granfalloons.* New York: Dell.

22

LIKE GOES WITH LIKE: THE ROLE OF REPRESENTATIVENESS IN PARANORMAL BELIEF

Thomas Gilovich and Kenneth Savitsky

I t was in 1983, at an infectious-disease conference in Brussels, that Barry Marshall, an internal-medicine resident from Perth, Australia, first staked his startling claim. He argued that the peptic ulcer, a painful crater in the lining of the stomach or duodenum, was not caused by a stressful lifestyle as everyone had thought. Instead, the malady that afflicts millions of adults in the United States alone was caused by a simple bacterium, and thus could be cured using antibiotics (Hunter 1993; Monmaney 1993; Peterson 1991; Wandycz 1993).

Although subsequent investigations have substantiated Marshall's claim (e.g., Hentschel et al. 1993), his colleagues initially were highly skeptical. Martin Blaser, director of the Division of Infectious Diseases at the Vanderbilt University School of Medicine, described Marshall's thesis as "the most preposterous thing I'd ever heard" (Monmaney 1993).

What made the idea so preposterous? Why were the experts so resistant to Marshall's suggestion? There were undoubtedly many reasons. For one, the claim contradicted what most physicians, psychiatrists, and psychologists knew (or thought they knew): Ulcers were caused by stress. As one author noted, "No physical ailment has ever been more closely tied to psychological turbulence" (Monmaney 1993, 64). In addition, science is necessarily and appropriately a rather conservative enterprise. Although insight, creativity, and even leaps of faith are vital to the endeavor, sound empirical evidence is the true coin of the realm. Much of the medical establishment's hesitation doubtless stemmed from the same healthy skepticism that readers of the *Skeptical Inquirer* have learned to treasure. After all, Marshall's results at the time were suggestive at best—no cause-effect relationship had yet been established.

But there may have been a third reason for the reluctance to embrace Marshall's contention, a reason we explore in this article. The belief that ulcers derive

---

This article first appeared in *Skeptical Inquirer* 20, no. 2 (March/April 1996): 34–40.

from stress is particularly seductive—for physicians and laypersons alike—because it flows from a general tendency of human judgment, a tendency to employ what psychologists Amos Tversky and Daniel Kahneman have called the "representativeness heuristic" (Kahneman and Tversky 1972, 1973; Tversky and Kahneman 1974, 1982). Indeed, we believe that judgment by representativeness plays a role in a host of erroneous beliefs, from beliefs about health and the human body to handwriting analysis and astrology (Gilovich 1991). We consider a sample of these beliefs in this article.

THE REPRESENTATIVENESS HEURISTIC

Representativeness is but one of a number of heuristics that people use to render complex problems manageable. Heuristics are often described as judgmental shortcuts that generally get us where we need to go—and quickly—but at the cost of occasionally sending us off course. Kahneman and Tversky liken them to perceptual cues, which generally enable us to perceive the world accurately, but occasionally give rise to misperception and illusion. Consider their example of using clarity as a cue for distance. The clarity of an object is one cue people use to decide how far away it is. The cue typically works well because the farther away something is, the less distinct it appears. On a particularly clear day, however, objects can appear closer than they are, and on hazy days they can appear farther away. In some circumstances, then, this normally accurate cue can lead to error.

Representativeness works much the same way. The representativeness heuristic involves a reflexive tendency to assess the similarity of objects and events along salient dimensions and to organize them on the basis of one overarching rule: "Like goes with like." Among other things, the representativeness heuristic reflects the belief that a member of a given category ought to resemble the category prototype, and that an effect ought to resemble the cause that produced it. Thus, the representativeness heuristic is often used to assess whether a given instance belongs to a particular category, such as whether an individual is likely to be an accountant or a comedian. It is also used in assigning causes to effects, as when deciding whether a meal of spicy food caused a case of heartburn or determining whether an assassination was the product of a conspiracy.[1]

Note that judgment by representativeness often works well. Instances often resemble their category prototypes and causes frequently resemble their effects. Members of various occupational groups, for example, frequently do resemble the group prototype. Likewise, "big" effects (such as the development of the atomic bomb) are often brought about by "big" causes (such as the Manhattan Project).

Still, the representativeness heuristic is only that—a heuristic or shortcut. As with all shortcuts, the representativeness heuristic should be used with caution. Although it can help us to make some judgments with accuracy and ease, it can also lead us astray. Not all members fit the category prototype. Some comedians are shy or taciturn, and some accountants are wild and crazy. And although causes are frequently representative of their effects, this relationship does not always hold: Tiny viruses give rise to devastating epidemics like malaria or AIDS; and splitting the nucleus of an atom releases an awesome amount of energy. In some cases, then, representativeness yields inaccuracy and error. Or even superstition. A nice example is provided by craps shooters, who roll the dice gently to coax a low number, and more vigorously to encourage a high one (Hanslin 1967). A small effect (low number) requires a small cause (gentle roll), and a big effect (high number) requires a big cause (vigorous roll).

How might the belief in a stress-ulcer link derive from the conviction that like goes with like? Because the burning feeling of an ulcerated stomach is not unlike the gut-wrenching, stomach-churning feeling of extreme stress (albeit more severe), the link seems natural: Stress is a representative cause of an ulcer.[2] But as Marshall suggested (and subsequent research has borne out), the link may be overblown. Stress alone does not appear to cause ulcers (Glavin and Szabo 1992; Soll 1990).

REPRESENTATIVENESS AND THE CONJUNCTION FALLACY

One of the most compelling demonstrations of how the representativeness heuristic can interfere with sound judgment comes from a much-discussed experiment in which participants were asked to consider the following description (Tversky and Kahneman 1982, 1983):

> Linda is 31 years old, single, outspoken, and very bright. She majored in philosophy. As a student, she was deeply concerned with issues of discrimination and social justice, and also participated in anti-nuclear demonstrations.
>
> Now, based on the above description, rank the following statements about Linda, from most to least likely:
> a. Linda is an insurance salesperson.
> b. Linda is a bank teller.
> c. Linda is a bank teller and is active in the feminist movement.

If you are like most people, you probably thought it was more likely that "Linda is a bank teller and is active in the feminist movement" than that "Linda is a bank teller." It is easy to see why: A feminist bank teller is much more representative of the description of Linda than is "just" a bank teller. It reflects the political activism, social-consciousness, and left-of-center politics implied in the description.

It may make sense, but it cannot be. The category "bank teller" subsumes the category "is a bank teller and is active in the feminist movement." The latter therefore cannot be more likely than the former. Anyone who is a bank teller and is active in the feminist movement is automatically also a bank teller. Indeed, even if one thinks it is impossible for someone with Linda's description to be solely a bank teller (that is, one who is not a feminist), being a bank teller is still *as* likely as being both. This error is referred to as the "conjunction fallacy" because the probability of two events co-occurring (i.e., their conjunction) can never exceed the individual probability of either of the constituents (Tversky and Kahneman 1982, 1983; Dawes and Mulford 1993).

Such is the logic of the situation. The psychology we bring to bear on it is something else. If we start with an unrepresentative outcome (being a bank teller) and then add a representative element (being active in the feminist movement), we create a description that is at once more psychologically compelling but objectively less likely. The rules of representativeness do not follow the laws of probability. A detailed description can seem compelling precisely because of the very details that, objectively speaking, actually make it less likely. Thus, someone may be less concerned about dying during a trip to the Middle East than about dying in a terrorist attack while there, even though the probability of death due to a *particular* cause is obviously lower than the probability of death due to the set of all possible causes. Likewise, the probability of global economic collapse can seem remote until one sketches a detailed scenario in which such a collapse follows, say, the destruction of the oil fields in the Persian Gulf. Once again, the additional details make the outcome less likely at the same time that they make it more psychologically compelling.

REPRESENTATIVENESS AND CAUSAL JUDGMENTS

Most of the empirical research on the representativeness heuristic is similar to the work on the conjunction fallacy in that the judgments people make are compared to a normative standard—in this case, to the laws of probability. The deleterious effect of judgment by representativeness is thereby established by the failure to meet such a standard. Previous work conducted in this fashion has shown, for example, that judgment by representativeness leads people to commit the "gambler's fallacy," to overestimate the reliability of small samples of data, and to be insufficiently "regressive" in making predictions under conditions of uncertainty.

The ulcer example with which we began this article does not have this property of being obviously at variance with a clear-cut normative standard. The same is true of nearly all examples of the impact of representativeness on causal judgments: It can be difficult to establish with certainty that a judgmental error has been made. Partly for this reason, there has been less empirical research on representativeness and causal judgments than on other areas, such as representativeness and the conjunction fallacy. This is not because representativeness is thought to have little impact on causal judgments, but because without a clear-cut normative standard it is simply more difficult to conduct research in this domain. The research that has been conducted, furthermore, is more suggestive than definitive. Nonetheless, the suggestive evidence is rather striking, and it points to the possibility that representativeness may exert at least as much influence over causal judgments as it does over other, more exhaustively researched types of judgments. To see how much, we discuss some examples of representativeness-thinking in medicine, in pseudoscientific systems, and in psychoanalysis.

REPRESENTATIVENESS AND MEDICAL BELIEFS

One area in which the impact of representativeness on causal judgments is particularly striking is the domain of health and medicine. Historically, people have often assumed that the symptoms of a disease should resemble either its cause or its cure (or both). In ancient Chinese medicine, for example, people with vision problems were fed ground bat in the mistaken belief that bats had particularly keen vision and that some of this ability might be transferred to the recipient (Deutsch 1977). Evans-Pritchard (1937) noted many examples of the influence of representativeness among the African Azande (although he discussed them in the context of magical-thinking, not representativeness). For instance, the Azande used the ground skull of the red bush monkey to cure epilepsy. Why? The cure should resemble the disease, so the herky-jerky movements of the monkey make the essence of monkey appear to be a promising candidate to settle the violent movements of an epileptic seizure. As Evans-Pritchard (quoted in Nisbett and Ross 1980, 116) put it:

> Generally the logic of therapeutic treatment consists in the selection of the most prominent external symptoms, the naming of the disease after some object in nature it resembles, and the utilization of the object as the principal ingredient in the drug administered to cure the disease. The circle may even be completed by belief that the external symptoms not only yield to treatment by the object which resembles them but are caused by it as well.

Western medical practice has likewise been guided by the representativeness heuristic. For instance, early Western medicine was strongly influenced by what was known as the "doctrine of signatures," or the belief that "every natural substance which possesses any medicinal virtue indicates by an obvious and well-marked external character the disease for which it is a remedy, or the object for which it should be employed" (quoted in Nisbett and Ross 1980, 116). Thus, physicians prescribed the lungs of the fox (known for its endurance) for asthmatics, and the yellow spice turmeric for jaundice. Again, disease and cure are linked because they resemble one another.

Or consider the popularity of homeopathy, which derives from the eighteenth century work of the German physician Samuel Hahnemann (Barrett 1987). One of the bedrock principles of homeopathy is Hahnemann's "law of similars," according to which the key to discovering what substance will cure a particular disorder lies in noting the effect that various substances have on healthy people. If a substance causes a particular reaction in an unafflicted person, then it is seen as a likely cure for a disease characterized by those same symptoms. As before, the external symptoms of a disease are used to identify a cure for the disease—a cure that manifests the same external characteristics.

Of course, there are instances in which substances that cause particular symptoms *are* used effectively as part of a therapeutic regimen to cure, alleviate, or prevent those very symptoms. Vaccines deliver small quantities of disease-causing viruses to help individuals develop immunities. Likewise, allergy sufferers sometimes receive periodic doses of the exact substance to which they are allergic so that they will develop a tolerance over time. The problem with the dubious medical practices described above is the *general* assumption that the symptoms of a disease should resemble its cause, its cure, or both. Limiting the scope of possible cures to those that are representative of the disease can seriously impede scientific discovery. Such a narrow focus, for example, would have inhibited the discovery of the two most significant developments of modern medicine: sanitation and antibiotics.

Representativeness-thinking continues to abound in modern "alternative" medicine, a pursuit that appears to be gaining in perceived legitimacy (Cowley, King, Hager, and Rosenberg 1995). An investigation by Congress into health fraud and quackery noted several examples of what appear to be interventions inspired by the superficial appeal of representativeness (U.S. Congress, House Subcommittee on Health and Long-Term Care 1984). In one set of suggested treatments, patients are encouraged to eat raw organ concentrates corresponding to the dysfunctional body part: e.g., brain concentrates for mental disorders, heart concentrates for cardiac conditions, and raw stomach lining for ulcers. Similarly, the fingerprints of representativeness are all over the practice of "rebirthing," a New Age therapeutic technique in which individuals attempt to reenact their own births in an effort to correct personality defects caused by having been born in an "unnatural" fashion (Ward 1994). One person who was born breech (i.e., feet first) underwent the rebirthing procedure to cure his sense that his life was always going in the wrong direction and that he could never seem to get things "the right way round." Another, born Caesarean, sought the treatment because of a lifelong difficulty with seeing things to completion, and always relying on others to finish tasks for her. As one author quipped, "God knows what damage forceps might inflict . . . a lifelong neurosis that you're being dragged where you don't want to go?" (Ward 1994, 90).

A more rigorous examination of the kind of erroneous beliefs about health and the human body that can arise from the appeal of representativeness has dealt with the adage, "You are what you eat." Just how far do people take this idea? In certain respects, the saying is undeniably true: Bodies are composed to a large extent of the molecules that were once ingested as food. Quite literally, we are what we have eaten. Indeed, there are times when we take on the character of what we ingest: People gain weight by eating fatty foods, and a person's skin can acquire an orange tint from the carotene found in carrots and tomatoes. But the notion that we develop the characteristics of the food we eat sometimes goes beyond such examples to almost magical extremes. The Hua of Papua New Guinea, for example, believe that individuals will grow quickly if they eat rapidly growing food (Meigs 1984, cited by Nemeroff and Rozin 1989).

But what about a more "scientifically minded" population? Psychologists Carol Nemeroff and Paul Rozin (1989) asked college students to consider a hypothetical culture known as the "Chandorans," who hunt wild boar and marine turtles. Some of the students learned that the Chandorans hunt turtles for their shells, and wild boar for their meat. The others heard the opposite: The tribe hunts turtles for their meat, and boar for their tusks.

After reading one of the two descriptions of the Chandorans, the students were asked to rate the tribe members on numerous characteristics. Their responses reflected a belief that the characteristics of the food that was eaten would "rub off" onto the tribe members. Boar-eaters were thought to be more aggressive and irritable than their counterparts—and more likely to have beards! The turtle-eaters were thought to live longer and be better swimmers.

However educated a person may be (the participants in Nemeroff and Rozin's experiment were University of Pennsylvania undergraduates), it can be difficult to get beyond the assumption that like goes with like. In this case, it leads to the belief that individuals tend to acquire the attributes of the food they ingest. Simple representativeness.

REPRESENTATIVENESS AND PSEUDOSCIENTIFIC BELIEFS

A core tenet of the field of astrology is that an individual's personality is influenced by the astrological sign under which he or she was born (Huntley 1990). A glance at the personality types associated with the various astrological signs reveals an uncanny concordance between the supposed personality of someone with a particular sign and the characteristics associated with the sign's namesake (Huntley 1990; Howe 1970; Zusne and Jones 1982). Those born under the sign of the goat (Capricorn) are said to be tenacious, hardworking, and stubborn; whereas those born under the lion (Leo) are proud, forceful leaders. Likewise, those born under the sign of Cancer (the crab) share with their namesake a tendency to appear hard on the outside; while inside their "shells" they are soft and vulnerable. One treatment of astrology goes so far as to suggest that, like the crab, those born under the sign of Cancer tend to be "deeply attached to their homes" (Read et al. 1978).

What is the origin of these associations? They are not empirically derived, as they have been shown time and time again to lack validity (e.g., Carlson 1985; Dean 1987; for reviews see Abell 1981; Schick and Vaughn 1995; Zusne and Jones 1982). Instead, they are conceptually driven by simple, representativeness-based assessments of the personalities that *should* be associated with various astrological signs. After all, who is more likely to be retiring and modest than a Virgo (the virgin)? Who better to be well balanced, harmonious, and fair than a Libra (the scales)? By taking advantage of people's reflexive associations, the system gains plausibility among those disinclined to dig deeper.

And it doesn't stop there. Consider another elaborate "scientific" system designed to assess the "secrets" of an individual's personality—graphology, or handwriting analysis. Corporations pay graphologists sizable fees to help screen job applicants by developing personality profiles of those who apply for jobs (Neter and Ben-Shakhar 1989). Graphologists are also called upon to provide "expert" testimony in trial proceedings, and to help the Secret Service determine if any real danger is posed by threatening letters to government officials (Scanlon and Mauro 1992). How much stock can we put in the work of handwriting analysts?

Unlike astrology, graphology is not worthless. It has been, and continues to be, the subject of careful empirical investigation (Nevo 1986), and it has been shown that people's handwriting can reveal certain things about them. Particularly shaky writing can be a clue that an individual suffers from some neurological disorder that causes hand tremors; whether a person is male or female is often apparent from his or her writing. In general, however, what handwriting analysis can determine most reliably tends to be things that can be more reliably ascertained through other means. As for the "secrets" of an individual's personality, graphology has yet to show that it is any better than astrology.

This has not done much to diminish the popularity of handwriting analysis, however. One reason for this is that graphologists, like astrologers, gain some surface plausibility or "face validity" for their claims by exploiting the tendency for people to employ the representativeness heuristic. Many of their claims have a superficial "sensible" quality, rarely violating the principle that like goes with like. Consider, for instance, the "zonal theory" of graphology, which divides a person's handwriting into the upper, middle, and lower regions. A person's "intellectual," "practical," and "instinctual" qualities supposedly correspond to the different regions (Basil 1989). Can you guess which is which? Could our "lower" instincts be reflected anywhere other than the lower region, or our "higher" intellect anywhere other than the top?

The list of such representativeness-based "connections" goes on and on. Handwriting slants to the left? The person must be holding something back, repressing his or her true emotions. Slants to the right? The person gets carried away by his or her feelings. A signature placed far below a paragraph suggests that the individual wishes to distance himself or herself from what was written (Scanlon and Mauro 1992). Handwriting that stays close to the left margin belongs to individuals attached to the past, whereas writing that hugs the right margin comes from those oriented toward the future.

What is ironic is that the very mechanism that many graphologists rely upon to argue for the persuasive value of their endeavor—that the character of the handwriting resembles the character of the person—is what ultimately betrays them: They call it "common sense"; we call it judgment by representativeness.

REPRESENTATIVENESS AND PSYCHOANALYSIS

Two prominent social psychologists, Richard Nisbett and Lee Ross, have argued that "the enormous popularity of Freudian theory probably lies in the fact that, unlike all its competitors among contemporary views, it encourages the layperson to do what comes naturally in causal explanation, that is, to use the representativeness heuristic" (Nisbett and Ross 1980, 244). Although this claim would be difficult to put to empirical test, there can be little doubt that much of the interpretation of symbols that lies at the core of psychoanalytic theory is driven by representativeness. Consider the interpretation of dreams, in which the images a client reports from his or her dreams are considered indicative of underlying motives. An infinite number of potential relationships exist between dream content and underlying psychodynamics, and it is interesting that virtually all of the "meaningful" ones identified by psychodynamically oriented clinicians are ones in which there is an obvious fit or resemblance between the reported image and inner dynamics. A man who dreams of a snake or a cigar is thought to be troubled by his penis or his sexuality. People who dream of policemen are thought to be concerned about their fathers or authority figures. Knowledge of the representativeness heuristic compels one to wonder whether such connections reflect something important about the psyche of the client, or whether they exist primarily in the mind of the therapist.

One area of psychodynamic theorizing in which the validity of such superficially plausible relationships has been tested and found wanting is the use of projective tests. The most widely known projective test is the Rorschach, in which clients report what they "see" in ambiguous blotches of ink on cards. As in all projective tests, the idea is that in responding to such an unstructured stimulus, a person must "project," and thus reveal, some of his or her inner dynamics. Countless studies, however, have failed to produce evidence that the test is valid—that is, that the assessments made about people on the basis of the test correspond to the psychopathological conditions from which they suffer (Burros 1978).[3]

The research findings notwithstanding, clinicians frequently report the Rorschach to be extremely helpful in clinical practice. Might representativeness contribute to this paradox of strongly held beliefs coexisting with the absence of any real relationship? You be the judge. A person who interprets the whole Rorschach card, and not its specific details, is considered by clinicians to suffer from a need to form a "big picture," and a tendency toward grandiosity, even paranoia. In contrast, a person who refers only to small details of the ink blots is considered to have an obsessive personality—someone who attends to detail at the expense of the more important holistic aspects (Dawes 1994). Once again, systematic research has failed to find evidence for these relationships, but the sense of representativeness gives them some superficial plausibility.

CONCLUSION

We have described numerous erroneous beliefs that appear to derive from the overuse of the representativeness heuristic. Many of them arise in domains in which the reach for solutions to important problems exceeds our grasp—such as the attempt to uncover (via astrology or handwriting analysis) simple cues to the complexities of human motivation and personality. In such domains in which no simple solutions exist, and yet the need or desire for such solutions remains strong, people often let down their guard. Dubious cause-effect links are then uncritically accepted because they satisfy the principle of like goes with like.

Representativeness can also have the opposite effect, inhibiting belief in valid claims that violate the expectation of resemblance. People initially scoffed at Walter Reed's suggestion that malaria was carried by the mosquito. From a representativeness standpoint, it is easy to see why: The cause (a tiny mosquito) is not at all representative of the result (a devastating disease). Reed's claim violated the notion that big effects should have big causes, and thus was difficult to accept (Nisbett and Ross 1980). Although skepticism is a vital component of critical thought, it should not be based on an excessive adherence to the principle that like goes with like.

Indeed, it is often those discoveries that violate the expected resemblance between cause and effect that are ultimately hailed as significant breakthroughs, as with the discovery of *Helicobacter pylori*, as the ulcer-causing bacterium is now named. As one author put it, "The discovery of *Helicobacter* is no crummy little shift. It's a mindblower—tangible, reproducible, unexpected, and, yes, revolutionary. Just the fact that a bug causes peptic ulcers, long considered the cardinal example of a psychosomatic illness, is a spear in the breast of New Age medicine" (Monmaney 1993, 68). Given these stakes, one might be advised to avoid an over-reliance on the shortcut of representativeness, and instead to devote the extra effort needed to make accurate judgments and decisions. (But not too much effort—you wouldn't want to give yourself an ulcer.)

NOTES

We thank Dennis Regan for his helpful comments on an earlier draft of this article.

1. The reason that the heuristic has been dubbed "representativeness" rather than, say, "resemblance" or "similarity" is that it also applies in circumstances in which the assessment of "fit" is not based on similarity. For example, when assessing whether a series of coin flips was produced by tossing a fair coin, people's judgments are influenced in part by whether the sequence is representative of one produced by a fair coin. A sequence of five heads and five tails is a representative outcome, but a sequence of nine heads and one tail is not. Note, however, that a fifty-fifty split does not make the sequence "similar" to a fair coin, but it does make it representative of one.

2. Some theories of the link between stress and ulcers are even more tinged with representativeness. Since the symptoms of an ulcer manifest themselves in the stomach, the cause "should" involve something that is highly characteristic of the stomach as well, such as hunger and nourishment. Thus, one theorist asserts, "The critical factor in the development of ulcers is the frustration associated with the wish to receive love—when this wish is rejected, it is converted into a wish to be fed," leading ultimately "to an ulcer." Echoing such ideas, James Masterson writes in his book *The Search for the Real Self* that ulcers affect those who are "hungering for emotional supplies that were lost in childhood or that were never sufficient to nourish the real self" (both quoted in Monmaney 1993).

3. Actually, a nonprojective use of the Rorschach, called the Exner System, has been shown to have some validity (Exner 1986). The system is based on the fact that some of the inkblots *do* look like various objects, and a person's responses are scored for the number and proportion that fail to reflect this correspondence. Unlike the usual Rorschach procedure, which is subjectively scored, the Exner system is a standardized test.

REFERENCES

Abell, G. O. 1981. "Astrology." In *Science and the Paranormal: Probing the Existence of the Supernatural*, edited by G. O. Abell and B. Singer. New York: Charles Scribner's Sons.

Barrett, S. 1987. Homeopathy: Is it medicine? *Skeptical Inquirer* 12(1) (Fall): 56–62.

Basil, R. 1989. Graphology and personality: Let the buyer beware. *Skeptical Inquirer* 13 (3) (Spring): 241–43.

Burros, O. K. 1978. *Mental Measurement Yearbook*. 8th ed. Highland Park, N.J.: Gryphon Press.

Carlson, S. 1985. A double-blind test of astrology. *Nature* 318: 419–25.

Cowley, G., P. King, M. Hager, and D. Rosenberg. 1995. Going mainstream. *Newsweek* (June 26): 56–57.

Dawes, R. M. 1994. *House of Cards: Psychology and Psychotherapy Built on Myth*. New York: Free Press.

Dawes, R. M., and M. Mulford. 1993. Diagnoses of alien kidnappings that result from conjunction effects in memory. *Skeptical Inquirer* 18(1) (Fall): 50–51.

Dean, G. 1987. Does astrology need to be true? Part 2: The answer is no. *Skeptical Inquirer* 11(3) (Spring): 257–73.

Deutsch, R.M. 1977. *The New Nuts among the Berries: How Nutrition Nonsense Captured America*. Palo Alto, Calif.: Ball Publishing.

Evans-Pritchard, E. E. 1937. *Witchcraft, Oracles and Magic among the Azande*. Oxford: Clarendon.

Exner, J. E. 1986. *The Rorschach: A Comprehensive System*. 2d ed. New York: John Wiley.

Gilovich, T. 1991. *How We Know What Isn't So: The Fallibility of Human Reason in Everyday Life*. New York: The Free Press.

Glavin, G. B., and S. Szabo. 1992. Experimental gastric mucosal injury: Laboratory models reveal mechanisms of pathogenesis and new therapeutic strategies. *FASEB Journal* 6: 825–31.

Hanslin, J. M. 1967. Craps and magic. *American Journal of Sociology* 73: 316–30.

Hentschel, E., G. Brandstatter, B. Dragosics, A. M. Hirschel, H. Nemec, K. Schutze, M. Taufer, and H. Wurzer. 1993. Effect of ranitidine and amoxicillin plus metronidazole on the eradication of Helicobacter pylori and the recurrence of duodenal ulcer. *New England Journal of Medicine* 328: 308–12.

Howe, E. 1970. Astrology. In *Man, Myth, and Magic: An Illustrated Encyclopedia of the Supernatural*, edited by R. Cavendish. New York: Marshall Cavendish.

Hunter, B. T. 1993. Good news for gastric sufferers. *Consumer's Research* 76 (October): 8–9.

Huntley, J. 1990. *The Elements of Astrology*. Shaftesbury, Dorset, Great Britain: Element Books.

Kahneman, D., and A. Tversky. 1972. Subjective probability: A judgment of representativeness. *Cognitive Psychology* 3: 430–54.

———. 1973. On the psychology of prediction. *Psychological Review* 80: 237–51.

Meigs, A. S. 1984. *Food, Sex, and Pollution: A New Guinea Religion*. New Brunswick, N.J.: Rutgers University Press.

Monmaney, T. 1993. Marshall's hunch. *New Yorker* 69 (September 20): 64–72.

Nemeroff, C., and P. Rozin. 1989. 'You are what you eat': Applying the demand-free 'impressions' technique to an unacknowledged belief. *Ethos* 17: 50–69.

Neter, E., and G. Ben-Shakhar. 1989. The predictive validity of graphological inferences: A meta-analytic approach. *Personality and Individual Differences* (10): 737–45.

Nevo, B., ed. 1986. *Scientific Aspects of Graphology: A Handbook*. Springfield, Ill.: Charles C. Thomas.

Nisbett, R., and L. Ross. 1980. *Human Inference: Strategies and Shortcomings of Social Judgment.* Englewood Cliffs, N.J.: Prentice-Hall.

Peterson, W. L. 1991. Helicobacter pylori and peptic ulcer disease. *New England Journal of Medicine* 324: 1043–48.

Read, A. W. et al., eds. 1978. *Funk and Wagnall's New Comprehensive International Dictionary of the English Language.* New York: Publishers Guild Press.

Scanlon, M., and J. Mauro. 1992. The lowdown on handwriting analysis: Is it for real? *Psychology Today* (November/December): 46–53; 80.

Schick, T., and L. Vaughn. 1995. *How to Think about Weird Things: Critical Thinking for a New Age.* Mountain View, Calif.: Mayfield Publishing Company.

Soll, A. H. 1990. Pathogenesis of peptic ulcer and implications for therapy. *New England Journal of Medicine* 322: 909–16.

Tversky, A., and D. Kahneman. 1974. Judgment under uncertainty: Heuristics and biases. *Science* 185: 1124–31.

———. 1982. "Judgments of and by Representativeness." In *Judgment under Uncertainty: Heuristics and Biases,* edited by D. Kahneman, P. Slovic, and A. Tversky. Cambridge: Cambridge University Press.

———. 1983. Extensional versus intuitive reasoning: The conjunction fallacy in probability judgment. *Psychological Review* 90: 293–315.

U.S. Congress. 1984. *Quackery: A $10 Billion Scandal: A Report by the Chairman of the (House) Subcommittee on Health and Long-Term Care.* Washington, D.C.: United States Government Printing Office.

Wandycz, K. 1993. The H. pylori factor. *Forbes* 152 (August 2): 128.

Ward, R. 1994. Maternity ward. *Mirabella* (February): 89–90.

Zusne, L. and W. H. Jones 1982. *Anomalistic Psychology.* Hillsdale, N.J.: Lawrence Erlbaum Associates.

23

THE PERSISTENT POPULARITY OF THE PARANORMAL

James Lett

(P) aranormal beliefs are firmly entrenched in our society. An overwhelming majority of all Americans hold at least some beliefs that are nonrational, nonscientific, and nonsensical (Gallup and Newport 1991), and the level of paranormal belief in the United States actually seems to be increasing (Frazier 1987). The irony, of course, is that no culture in the history of humanity has ever possesses a greater store of objective, accurate, and reliable knowledge about the universe.

That irony is well appreciated by the physicist Victor Stenger, who offers a thorough critique of transcendental claims in his recent book *Physics and Psychics.* Stenger (1990, 298) writes that he is "astonished that so many people in a modern nation like the United States still take the paranormal seriously," and then he goes on to say that he "shudder[s] at what this fact implies about the general state of scientific education in America."

As a science educator, I share Professor Stenger's concerns about the deplorable state of scientific education in our country. Those concerns have led me to develop a course at my college called "Anthropology and the Paranormal," which attempts to teach students the fundamentals of critical thinking (Lett 1990). As an anthropologist, however, I am not at all astonished by the popularity of the paranormal, and I am aware that all of my best efforts, and all of the collective best efforts of all the world's scientists and skeptics, are not likely to appreciably diminish the level of paranormal belief.

To understand why that's true, you simply have to look at the underlying reasons for the popularity of the paranormal. Those reasons are well understood by anthropologists and widely appreciated by many other skeptics (see, for example, Singer and Benassi 1981; Kurtz 1984; Asimov 1986). In brief, the popularity of the paranormal in the United States can be attributed to four principal factors: the uncertainty of the public, the unreliability of the media, the inadequacy of the

This article first appeared in *Skeptical Inquirer* 16, no. 4 (Summer 1992): 381–88.

educational system, and the inaccuracy of the American worldview. Let me review each of the factors in turn, and then consider why they ensure the persistence of paranormal beliefs.

1. *The Uncertainty of the Public.* Anthropologists have documented the fact that religion is a cultural universal. Belief in the supernatural—in a reality that is not necessarily amenable to the five senses—occurs in every society on the planet. American culture is hardly unique in having a range of paranormal beliefs. What is remarkable about American culture at the moment is that paranormal beliefs seem to be proliferating, and that presents an interesting anthropological puzzle. Why are the proponents of some of the most preposterous and platitudinous forms of the paranormal able to attract such wide followings?

The answer is fairly simple. Since early in this century, when Bronislaw Malinowski first published the results of his ethnographic research in the southwestern Pacific, anthropologists have known that people are likely to turn to magical solutions to solve their problems when practical solutions are unavailable. In a classic study, Malinowski noticed that the inhabitants of the Trobriand Islands always used magical rituals to protect themselves when fishing in the rough, shark-infested waters of the open ocean, but never turned to magic when fishing in the calm, protected waters of the lagoon. He concluded that human beings are likely to seek reassurance in paranormal beliefs whenever they face situations where the outcome is uncertain or beyond their control. Subsequent anthropological research around the world has confirmed the obvious truth of Malinowski's principle.

It might seem, then, that the level of paranormal belief in the United States should be on the decline, given the scientific advances of the past century that have substantially increased our understanding and control of so many facets of reality. However, the evidence indicates that the average American's degree of certainty and security about life has actually decreased over the past generation or so, at the same time that the incidence of paranormal beliefs has increased. The reason? In a phrase, rapid and profound culture change.

Millions of Americans are increasingly bewildered and disoriented by the far-reaching changes that have occurred in virtually every aspect of daily life since World War II. The American family has undergone radical restructuring, sexual morality has been turned upside-down, gender roles have been substantially redefined, the birth rate has dropped, the divorce rate has jumped, the crime rate has soared, the cost of living has skyrocketed, the quality of American-made goods and services has plummeted, the average real income of American households has fallen, the international power and prestige of the country has declined, the impoverished underclass has mushroomed, and the American dream of home ownership has become unattainable for increasing numbers of middle-class citizens. The population of the planet is growing at an alarming rate, the world's natural resources are being readily depleted, the environment is suffering calamitous abuse, and the threat of nuclear annihilation still hangs over the whole of humanity.

Small wonder, then, that millions of Americans are suspicious of the present and fearful of the future. At the beginning of the past decade, the anthropologist Marvin Harris (1981) noted these trends in American life and linked them to a number of phenomena, including the rise of the Moonies and other nontraditional religious cults, the popularity of Carlos Castañeda's shamanic mysticism, the development of Scientology, the growth of Transcendental Meditation and other self-help strategies, and the emergence of the televangelists. What was true ten years ago is even more true today. For millions of Americans, life is a perplexing, frightening, and unsatisfying whirl, and they are looking for a way off the merry-go-round. They are looking for answers.

It should be no surprise that millions of Americans find their answers in paranormal beliefs. At a time when traditional religious institutions are declining in influence (because of their increasing irrelevance to the new facts of daily life), paranormal beliefs are emerging as the new folk religions. The reason, as Isaac Asimov (1986, 212) pithily puts it, is obvious: "Inspect every piece of pseudoscience and you will find a security blanket, a thumb to suck, a skirt to hold." Paul Kurtz (1986) argues that the tendency to embrace nonempirical explanations of puzzling phenomena is deeply rooted in the human psyche—he calls it the "transcendental temptation." I would agree. By their nature, human beings are meaning-seeking animals, but the sad conclusion of cross-cultural anthropological research is that most individual humans, and all human cultures, are content with the *illusion* of meaning. For most people, it matters not, apparently, whether their explanations are true or false; it only matters that they are emotionally satisfying. For millions of Americans, paranormal beliefs fulfill that criterion.

2. *The Unreliability of the Media.* Book and magazine publishers, television and movie producers, and newspaper and broadcast journalists are all sure about one thing when it comes to the paranormal: it sells. There is a huge market for stories (true or not) about ghosts, reincarnation, ESP, psychics, UFOs, ancient astronauts, Bigfoot, the Bermuda Triangle, and every other imaginable variety of the paranormal, and the media reap huge profits by producing a large quantity of uncritical material to supply that demand.

The entertainment media unashamedly fuel the public passion for the paranormal. They do so even when they themselves—writers, editors, and producers—recognize this nonsense for what it is. Time-Life Books, for example, is promoting a best-selling series of pro-paranormal books called "Mysteries of the Unknown," despite the fact that the editors know very well that nothing of what they are publishing is either mysterious or unknown. (*Time* magazine has frequently published skeptical stories about the paranormal, including, for example, a December 7, 1987, cover story about the New Age, and a June 13, 1988, feature article about magician/investigator James Randi.)

The news media, which are regulated to a certain extent by both the law and their own professional standards, supposedly offer the public the objective truth about paranormal phenomena. In practice, however, the news media are only slightly more skeptical than the entertainment media; far too frequently, they are just as credulous when it comes to the paranormal.

The infamous Tamara Rand hoax is widely recalled by skeptics (Frazier and Randi 1981), although it has undoubtedly been forgotten by the general public. Rand was the self-professed psychic who claimed to have accurately predicted the attempted assassination of Ronald Reagan in March 1981. Four days after John Hinckley shot the president, three of the four major television news networks broadcast a videotape of Rand's prediction, which she said was made more than two months before the assassination attempt. Investigation revealed that the tape had been made 24 hours after the shooting; after repeated requests from CSICOP, the networks retracted their stories.

The real indictment, however, lies not in the fact that the networks were slow to correct their error, but in the fact that they could ever have made such an error in the first place. How could the networks have reported such an improbable story without any preliminary investigation whatsoever? No such lack of skepticism would be applied to any political story by any self-respecting journalist. The unfortunate truth is that the lessons of the Tamara Rand incident made little impression upon the news media. They continue to apply the same uncritical standards in their reporting on the paranormal.

In his book *Superstition and the Press*, the distinguished journalism professor Curtis MacDougall (1983) compiled a voluminous list of preposterous paranormal claims that had been credulously reported by reputable newspapers in American cities. MacDougall died in 1985, but an equally energetic reader and clipper would be able to compile an annual update that would add substantially to MacDougall's 600-page book. (In fact, the Education Subcommittee of CSICOP maintains such a file—see Feder 1987–88.) The broadcast media, as a whole, are equally uncritical.

There are two reasons for the paranormal receiving such a prominent play in the news media. In the first place, journalists know that the public is greatly interested in stories about the paranormal, and there are major incentives—they are called ratings and circulation—to give the public what it wants. In the second place (and this is the more disturbing and more dangerous of the two factors), most journalists lack the knowledge, training, and experience to critically evaluate paranormal claims. Journalism professor Philip Meyer (1986, 39–40) admits that many journalists don't even care about the objective truth of the matter: "It doesn't make any difference what the facts are—if somebody with an impressive-sounding degree or title says something interesting, then it's a story."

As a cultural anthropologist, I have many things in common with journalists: both of us are interested in observing, recording, describing, and explaining human behavior. There are important differences between us, however, one of which is the way we define "objectivity." As a scientist, objectivity means, for me, being fair to the truth; for the journalist (and I speak from personal experience, having spent time as a television journalist), objectivity means being fair to everyone involved.

I've written about these issues before (Lett 1986, 1987), and my conclusion is that journalists are not likely to become skeptics. As long as the paranormal sells, journalists cannot afford to debunk it. They are, after all, competing with the entertainment media for the attention of the public. News coverage of the notorious shroud of Turin provides an illuminating example. On September 27, 1988, when it was publicly revealed that radiocarbon dating had proved that the shroud could not be the burial cloth of Christ, ABC News reported that scientists were nevertheless baffled as to how the image could have been made. Not true, of course—scientists know very well how the image could have been made (see Nickell 1983, 1989)—but that's what journalistic "objectivity" means.

I am aware that there are exceptions. Beginning on November 28, 1988, to name one outstanding instance, "Good Morning America" broadcast a three-part series on astrology by ABC News consumer editor John Stossel. Stossel was unabashedly skeptical in his report— he called astrology "total nonsense"— and he did an excellent job of clearly and forcefully debunking the paranormal claims of Hollywood astrologer Joyce Jillson. Nevertheless, such reports are by far the exception rather than the rule in broadcast news.

The real significance of the fact that the media are generally uncritical lies in the fact that the media are, in addition, highly pervasive and highly influential (see Kottak 1990). The media are an important agent of enculturation for all Americans, and they are a predominant source of information for Americans who have completed their formal schooling. Most adult Americans depend primarily upon the media for their knowledge and understanding of the world. The knowledge offered by the media is hardly reliable, however. On balance, the message of both the entertainment and the news media with regard to the paranormal is the same: namely, that there is, or at least very well may be, another dimension to ordinary reality that is well known to the average citizen but obscure to established scientists, who tend to be cloistered, conservative, and closed-minded. Obviously that message is highly misleading.

3. *The Inadequacy of the Educational System.* The scientific method has been responsible for an enormous and unprecedented increase in the fund of human knowledge in the past century, yet the American public-educational system fails to teach students the basic principles of that method. Beginning in grade school, and continuing, unfortunately, through high school and college, students are taught science as though it were a body of facts, a dry and static subject matter, rather than a technique for acquiring knowledge. The skills of evidential reasoning and logical analysis that are the heart of the scientific approach are rarely identified and examined; instead, students are given the task of memorizing the products of past scientific research.

These points are well appreciated by many science educators, of course, especially at the university level. Charles Vigue (1988, 326), for example, is a biology professor who helped develop an interdisciplinary course called "The Nature of Science," which explores such topics as "deductive and inductive logic," "scientific methodology," "causal and noncausal explanation," "the interpretation of data," and the like. Nevertheless, these topics are rarely addressed in precollegiate science classrooms, and they continue to be deemphasized in post-secondary curricula.

All of which means that most students finish their schooling without developing their abilities to think critically, reason logically, or evaluate properly. They have learned to respect the accomplishments of science, and they have learned to recognize the superficial trappings of science (such as lofty credentials, impressive jargon, lengthy citations, and professional conferences). Unfortunately, a little learning can be a dangerous thing, because that leaves students easy prey for pseudoscientists, who can often claim those same superficial trappings for themselves.

4. *The Inaccuracy of the American Worldview.* Every culture has a "worldview," a set of assumptions about the nature of reality, and every culture's worldview is a mix of sense and nonsense. The American worldview—to which all of us who are native members of U.S. society are enculturated from the moment of birth—includes a large number of erroneous assumptions.

Consider, for example, the following propositions: polygraphists can detect lies; graphologists can discern personality traits from handwriting samples; psychotherapists can diagnose, treat, and cure "mental illnesses" based on a scientific understanding of human personality processes; alcoholism is a disease (as are drug addiction, overeating, undereating, and any number of antisocial activities and proclivities, such as an apolitical motivation for political assassination); human beings have souls, which are in some way independent of their bodies; there is life after death; a benevolent, omnipotent, omniscient deity exists and is concerned about the personal fate of every individual on the planet.

Each of these assumptions is part of the American worldview; each is widely accepted by the great majority of Americans, each enjoys considerable prestige, and each is firmly institutionalized in the legal, social, political, economic, and educational systems of our culture—yet each is, by any reasonable measure, patently and completely untrue. None of them enjoys evidential support, and none of them can withstand logical and empirical scrutiny. Because these propositions are part of the culture's worldview, however, most Americans take them for granted and regard them as obvious givens. Indeed, as far as most Americans are concerned, anyone who would even think to question these basic assumptions would have to be lacking in common sense.

These dubious assumptions about the nature of reality can pose a fundamental obstacle to the rational investigation of paranormal phenomena. Using the methods of scientific investigation, for example, you can show someone that extrasensory perception does not exist (or at least that it has never been demonstrated), and the average American is likely to respond (as many of my students have) with the following syllogism: I know that God exists; the proposition that God exists cannot be substantiated by the methods of scientific investigation; therefore there must be something wrong with the methods of scientific investigation, at least when those methods are applied to the supernatural (or paranormal). That argument is perfectly valid; it is, of course, unsound, because the first premise is indefensible, but for psychological and political reasons it can be very difficult to explain that to students.

In short, when we try to decrease the incidence of paranormal beliefs in American society by appealing to the rationality of Americans, we are fighting an uphill battle, because we are dealing with an audience that has been persuaded that nonrational thinking is perfectly appropriate in some cases. Moreover, when we use rationality to attack beliefs like precognition and psychokinesis, we are implicitly attacking such beliefs as the efficacy of psychotherapy and the immortality of the soul. The degree of intellectual and emotional commitment to such beliefs is overwhelming, and it is a rare individual who will examine the merits of that commitment.

Collectively, I believe, these four factors—the uncertainty of the public the unreliability of the media, the inadequacy of the educational system, and the inaccuracy of the American worldview—explain the popularity of paranormal beliefs in American culture today. The stresses and strains of life account for the *appeal* of paranormal beliefs, the uncritical publication of extraordinary claims accounts for the *source* of paranormal beliefs, and the widespread inability to think critically, together with the cultural legitimization of uncritical thinking, accounts for the *persistence* of paranormal beliefs.

The only way to diminish the incidence of paranormal beliefs, then, is to address all four problems: to somehow reduce anxiety, revamp the media, reform education, and revise the culture's worldview. That's a daunting task, and the sad truth is that it will probably never be accomplished. We *might* someday succeed in inspiring the media to be more responsible on these subjects (although I doubt it), we *might* manage to restructure the educational system (although I suspect the possibility is remote), but we will *never* be able to appreciably reduce anxiety, because we will never be able to eliminate the sources of anxiety. As long as death, disease, disability, disaster, and disharmony exist, paranormal beliefs will flourish.

I know from personal experience that showing people the facts, and teaching them how to think critically, will not necessarily be enough to get them to abandon paranormal beliefs. In teaching my course on "Anthropology and the Paranormal," I'm often able to persuade my students that the evidence doesn't support their belief in the Loch Ness monster or levitation or spontaneous human combustion, but I rarely succeed in persuading students that the evidence doesn't support their belief in life after death. No amount of training in evidential reasoning will be sufficient to dissuade most people from beliefs to which they have a strong emotional commitment.

Nevertheless, I am not willing to capitulate and to allow irrationality to go unchallenged. I agree in principle with Carl Sagan (1986, 227) that "the best antidote for pseudoscience...is science," and I believe that the wide-ranging efforts of CSICOP represent our best hope for an effective response to paranormal beliefs. We should encourage scientists to share their knowledge with the general public, just as we should encourage journalists and other members of the mass media to exercise responsible judgment in the publication of nonfiction (and everything proffered as nonfiction). We should insist that instruction in critical-thinking skills be incorporated into educational curricula at all grade levels. We should, in addition, encourage the deinstitutionalization of pseudoscience in our culture (for example, I would support efforts to make it illegal for psychotherapists to give expert testimony in court).

What we shouldn't do, however, is delude ourselves. Skepticism, unfortunately, isn't a drug that will cure irrationality, or even a vaccine that will prevent it; at best, skepticism is only an antidote to an insidious poison. The problem, however, is that the antidote of skepticism isn't effective for everyone who's exposed to the poison. It's only effective for people who have the qualities of heart and mind required to face the anxieties and uncertainties of life without the quick comfort and simple answers offered by paranormal beliefs. As long as those qualities remain in short supply among the general public, paranormal beliefs will continue to be prevalent.

REFERENCES

Asimov, Isaac. 1986. The perennial fringe. *Skeptical Inquirer* 10: 212–14.

Feder, Ken. 1987-88. Trends in popular media: Credulity still reigns. *Skeptical Inquirer* 12: 124–26.

Frazier, Kendrick. 1987. Mainlining of mysticism: Poll shows new popularity (News and Comment). *Skeptical Inquirer* 11: 333–34.

Frazier, Kendrick, and James Randi. 1981. Prediction after the fact: Lessons of the Tamara Rand hoax. *Skeptical Inquirer* 6(1): 4–7.

Gallup, George H., Jr., and Frank Newport. 1991. Belief in paranormal phenomena among adult Americans. *Skeptical Inquirer* 15: 137–46.

Harris, Marvin. 1981. *America Now: The Anthropology of a Changing Culture.* New York: Simon and Schuster.

Kottak, Conrad Phillip. 1990. *Prime-Time Society: An Anthropological Analysis of Television and Culture.* Belmont, Calif.: Wadsworth.

Kurtz, Paul. 1984. Debunking, neutrality, and skepticism in science. *Skeptical Inquirer* 8: 239–46.

———. 1986. *The Transcendental Temptation: A Critique of Religion and the Paranormal.* Amherst, N.Y.: Prometheus Books.

Lett, James. 1986. Anthropology and journalism. *Communicator* 40 (5): 33–35.

———. 1987. An anthropological view of television journalism. *Human Organization* 46 (4): 356–59.

———. 1990. A field guide to critical thinking: Six simple rules to follow in examining paranormal claims. *Skeptical Inquirer* 14: 153–60.

MacDougall, Curtis D. 1983. *Superstition and the Press.* Amherst, N.Y.: Prometheus Books.

Meyer, Philip. 1986. Ghostbusters: The press and the paranormal. *Columbia Journalism Review,* March/April, pp. 38–41.

Nickell, Joe. 1983. *Inquest on the Shroud of Turin.* Amherst, N.Y.: Prometheus Books (Updated paperback ed., 1987).

———. 1989. Unshrouding a mystery: Science, pseudoscience, and the cloth of Turin. *Skeptical Inquirer* 13: 296–99.

Sagan, Carl. 1986. Night walkers and mystery mongers: Sense and nonsense at the edge of science. *Skeptical Inquirer* 10(3): 218–28.

Singer, B., and V. A. Benassi. 1981. Occult beliefs. *American Scientist* 69: 49–55.

Stenger, Victor J. 1990. *Physics and Psychics: The Search for a World Beyond the Senses.* Amherst, N.Y.: Prometheus Books.

Vigue, Charles L. 1988. Teaching the nature of science. *Skeptical Inquirer* 12: 325–26.

24

EYEWITNESS TESTIMONY AND THE PARANORMAL

Richard Wiseman, Matthew Smith, and Jeff Wiseman

(M) uch of the evidence relating to paranormal phenomena consists of eyewitness testimony. However, a large body of experimental research has shown that such testimony can be extremely unreliable.

For example, in 1887 Richard Hodgson and S. John Davey held séances in Britain (in which phenomena were faked by trickery) for unsuspecting sitters and requested each sitter to write a description of the séance after it had ended. Hodgson and Davey reported that sitters omitted many important events and recalled others in incorrect order. Indeed, some of the accounts were so unreliable that Hodgson later remarked:

> The account of a trick by a person ignorant of the method used in its production will involve a misdescription of its fundamental conditions . . . so marked that no clue is afforded the student for the actual explanation (Hodgson and Davey 1887, 9).

In a partial replication of this work, Theodore Besterman (1932) in Britain had sitters attend a fake séance and then answer questions relating to various phenomena that had occurred. Besterman reported that sitters had a tendency to underestimate the number of persons present in the séance room, to fail to report major disturbances that took place (e.g., the movement of the experimenter from the séance room), to fail to recall the conditions under which given phenomena took place, and to experience the illusory movements of objects.

More recently, Singer and Benassi in the United States (1980) had a stage magician perform fake psychic phenomena before two groups of university students. Students in one group were told that they were about to see a magician; the other group, that they were about to witness a demonstration of genuine psychic ability. Afterward, all of the students were asked to note whether they believed the performer was

---

This article first appeared in *Skeptical Inquirer* 19, no. 6 (November/December 1995): 29–32.

a genuine psychic or a magician. Approximately two-thirds of *both* groups stated they believed the performer to be a genuine psychic. In a follow-up experiment the researchers added a third condition, wherein the experimenter stressed that the performer was *definitely* a magician. Fifty-eight percent of the people in this group still stated they believed the performer to be a genuine psychic!

These studies admirably demonstrate that eyewitness testimony of supposedly paranormal events can be unreliable. Additional studies have now started to examine some of the factors that might cause such inaccuracy.

Clearly, many supposedly paranormal events are difficult to observe simply because of their duration, frequency, and the conditions under which they occur. For example, ostensible poltergeist activity, séance phenomena, and UFO sightings often occur without warning, are over within a few moments, take place under poor lighting or weather conditions, or happen at a considerable distance from observers. In addition, some people have sight/hearing deficiencies, while others have observed these phenomena under the influence of alcohol, drugs, or when they are tired (especially if they have had to wait a relatively long time for the phenomena to occur).

It is also possible that observers' beliefs and expectations play an important role in the production of inaccurate testimony. Different people clearly have different beliefs and expectations prior to observing a supposed psychic—skeptics might expect to see some kind of trickery; believers may expect a display of genuine psi. Some seventy years ago Eric Dingwall in Britain (1921) speculated that such expectations may distort eyewitness testimony:

> The frame of mind in which a person goes to see magic and to a medium cannot be compared. In one case he goes either purely for amusement or possibly with the idea of discovering 'how it was done,' whilst in the other he usually goes with the thought that it is possible that he will come into direct contact with the other world (p. 211).

Recent experimental evidence suggests that Dingwall's speculations are correct. Wiseman and Morris (1995a) in Britain carried out two studies investigating the effect that belief in the paranormal has on the observation of conjuring tricks. Individuals taking part in the experiment were first asked several questions concerning their belief in the paranormal. On the basis of their answers they were classified as either believers (labeled "sheep") or skeptics (labeled "goats"). [Gertrude Schmeidler, City College, New York City, coined the terms *sheep* and *goats*.]

In both experiments individuals were first shown a film containing fake psychic demonstrations. In the first demonstration the "psychic" apparently bent a key by concentrating on it; in the second demonstration he supposedly bent a spoon simply by rubbing it.

After they watched the film, witnesses were asked to rate the "paranormal" content of the demonstrations and complete a set of recall questions. Wiseman and Morris wanted to discover if, as Hodgson and Dingwall had suggested, sheep really did tend to misremember those parts of the demonstrations that were central to solving the tricks. For this reason, half of the questions concerned the methods used to fake the phenomena. For example, the psychic faked the key-bending demonstration by secretly switching the straight key for a pre-bent duplicate by passing the straight key from one hand to the other. During the switch the straight key could not be seen. This was clearly central to the trick's method; and one of the "important" questions asked was whether the straight key had always remained in sight. A second set of "unimportant" questions asked about parts of the demonstration that were not related to the tricks' methods.

Overall, the results suggested that sheep rated the demonstrations as more "paranormal" than goats did, and that goats did indeed recall significantly more "important" information than sheep. There was no such difference for the recall of the "unimportant" information.

This is not the only study to investigate sheep/goat differences in observation and recall of "paranormal" phenomena. Jones and Russell in the United States (1980) asked individuals to observe a staged demonstration of extrasensory perception (ESP). In one condition the demonstration was successful (i.e., ESP appeared to occur) while in the other it was not. All individuals were then asked to recall the demonstration. Sheep who saw the unsuccessful demonstration distorted their memories of it and often stated that ESP had occurred. Goats tended to correctly recall the demonstration, even if it appeared to support the existence of ESP.

In addition, Matthew Smith in Britain (1993) investigated the effect that instructions (given prior to watching a film containing a demonstration of apparent psychic ability) had on the recall of the film. Individuals were split into two groups. One group was told that the film contained trickery; the other group was told that it contained genuine paranormal phenomena. The former group recalled significantly more information about the film than the latter group.

All of the above experiments were carried out in controlled laboratory settings. However, another recent study suggests that the same inaccuracies may exist in a more natural setting, namely, the séance room.

Many individuals have reported experiencing extraordinary phenomena during dark-room séances. Eyewitnesses claim that objects have mysteriously moved, strange sounds have been produced, or ghostly forms have appeared, and that these phenomena have occurred under conditions that render normal explanations practically impossible.

Believers argue that conditions commonly associated with a séance (such as darkness, anticipation, and fear) may act as a catalyst to produce these phenomena (Batcheldor 1966). Skeptics suggest that reports of séances are unreliable and that eyewitnesses are either fooling themselves or being fooled by fraudulent mediums.

The authors carried out an experiment in the United Kingdom to assess both the reliability of testimony relating to séance phenomena, and whether paranormal events could be produced in a modern séance. We carried out our experiment, titled "Manifestations," three times. Twenty-five people attended on each occasion. They were first asked to complete a short questionnaire, noting their age, gender, and whether they believed that genuine paranormal phenomena might sometimes take place during séances.

A séance room had been prepared. All of the windows and doors in the room had been sealed and blacked out, and twenty-five chairs had been arranged in a large circle. Three objects—a book, a slate, and a bell—had been treated with luminous paint and placed onto three of the chairs. A small table, the edges of which were also luminous, was situated in the middle of the circle. Two luminous maracas rested on the table.

Following a brief talk on the aims of the project, the participants were led into the darkened séance room. Richard Wiseman played the part of the medium. With the help of a torch, he showed each person to a chair, and, where appropriate, asked them to pick up the book, slate, or bell.

Next, he drew participants' attention to the table and maracas. Those participants who had picked up the other luminous objects were asked to make themselves known, and the "medium" collected the objects one by one and placed them on the table.

He then pointed out the presence of a small luminous ball, approximately 5 centimeters in diameter, suspended on a piece of rope from the ceiling. Finally, he took his place in the circle, extinguished the torch, and asked everybody to join hands.

The medium first asked the participants to concentrate on trying to move the luminous ball and then to try the same with the objects on the table. Finally, the participants were asked to concentrate on moving the table itself. The séance lasted approximately ten minutes.

Clearly, it was important that some phenomena occurred to assess the reliability of eyewitness testimony. The maracas were therefore "gimmicked" to ensure their movement during the séance. In the third séance the table was also similarly moved by trickery. Finally, we also used trickery to create a few strange noises at the end of each séance.

All of the ungimmicked objects were carefully placed on markers so that any movement would have been detectable. After leaving the séance room, the participants completed a short questionnaire that asked them about their experience of the séance.

No genuine paranormal phenomena took place during any of the séances. However, our questionnaire allowed us to assess the reliability of participants' eyewitness testimony.

Would participants remember which objects had been handled before the start of the séance? As the maracas were gimmicked, we had to ensure that they were not examined or handled by anyone. Nevertheless, one in five participants stated that they had been. This was an important inaccuracy, as observers are likely to judge the movement of an object more impressive if they think that the item has been scrutinized beforehand.

This type of misconception was not confined to the maracas. In the first two séances, the slate, bell, book, and table remained stationary. Despite this, 27 percent of participants reported movement of at least one of these. In the third séance the table was gimmicked so that it shifted four inches toward the medium, but participants' testimony was again unreliable, with one in four people reporting no movement at all.

An interesting pattern develops if the results are analyzed by separating the participants by belief. The ball, suspended from the ceiling, did not move at any time. Seventy-six percent of disbelievers were certain that it hadn't moved. In contrast, the same certainty among believers was only 54 percent. In addition, 40 percent of believers thought that at least one other object had moved, compared to only 14 percent of disbelievers. The answers to the question "Do you believe you have witnessed any genuine paranormal phenomena?" perhaps provide the most conclusive result for the believer/disbeliever divide. One in five believers stated that he or she had seen genuine phenomena. None of the disbelievers thought so. This would suggest that while we are all vulnerable to trickery, a belief or expectation of paranormal phenomena during séances may add to that vulnerability.

The results clearly show that it is difficult to obtain reliable testimony about the séance. Indeed, our study probably underestimated the extent of this unreliability as the séance lasted only ten minutes and participants were asked to remember what had happened immediately afterward.

Although a minority of participants believed that they had observed genuine paranormal phenomena, it does not seem unreasonable to assume that these individuals might be the most likely to tell others about their experience. Our results suggest that many of their reports would be fraught with inaccuracies and it might only take a few of the more distorted accounts to circulate before news that "genuine" paranormal phenomena had occurred became widespread.

In short, there is now considerable evidence to suggest that individuals' beliefs and expectations can, on occasion, lead them to be unreliable witnesses of supposedly paranormal phenomena. It is vital that investigators of the paranormal take this factor into account when faced with individuals claiming to have seen extraordinary events. It should be remembered, however, that such factors may hinder accurate testimony regardless of whether that testimony is for or against the existence of paranormal phenomena; the observations and memory of individuals with a strong need to *disbelieve* in the paranormal may be as biased as extreme believers. In short, the central message is that investigators need to be able to carefully assess testimony, regardless of whether it reinforces or opposes their own beliefs concerning the paranormal.

Accurate assessment of the reliability of testimony requires a thorough understanding of the main factors that cause unreliable observation and remembering. Research is starting to reveal more about these factors and the situations under which they do, and do not, occur. Indeed, this represents part of a general movement to increase the quality of the methods used to investigate psychic phenomena (Wiseman and Morris 1995b). Given the important role that eyewitness testimony plays in parapsychology, understanding observation is clearly a priority for future research.

REFERENCES

Batcheldor, K. J. 1966. Report on a case of table levitation and associated phenomena. *Journal of the Society for Psychical Research* 43: 339–56.

Besterman, T. 1932. The psychology of testimony in relation to paraphysical phenomena: Report of an experiment. *Proceedings of the Society for Psychical Research* 40: 363–87.

Dingwall, E. 1921. Magic and mediumship. *Psychic Science Quarterly* 1(3): 206–19.

Hodgson, R., and S. J. Davy. 1887. The possibilities of mal-observation and lapse of memory from a practical point of view. *Proceedings of the Society for Psychical Research* 4: 381–495.

Jones, W. H., and D. Russell. 1980. The selective processing of belief disconfirming information. *European Journal of Social Psychology* 10: 309–12.

Smith, M. D. 1993. The effect of belief in the paranormal and prior set upon the observation of a 'psychic' demonstration. *European Journal of Parapsychology* 9: 24–34.

Singer, B., and V. A. Benassi. 1980. Fooling some of the people all of the time. *Skeptical Inquirer* 5, no. 2 (Winter): 17–24.

Wiseman, R. J., and R. L. Morris. 1995a. Recalling pseudo-psychic demonstrations. *British Journal of Psychology* 86: 113–25.

———. 1995b. *Guidelines for Testing Psychic Claimants.* Amherst, N.Y.: Prometheus Books.

25

PSYCHIC EXPERIENCES: PSYCHIC ILLUSIONS

Susan Blackmore

(**W**) hy do so many people believe in psychic phenomena? Because they have psychic experiences. And why do they have psychic experiences? Because such experiences are an inevitable consequence of the way we think. I suggest that, like visual illusions, they are the price we pay for a generally very effective relationship with a massively complex world.

The latest Gallup poll (Gallup and Newport 1991) shows that about a third of Americans believe in telepathy and about a quarter claim to have experienced it themselves. Rather fewer have experienced clairvoyance or psychokinesis (PK), but still the numbers are very high and have not been decreasing over the years. Previous surveys have found similar results and also that the most common reason for belief in the paranormal is personal experience (Palmer 1979; Blackmore 1984).

A "psychic experience" is here defined as any experience interpreted by the experient as requiring a psychic or paranormal interpretation. The question of whether such a hypothesis is required is not addressed. Rather we are attempting to understand how such experiences come about even if no genuinely paranormal phenomena occur. It should be noted that many experimental studies of psi (such as guessing long strings of targets) do not produce psychic experiences in this sense, although they may produce evidence of the paranormal. Others (such as ganzfeld studies and remote viewing, perhaps) do, but the experience is a separate issue from the question of statistical significance or evidence for psi. We are here concerned with experience and belief, not the evidence for psi.

My hypothesis is that psychic experiences are comparable to visual illusions. The experience is real enough, but its origin lies in internal processes, not peculiarities in the observable world. Like visual illusions they arise from cognitive processes that are usually appropriate but under certain circumstances give rise to the wrong answer. In other words, they are a price we pay for using efficient heuristics.

This article first appeared in *Skeptical Inquirer* 16, no. 4 (Summer 1992): 367–76.

In the case of vision, illusions arise when, for example, depth is seen in two-dimensional figures and constancy mechanisms give the answer that would be correct for real depth. The equivalent in the case of psychic experiences may be the illusion that a cause is operating and an explanation is required when in fact none is. In other words, psychic experiences are illusions of causality. I shall discuss five types of illusion.

1. ILLUSIONS OF CONNECTION

Experiences of telepathy, clairvoyance, and precognition imply a coincidence that is "too good to be just chance." This is so whether the experience involves dreaming about a person's death and that person dies within a few hours, feeling the urge to pick up one's partner from the station and in fact he was stranded and needed help, or betting on a horse that later wins a race.

Some people's response to such events is to say, "That was just a chance coincidence", while others' is to say, "That cannot be chance." In the latter case the person will then look for a causal explanation for the coincidence. If none can be found, a "cause," such as ESP, may be invoked. Alternatively, some kind of non-causal but meaningful connection may be sought, such as Carl Jung's "acausal connecting principle" (Jung 1973).

There are two possible types of error that may be made here. First, people may treat connected events as chance coincidences, thereby missing real connections between events and failing to look for explanations. Second, they may treat chance events as connected and seek for explanations where none are required. In the real world of inadequate information and complex interactions one would expect errors of both types to occur. It is the latter type that, I suggest, give rise to experiences of ESP.

This is comparable to classical signal-detection theory. Figure 1 shows two distributions. For any given stimulus strength there could be just noise or noise plus a signal. At low signal-to-noise ratios, it is not possible to be a perfect detector. Mistakes are inevitable and may be either in missing a true signal or in thinking there is a signal when there is not. I am suggesting that believers in the paranormal (called "sheep" in psychological parlance) are more likely to make the latter kind of error than are disbelievers (called "goats"). In signal-detection theory, this is described in terms of a variable criterion. As the payoffs change, people may use a different criterion, making more of one kind of error and fewer of another. Their sensitivity (d') may not change when their criterion does (see Figure 2). It is not a question of right and wrong but of which kind of error you would rather make, given you have to make some.

One prediction of this approach is that those people who more frequently look for explanations of chance coincidences are more likely to have psychic experiences. Therefore, sheep should be those who underestimate the probability of chance coincidences.

It has long been known that probability judgments can be extremely inaccurate. Kahneman and Tversky (1973) have explored some of the heuristics, such as

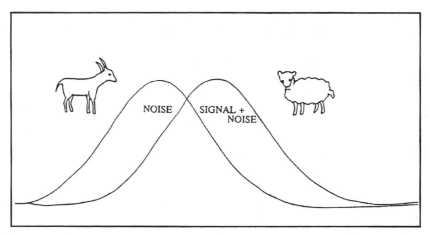

Figure 1. Sheep (believers in the paranormal) can be seen as more likely than goats (disbelievers) to decide that a connection Is meaningful, a series of events nonrandom, or a form present in ambiguous images.

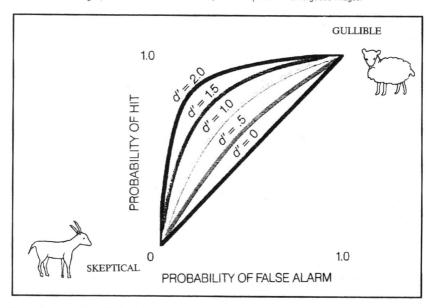

Figure 2. In an analogy with signal-detection theory, sheep and goats might have the same sensitivity (d′) but differ in criterion.

"representativeness" and "availability," that people use to make judgments and that can give rise to serious errors. In addition, people have great confidence in erroneous judgments, even in the face of contrary evidence (Einhorn and Hogarth 1978). Falk and collaborators have investigated what makes people find coincidences surprising (Falk 1982; Falk and McGregor 1983). Adding specific but superfluous details can make coincidences seem more surprising, and things that happen to subjects themselves seem more surprising to them than the same things happening to other people. Diaconis and Mosteller (1989) have reviewed some ways of studying the psychology of coincidences and have provided models for calculating probabilities.

There is, however, little research relating these misjudgments to belief in the paranormal or to having psychic experiences. Blackmore and Troscianko (1985) found that sheep performed worse than goats on a variety of probability tasks. For example, in questions testing for responsiveness to sample size, sheep did significantly worse than goats. The well-known birthday question was asked: How many people would you need to have at a party to have a 50:50 chance that two have the same birthday? (See Diaconis and Mosteller 1989 for a general model for this type of question.) As predicted, goats got the answer right significantly more often than sheep.

Subjects also played a coin-tossing computer game and were asked to guess how many hits they would be likely to get by chance. The correct answer, 10 hits in 20 trials, seems to be rather obvious. However, the sheep gave a significantly lower mean estimate of only 7.9, while goats gave a more accurate estimate of 9.6.

Further research is called for here. It would be interesting to test whether sheep and goats differ in the probability they assign to various kinds of coincidences happening both in laboratory tests and in assessing probabilities of real-world events.

2. ILLUSIONS OF CONTROL

Where the coincidence is between a person's own action and an event external to them, the same effect may be at work but the assumed cause will be personal control; or in the context of psi, it will be PK. This has been called the "illusion of control" by Langer (1975). Sheep have been found to show a greater illusion of control than goats in a psi task (Ayeroff and Abelson 1976, Jones et al. 1977; Benassi et al. 1979).

One might argue that if PK occurs then the perception of personal control in such tasks is not an illusion. This is less likely, given that no PK was found in these experiments. However, to rule out this as an explanation for the difference, Blackmore and Troscianko (1985) used a covert psi task. There was no evidence of PK and a greater illusion of control for sheep than for goats.

3. ILLUSIONS OF PATTERN AND RANDOMNESS

Pattern and randomness cannot be unambiguously distinguished. In a long enough series of events, any combination or string of events is likely to occur by chance. However, the process of extracting pattern from noise is central to all sensory processes. As in the case of coincidences, two kinds of error can occur. One is the failure to detect patterns that are there; the second is the tendency to see patterns that are not there. We are arguing that the second type of error will make people search for a cause and that, since there is no cause, they may turn to paranormal explanations.

This predicts that people who make this type of error are more likely to have psychic experiences (or experiences they interpret as psychic) and hence to believe in the paranormal.

It has long been known that people are bad at judging randomness. In particular, when asked to generate a string of random numbers (subjective random generation, or SRG), people typically give far fewer repetitions of the same digit than would be expected by chance (see reviews by Budescu 1987 and Wagenaar 1972). This is related to the "Gambler's Fallacy," whereby some people think that a long string of reds must be followed by black. ESP experiments are often equivalent to SRG and show the same bias.

Blackmore and Troscianko (1985) found no differences between sheep and goats in SRG for strings of digits 1 to 5 or in the ability of sheep and goats to discriminate random sequences from biased ones. However, Brugger, Landis, and Regard (1990) did. They argued that the same variables affect ESP scoring and SRG in the same direction—variables like task duration, stimulant and depressant drugs, and age. They even suggest that many laboratory ESP findings may be explained by correspondences between target sequences and human biases. Although there is some evidence for this in studies giving immediate feedback (Gatlin 1979; Tart 1979), this cannot easily explain results obtained without feedback and with adequate target randomization.

They tested the relationship to belief in the paranormal in three experiments. SRG was studied in a telepathy experiment with five symbols to choose from. Sheep produced significantly fewer repetitions than goats did. Subjects intermediate in belief gave intermediate repetitions. There was no evidence of ESP occurring and no sheep-goat effect (i.e., sheep did not do better at the ESP test).

In a second experiment, SRG was studied in mimicking the roll of dice (6 choices). The same effect was found. Third, subjects were shown dice sequences with different numbers of repetitions and asked which was more likely to appear first by chance. Of course all strings were equally likely to occur, but subjects tended to choose the string with fewer repetitions. Sheep did so more than goats, and the intermediate group was in between. These results appear to be highly consistent and to show the expected greater bias in sheep.

To test this further, Katherine Galaud, at Bristol University, carried out an experiment to compare SRG for different numbers of choices. It might be argued that most people can predict or calculate likely sequences when only two choices are involved but that the real world typically involves multiple choices and low probabilities. Perhaps SRG would be even less random when more choices are possible. Furthermore, differences between sheep and goats may be more extreme where more choices are available. This experiment studied the variation in results with different numbers of choices available.

One hundred twenty students were given the Belief in the Paranormal Scale (BPS) (Jones, Russell, and Nickel 1977), a randomness questionnaire, and a probability questionnaire. The probability questionnaire consisted of three questions based on the "taxi problem" (Kahneman and Tversky 1972) manipulated to give correct answers of 20, 40, and 80 percent. The randomness questionnaire asked subjects to generate strings of random numbers, choosing from the digits 1 to 2, 1 to 4, or 1 to 8, with expected numbers of repetitions being 12, 6, and 3, respectively. No differences were found between sheep, goats, and intermediates (Blackmore, Galaud, and Walker 1994).

There are two differences between this experiment and Brugger's that might account for the different results. One is that Brugger et al. timed the generation of digits with a metronome. It could be that, given time to think about randomness, people can to some extent compensate for their biases and that untimed and unpressured responses like those in the present experiment cannot reveal them. However, it could also be argued that in real-life situations there is not usually time pressure. Another difference is that they used only one question on ESP to divide subjects into sheep, goats, and intermediates. Further experiments now under way at Bristol are trying to find out if these factors are responsible.

4. ILLUSIONS OF FORM

Object recognition can entail the same two types of error. A conservative approach means missing interesting forms that are there. A less cautious approach means seeing things that are not. Possibly, those people who are more likely to see forms when none is present are also more likely to see apparitions or ghosts or to seek paranormal explanations when none is required.

In a second experiment at Bristol, carried out by Catherine Walker (Blackmore, Galaud, and Walker 1994), we tested this and a related question. If sheep are more willing to see forms in noisy displays, is this an error compared with goats, or are goats more likely to miss forms that are present. This is the familiar question of accuracy versus criterion. Sheep might simply have a lower criterion for seeing forms than goats, with the same accuracy for discriminating forms, or they may actually make more errors altogether.

Fifty subjects were given the Belief in the Paranormal Scale and tested on an object-identification task. The stimuli consisted of four sets of seven pictures each; ranging from barely identifiable blobs to clear outline shapes (see Figure 3). The final shapes were two leaves, a bird, a fish and an axe. They were presented for 10 milliseconds each, with a mask of black dots on a white background shown between presentations. The four least identifiable stimuli were shown first, progressing through the series with the four at each level being randomized for order. The subjects were asked whether they could see any shape; and, if they could, what shape it was.

It was predicted that sheep would report seeing forms earlier in the series than goats but would not be any more accurate in identifying the forms. In other words, they would have a lower criterion for identification. This is exactly what was found. BPS scores did not correlate with the number of pictures correctly identified but did correlate closely with the number of incorrect identifications and the tendency to say there was a shape but not identify it. In other words, the sheep were more likely to make wrong guesses but were no worse at detecting the

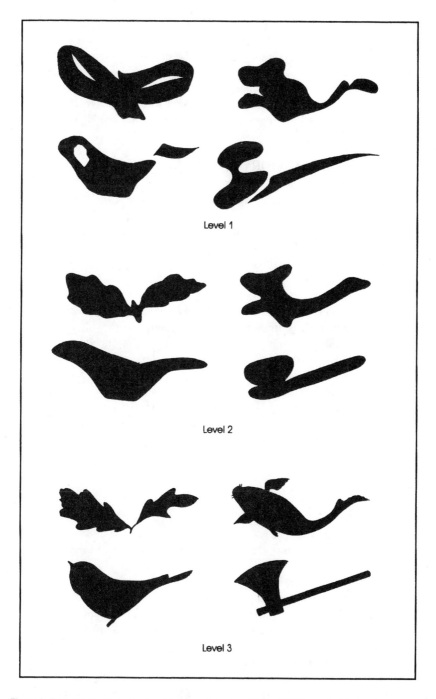

Level 1

Level 2

Level 3

Figure 3. Examples of stimuli taken from Blackmore, Galaud, and Walker 1994. For each of the four forms there are seven levels of detectability. Three levels are shown here.

pictures that were there. Goats, although willing to say there was a shape, were less willing than sheep to guess at identifying it.

This confirms that sheep are more likely to claim to see identifiable forms in ambiguous stimuli, but there are many possible reasons for this. For example, creativity may correlate with belief in the paranormal and with tendency to see forms. Whatever the origins of the tendency, the findings fit with the idea that paranormal belief may be encouraged in those who more often see form in ambiguity.

5. ILLUSIONS OF MEMORY

In addition to all the processes above, selective memory may make coincidences appear to occur more often than they do in fact occur. Hintzman, Asher, and Stern (1978) demonstrated selective remembering of meaningfully related events. Fischhoff and Beyth (1975) showed that people misremember their previous predictions to conform with what actually happened.

We might predict that people who are particularly prone to such memory effects are more likely to seek paranormal explanations and therefore to have psychic experiences and believe in the paranormal. If so, these effects would be greater for sheep than for goats, but this has not been tested.

The popularity of fortune-tellers may also depend to some extent on selective memory. Selective recall of meaningful coincidences and true statements about the person will add to the Barnum effect, or the tendency to accept certain kinds of personality readings as true of oneself but not of others (Dickson and Kelly 1985). If this is so we would expect the people who frequent fortune-tellers to be more prone to this kind of selective memory. Again this has not been tested, but a project is now underway at Bristol to investigate it.

CONCLUSIONS

Five types of psychic illusion have been explored. They may be the basis for many spontaneous psychic experiences that generate belief in the paranormal. The tendency for sheep to show many of these effects to a greater extent than goats tends to confirm this hypothesis.

This conclusion does not apply to many kinds of psi experiments, especially those giving no feedback and using sound randomization techniques. It therefore has no bearing on the issue of whether any laboratory experiments provide evidence for psi. Also in life outside the lab these processes may operate to produce psychic experiences and belief in the paranormal quite independently of whether genuinely paranormal phenomena ever occur.

These findings are therefore not so much evidence against the occurrence of paranormal phenomena as a suggestion that we should expect to find a high incidence of psychic experiences and widespread belief in the paranormal whether or not psychic phenomena ever occur.

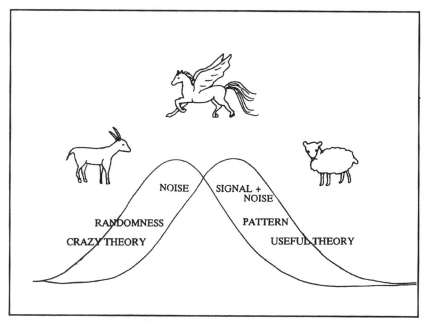

Figure 4. The true skeptic is not entrenched at one end of the spectrum.

THE NATURE OF SKEPTICISM

The whole basis of this approach is that human beings, in trying to make sense of their world, must make mistakes. On the one hand, they miss things that are there and, on the other, invent things that are not. This applies as much to simple signals as to complex correlations and to scientific theories as well as to perceptual ones. I have tried to show some of these in Figure 4.

In everyday life the equivalent of the sheep is someone who will see something interesting in everything. The problem is that they may be seeing things that are not there. The equivalent of the goat is someone who needs lots of evidence before seeing or experiencing anything. They are likely to miss out on a lot of fun!

Similarly, in science the equivalent of the sheep is someone who enjoys every crazy theory and follows every faint lead. The problem is that they may easily be following a false lead. The equivalent of the goat is someone who takes no interest in wacky theories and sticks only to the conventional. They may be safe but are likely to miss the really exciting new theory when it comes along.

You takes your choice and with it the consequences—fun or boredom, fear of failure or love of novelty. But what of skepticism? I do not think the true skeptic is the goat. The true skeptic does not always stick to one end of the spectrum but can shift criteria as the circumstances demand. The true skeptic is as skeptical of the goat who denies everything as of the sheep who embraces everything (as is John Palmer's [1986] "progressive skeptic"). True skeptics can drop their fear of looking silly or curtail their love of the novel as appropriate; can apply caution or stick their neck out according to their understanding of the issues. The true skeptic is not the ultimate goat but something more like a flying horse.

REFERENCES

Ayeroff, F., and R. P. Abelson. 1976. ESP and ESB: Belief in personal success at mental telepathy. *Journal of Personality and Social Psychology* 34: 240–47.

Benassi, V. A., P. D. Sweeney, and G. E. Drevno. 1979. Mind over matter: Perceived success at psychokinesis. *Journal of Personality and Social Psychology* 37: 1377–86.

Blackmore, S. J. 1984. A postal survey of OBEs and other experiences. *Journal of the Society for Psychical Research* 52: 225–44.

Blackmore, S. J., K. Galaud, and C. Walker. 1994. "Psychic Experiences as Illusions of Causality." In *Research in Parapsychology 1991,* edited by E. W. Cook and D. Delanoy, 89–93. Metuchen, N.J.: Scarecrow Press.

Blackmore, S. J., and T. Troscianko. 1985. Belief in the paranormal: Probability judgements, illusory control, and the "chance baseline shift." *British Journal of Psychology* 76: 459–68.

Brugger, P., T. Landis, and M. Regard. 1990. A "sheep-goat effect" in repetition avoidance: Extra-sensory perception as an effect of subjective probability? *British Journal of Psychology* 81: 455–68.

Budescu, D. V. 1987. A Markov model for generation of random binary sequences. *Journal of Experimental Psychology: Human Perception and Performance* 13: 25–39.

Diaconis, P., and F. Mosteller. 1989. Methods for studying coincidences. *Journal of the American Statistical Association* 84: 853–61.

Dickson, D. H., and I. W. Kelly. 1985. The Barnum effect in personality assessment: A review of the literature. *Psychological Reports* 57: 367–82.

Einhorn, H. J., and R. M. Hogarth. 1978. Confidence in judgement: Persistence of the illusion of validity. *Psychological Review* 85: 395–416.

Falk, R. 1982. On coincidences. *Skeptical Inquirer* 6 (1): 18-31.

Falk, R., and D. MacGregor. 1983. "The Surprisingness of Coincidences." In *Analysing and Aiding Decision Processes,* edited by P. Humphreys, O. Svenson, and A. Vari, 489–502. New York: North Holland.

Fischhoff, B. and R. Beyth. 1975. "I knew it would happen": Remembered probabilities of once-future things. *Organizational Behaviour and Human Performance* 13: 116.

Gallup, G. H., and F. Newport. 1991. Belief in paranormal phenomena among adult Americans. *Skeptical Inquirer* 15: 137–46.

Gatlin, L. L. 1979. A new measure of bias in finite sequences with applications to ESP data. *Journal of the American Society for Psychical Research* 73: 29–43.

Hintzman, D. L., S. J. Asher, and L. D. Stern. 1978. Incidental retrieval and memory for coincidences. In *Practical Aspects of Memory,* edited by M. M. Gruneberg, P. E. Morris, and R. N. Sykes, 61–66. New York: Academic Press.

Jones, W. H., D. W. Russell, and T. W. Nickel. 1977. Belief in the Paranormal Scale: An objective instrument to measure beliefs in magical phenomena and causes. *JSAS Catalog of Selected Documents in Psychology* 7:100 (Ms. No. 1577).

Jung, C. G. 1973. *Synchronicity: An Acausal Connecting Principle.* Princeton, N.J.: Princeton University Press.

Kahneman, D., and A. Tversky. 1972. On prediction and judgment. *Oregon Research Institute Bulletin* 12.

———. 1973. On the psychology of prediction. *Psychological Review* 80: 237–51.

Langer E. J. 1975. The illusion of control. *Journal of Personality and Social Psychology* 32: 311–28.

Palmer, J. 1979. A community mail survey of psychic experiences. *Journal of the American Society for Psychical Research* 73: 221–52.

Palmer, J. 1986. Progressive skepticism. *Journal of Parapsychology* 50: 29–42.

Tart, C. T. 1979. Randomicity, predictability and mathematical inference strategies in ESP feedback experiments. *Journal of the American Society for Psychical Research* 73: 44–60.

Wagenaar, W. A. 1972. Generation of random sequences by human subjects: A critical survey of the literature. *Psychological Bulletin* 77: 65–72.

26

BELIEF IN ASTROLOGY: A TEST OF THE BARNUM EFFECT

Christopher C. French, Mandy Fowler, Katy McCarthy, and Debbie Peers

(T) here is no empirical support for the claims of traditional astrological theory (see Culver and Ianna 1988; Dean and Mather 1977; Eysenck and Nias 1982; Gauquelin 1979; Jerome 1977; Kelly 1979; Startup 1984). Despite this, the level of belief in astrology in the general population is high and shows no sign of declining. Most people who have their horoscopes cast perceive those horoscopes to be an accurate description of their personalities. Why should this be?

Several factors have been suggested as playing a role in forming and maintaining a belief in the validity of horoscopes (Dean 1987; Tyson 1982). One of the most well known is the so-called Barnum effect, the tendency for people to accept vague, ambiguous, and general statements as descriptive of their unique personalities.

There are two differing reasons given in the literature for naming this inclination to believe "the Barnum effect," although both are based on quotations from P. T. Barnum. The first is that the famous circus-owner maintained that his secret of success was always to have a little something for everyone. Likewise, the typical astrological personality profile consists of a collection of statements carefully selected to enable everyone to see something of themselves in the description. The second, more cynical reason is that Barnum's most infamous phrase was, of course, "There's a sucker born every minute."

The best way to appreciate the force of the Barnum effect is to actually read a typical Barnum profile. Try it:

> You have a great need for other people to like you and admire you. You have a tendency to be critical of yourself. You have a great deal of unused capacity which you have not used to your advantage. While you have some personality weaknesses, you are generally able to compensate for them. Your sexual adjustment has presented problems for you. Disciplined and self-controlled outside, you tend to be worrisome and insecure inside. At times you have serious doubts

This article first appeared in *Skeptical Inquirer* 15, no. 2 (Winter 1991): 166–72.

as to whether you have made the right decision or done the right thing. You prefer a certain amount of change and variety and become dissatisfied when hemmed in by restrictions and limitations. You pride yourself on being an independent thinker and do not accept others' statements without satisfactory proof. You have found it unwise to be too frank in revealing yourself to others. At times you are extroverted, affable and sociable, while at other times you are introverted, wary and reserved. Some of your aspirations tend to be pretty unrealistic. Security is one of your major goals in life.

Typically, a naive subject reading the personality description above would be impressed by its accuracy if told that the description was based upon his or her horoscope. This profile was actually first used in a study some 40 years ago (Forer 1949), but its appeal is as strong today as it was then.

It is important to realize that the Barnum effect does not apply only to personality descriptions supposedly based upon horoscopes. The effect is found if the profile is said to be based upon any form of personality assessment, including palmistry, objective psychological tests, projective tests, personal interview, graphology, or Tarot cards. A considerable amount of research has been done on the psychological factors that influence the Barnum effect. Although detailed review of these studies is beyond the scope of this discussion (see Dickson and Kelly 1985; Furnham and Schofield 1987; Snyder, Shenkel, and Lowery 1977), it is clear that the effect is an important factor in the acceptance of horoscopes (see, e.g., studies by Rosen 1975; Snyder 1974; Snyder, Larsen, and Bloom 1976).

However, not all statements in horoscopes are Barnum-type statements. For example, the typical "Aries" is said to be bold, energetic, assertive, selfish, insensitive, and aggressive. Surely, not everyone would see themselves as fitting this description. But, as Sundberg (1955 pointed out, Barnum profiles consist of a variety of statements:

> Vague, e.g., "You enjoy a certain amount of change and variety in life", double-headed, e.g., "You are generally cheerful and optimistic but get depressed at times", modal characteristics of the subject's group, e.g., "You find that study is not always easy", favorable, e.g., "You are forceful and well-liked by others."

The typical horoscope is a mix of general statements and rather more specific ones. People tend to be impressed by the specific details that appear to fit (and pay less attention to those that do not), while the general Barnum-type statements provide readily acceptable "padding."

It seemed possible to us that different psychological mechanisms might be required to explain the formation and maintenance of belief in strong believers compared with moderate believers. One possibility was suggested by Goldberg (1979). Some "Virgos" actually will, by chance alone, have the personality characteristics typically associated with that sun-sign, and similarly for all of the other sun-signs. Such people will be constantly amazed at the accuracy of horoscopes based upon this information and are far more likely to take their interest in the subject further than those who feel that the personality descriptions typical for *their* sun-signs are not appropriate—such as a timid "Aries." The believer is likely to buy popular books on astrology and be attracted to others with an interest in astrology, and some of these others will by coincidence be typical examples of their sun-signs, providing for the believer seemingly incontrovertible proof that astrology is valid. We shall henceforth refer to this model as the "Coincidence Hypothesis," as the original match between the typical sun-sign profile and the individual's personality is totally coincidental.

The scenario above, although speculative, seemed plausible to us and led to some testable hypotheses. If different mechanisms are responsible for producing different levels of belief in the way described, then we would predict that strong believers would show *less* acceptance of the Barnum-type profile than moderate believers, for the following reasons. Strong believers would be likely to have more knowledge of the typical characteristics associated with sun-signs, particularly the believers' own signs. Therefore they would be more impressed by reading a description that corresponded to this typical pattern and contained reference to specific expected traits than by the more general Barnum-type description. Furthermore, strong believers would rate horoscopes cast on the basis of their birth details (henceforth referred to as "genuine" horoscopes) as more accurate than randomly selected horoscopes ("false" horoscopes). Moderate believers, on the other hand, would not be as inclined to look for the typical profile because they are unlikely to possess detailed knowledge of typical sun-sign profiles. Moderate believers are likely to be more impressed by the carefully selected Barnum-type statements and to be less able to distinguish between genuine and false horoscopes. All of these effects would be relative, of course, since even moderate believers may have some knowledge of their typical sun-sign profiles.

An alternative hypothesis maintains simply that the Barnum effect will be equally strong for everyone and that some other (unspecified) factor is required to account for differences in belief between strong and moderate believers (see Dean 1987, for possibilities). In this case, one would argue that the effectiveness of horoscopes is due largely to the Barnum effect and that genuine horoscopes are effective only to the extent that they incidentally capitalize on the effect. We would predict a different pattern of results on the basis of this hypothesis. Both genuine and Barnum-type horoscopes would be judged as accurate by all believers, but the Barnum profile ought to be judged as more accurate, as the careful selection of statements would maximize the effect. Furthermore, there would be no difference in the perceived accuracy of genuine and false horoscopes. We shall refer to this hypothesis simply as the "Barnum Hypothesis."

A third hypothesis, which has already been thoroughly discredited, can also be outlined. The "Astrological Hypothesis" would maintain that the position of the stars and planets at birth really does influence the formation of personality as outlined in traditional astrology. If this were so, everyone, regardless of degree of belief, ought to rate genuine horoscopes as more accurate than either false or Barnum-type horoscopes.

In order to test these hypotheses, data were collected from 52 subjects, most of whom were attending a sixth-form college. Ages ranged from 16 to 35, with a mean

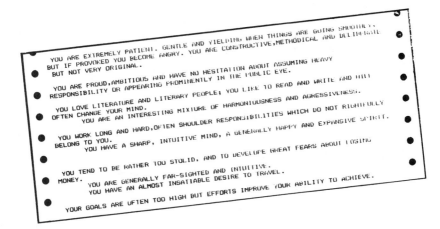

Figure 1: A typical example of the computerized horoscope used in this study.

age of 18. Thirty-five of the subjects were female. Subjects were told that the study was an assessment of three different computer programs for casting horoscopes and were initially asked to provide information on their date, time, and place of birth as well as information relating to their belief in and knowledge of astrology.

Several days after collection of the initial data, each subject was presented with a booklet containing a "genuine" horoscope, a randomly selected ("false") horoscope, and a Barnum-type horoscope. The order of the horoscopes was counterbalanced across subjects. The genuine horoscopes were cast using a modified version of the HOROSCOPICS program (Copyright 1983, Patched by PAS Inc., 306 S. Homewood Ave., Pittsburgh, PA 15208) run on an IBM XT personal computer. The program casts a horoscope on the basis of the date of birth only and produces a profile consisting of a dozen statements (a typical example of the output is shown in Figure 1). The program was modified to produce output consisting solely of the personality profile, omitting the astrological data upon which the interpretation was based. The false horoscopes were randomly selected horoscopes from the pool of genuine horoscopes, so that the two pools were in fact identical. The Barnum horoscope was the same as the one presented earlier except that one sentence ("Your sexual adjustment has presented problems for you") was omitted in order to equate the number of statements in each horoscope. The horoscopes were all presented on computer print-out paper with the same layout. Subjects were asked to read and rate each horoscope before considering the next one.

Of the 52 subjects, 7 stated that they believed in astrology "strongly," 31 "moderately," and 14 "not at all." There was a striking difference in distribution between male and female subjects. All 7 strong believers were female, as were 25 out of 31 moderate believers. Only 3 of 14 disbelievers were female (chi-square = 19.3, df = 2, $p < 0.0001$). This finding is in line with previous research showing that more women than men believe in astrology and are interested in their horo-

Table 1

**Means (and Standard Deviations) for Accuracy Ratings (on a Scale of 1 to 5)
for Barnum, False, and Genuine Horoscopes**

	Barnum	False	Genuine
Strong Belief (N= 7)	4.14 (1.07)	3.29 (1.11)	3.29 (0.76)
Moderate Belief (N = 31)	3.77 (0.85)	3.00 (1.06)	3.16 (0.90)
No Belief (N = 14)	3.29 (1.49)	3.07 (1.14)	2.79 (1.25)
Entire Group (N = 52)	3.69 (1.09)	3.06 (1.07)	3.08 (0.99)

Table 2

**Means (and Standard Deviations) for Applicability Ratings
for Barnum, False, and Genuine Horoscopes**

	Barnum	False	Genuine
Strong Belief (N = 7)	2.96 (1.08)	2.33 (0.96)	2.42 (0.94)
Moderate Belief (N = 31)	3.29 (1.11)	2.86 (0.90)	2.57 (0.79)
No Belief (N = 14)	2.97 (1.05)	2.26 (0.93)	2.48 (0.85)
Entire Group (N = 52)	2.79 (1.19)	2.21 (1.05)	2.21 (1.19)

scopes (e.g., DeFrance, Fischler, Morin, and Petrossian 1971; Gallup 1975; Sobal and Emmons 1982; Wuthnow 1976).

As would be expected, level of belief was significantly correlated with self-reported frequency of reading newspaper horoscopes, with self-assessed knowledge of astrology, and with self-assessed influence of astrology on subjects' everyday lives. The initial questionnaire also included a question asking subjects to write down their sun sign, ascendant, and moon sign, if known. This allowed for a maximum score of three on this rather crude measure of astrological knowledge, which was also found to correlate significantly with belief.

The second questionnaire asked subjects to rate how accurate they felt each horoscope was on a scale from one (not at all accurate) to five (completely accurate). The mean ratings of each group are shown in Table 1.

In order to test the experimental hypotheses, data were subjected to a two-way analysis of variance with type of horoscope and level of belief as factors. The only significant effect was related to type of horoscope ($F(2, 98) = 4.95$, $p < 0.01$), and reflected the fact that the Barnum horoscope was rated as much more accurate than the other two. No significant difference was found between the ratings of the genuine and false horoscopes and no interaction was found between type of horoscope and level of belief.

It might be objected that the reason that people so readily accept the Barnum profile is that the statements in it actually *do* apply to everyone. If so, then people are behaving quite rationally in rating its accuracy so highly. It is therefore important to show that people rate the Barnum profile as highly accurate while at the same time not realizing its general applicability. Therefore we asked subjects to rate how general they found the horoscopes on a scale from one to four (1 = very general) 2 = quite general; 3 = quite applicable to you personally; 4 = very applicable to you personally). Mean ratings are presented in Table 2.

Once again, these data were analyzed using a two-way analysis of variance with type of horoscope and level of belief as factors and, once again, the only significant effect was that the Barnum profile was rated as *more* applicable than the other two ($F(2, 98) = 5.35$, $p < 0.01$). The genuine and false horoscopes did not differ in applicability ratings.

A final question on the questionnaire asked subjects, for each horoscope, if they felt that it constituted evidence for astrological belief. Twenty subjects out of 52 felt that the Barnum profile constituted such evidence, whereas only 12 and 11, respectively, felt this way about the false and genuine horoscopes. These proportions are significantly different (Cochran's Q = 10.43, df = 2, $p < 0.01$).

There can be no doubt that this experiment offers strong support for the Barnum Hypothesis and no support whatsoever for the Coincidence Hypothesis or the Astrological Hypothesis, at least for the sample under study. It might be objected that the group of strong believers was small in comparison with the other two groups; but as examination of the tables reveals, there was no sign of a trend in favor of either of the latter hypotheses. In fact, this group tended to be more influenced by the Barnum profile than the other two groups, although this effect did not reach statistical significance. The nonbelievers tended to be least influenced by the Barnum effect. Perhaps with larger samples these effects would have reached significance.

To summarize, the Barnum profile was rated as most accurate and most personally applicable by all groups, as predicted by the Barnum Hypothesis. Furthermore, a significantly greater number of subjects felt that the Barnum profile, compared with the other two horoscopes, constituted evidence in favor of astrology. No interaction was found between level of belief and type of horoscope, thus failing to support the Coincidence Hypothesis. No support was found for the Astrological Hypothesis. No group was able to differentiate the genuine from false horoscopes, which were both rated as less accurate and less applicable than the Barnum profile.

For this sample, then, the Barnum effect offered the best explanation of belief in astrology. However, we may need to be cautious in generalizing these results too widely. The strong believers did claim and demonstrate more knowledge of astrology than the other two groups, but their level of knowledge was still not very great, as might be expected in a study where the average age of the subjects was 18. A more refined version of the Coincidence Hypothesis would recognize the many levels of astrological knowledge attainable and the possibly complex interactions that this could produce in studies like this one when applied to different subject groups.

At the lowest level are those who profess no knowledge of astrology. Next are those who have some vague notion of their sun-sign and its associated characteristics, followed by those who may take astrology seriously enough to buy popular books on the subject. The Coincidence Hypothesis as outlined earlier in this article applies to these three levels. However, beyond sun-sign astrology we have what Dean (1986–87) refers to as "the real thing," involving a consultation between a professional astrologer and a client in which the astrologer's interpretation is based upon as many as 40 interacting chart factors, of which sun-sign is only one. As Eysenck and Nias (1982) discuss, it is possible that those who are very knowledgeable would not be so influenced by consideration of the sun-sign, recognizing instead that "real" astrology is a much more complex enterprise.

The fact remains, however, that most people who profess a belief in astrology, whether strong or moderate, do not possess much knowledge of the subject. It is of great interest to understand the factors that produce belief in such individuals, and the current study strongly suggests that we need look no further than the basic Barnum effect.

POSTSCRIPT BY CHRISTOPHER C. FRENCH:

Since the publication of our original study, I have carried out a number of further (unpublished) studies in collaboration with my students in an attempt to produce evidence in support of the Coincidence Hypothesis. In these later studies we tended to focus upon knowledge of astrology rather than belief in astrology, on the assumption that the former would be more directly relevant even though the two measures are highly correlated. For example, in one study we compared a group of students attending an astrology course with non-astrology students. As expected, the former group were much more knowledgeable regarding astrology than the latter group (and also more knowledgeable than the participants in our original study). The results, however, were very similar to those of our original study, with the Barnum Hypothesis receiving further strong support. It is somewhat surprising that believers were not found to be more susceptible to the Barnum effect than nonbelievers, but a similar finding is reported by Tobacyk, Milford, Springer, and Tobacyk (1988). They reported that believers in the paranormal (not specifically astrology) were no more susceptible than nonbelievers.

At least one study, however, has reported differences between believers in astrology and skeptics. Glick, Gottesman, and Jolton (1989) provided each participant with a Barnum-type profile that was either positive or negative. Additionally, the profiles were either said to be astrologically derived or else simply to be a set of statements that might or might not describe the participant. Overall, believers were more likely than skeptics to accept any description as accurate. Furthermore, although positive descriptions were usually rated as being more accurate than negative, believers accepted a negative description as readily as a positive one. Interestingly, all subjects rated the "astrological" profile as more accurate than the "non-astrological" even though they were in fact identical. It is clear from all of these studies that the Barnum effect is a major factor in the acceptance of horoscopes.

REFERENCES

Culver, R., and P. Ianna. 1988. *Astrology: True or False?* Amherst, N.Y.: Prometheus Books.

Dean, G. 1986–87. Does astrology need to be true? Part 1: A look at the real thing. *Skeptical Inquirer* 11: 166–84.

———. 1987. Does astrology need to be true? Part 2: The answer is no. *Skeptical Inquirer* 11: 257–73.

Dean, G., and A. Mather. 1977. *Recent Advances in Natal Astrology: A Critical Review 1900–1976.* Subiaco, Western Australia: Analogic.

DeFrance, P., C. Fischler, E. Morin, and L. Petrossian. 1971. *Le Retour des Astrologues.* Paris: Cahiers du Club Nouvel Observateur.

Dickson, D. H., and I. W. Kelly. 1985. The 'Barnum effect' in personality assessment: A review of the literature. *Psychological Reports* 57: 367–82.

Eysenck, H. J., and D. K. B. Nias. 1982. *Astrology: Science or Superstition?* London: Temple Smith.

Forer, B. R. 1949. The fallacy of personal validation: A classroom demonstration of gullibility. *Journal of Abnormal and Social Psychology* 44: 118–23.

Furnham, A., and S. Schofield. 1987. Accepting personality test feedback: A review of the Barnum effect. *Current Psychological Research & Reviews* 6: 162–78.

Gallup, G. H. 1975. Thirty-two million Americans express belief in astrology. *Gallup Poll.* Institute of Public Opinion. October 19.

Gauquelin, M. 1979. *Dreams and Illusions of Astrology.* Amherst, N.Y.: Prometheus Books.

Glick, P., D. Gottesman, and J. Jolton. 1989. The fault is not in the stars: Susceptibility of skeptics and believers in astrology to the Barnum effect. *Personality and Social Psychology Bulletin* 15: 572–83.

Goldberg, S. 1979. Is astrology a science? *The Humanist* 39(2): 9–16.

Jerome, L. 1977. *Astrology Disproved.* Amherst, N.Y.: Prometheus Books.

Kelly, I. W. 1979. Astrology and science: A critical examination. *Psychological Reports* 44: 1231–40.

Rosen, G. M. 1975. Effects of source prestige on subjects' acceptance of the Barnum effect: Psychologist versus astrologer. *Journal of Consulting and Clinical Psychology* 43: 95.

Snyder, C. R. 1974. Why horoscopes are true: The effect of specificity on acceptance of astrological interpretations. *Journal of Consulting and Clinical Psychology* 30: 577–80.

Snyder, C. R., D. K. Larsen, and L. J. Bloom. 1976. Acceptance of general personality interpretations prior to and after receiving diagnostic feedback supposedly based on psychological, graphological, and astrological assessment procedures. *Journal of Consulting and Clinical Psychology* 32: 258–65.

Snyder, C. R., R. J. Shenkel, and C. R. Lowery. 1977. Acceptance of personality interpretations: The "Barnum effect" and beyond. *Journal of Consulting and Clinical Psychology* 45: 104–14.

Sobal, J., and C. F. Emmons. 1982. Patterns of belief in religious, psychic, and other paranormal phenomena. *Zetetic Scholar* 9: 7–17

Startup, M. J. 1984. *The Validity of Astrological Theory as Applied to Personality, with Special Reference to the Angular Separation Between Planets.* Ph.D. thesis, Goldsmiths' College, London University; 350 references.

Sundberg, N. D. 1955. The acceptance of "fake" versus "bona fide" personality test interpretations. *Journal of Abnormal and Social Psychology* 50: 145–47.

Tobacyk, J., G. Milford, T. Springer, and Z. Tobacyk. 1988. Paranormal beliefs and the

Barnum effect. *Journal of Personality Assessment* 52: 737–39.Tyson, G. A. 1982. Why people perceive horoscopes as being true: A review. *Bulletin of the British Psychological Society* 35: 186–88.

Wuthnow, R. 1976. Astrology and marginality. *Journal for the Scientific Study of Religion* 15: 157–68.

Address correspondence to Dr. Christopher C. French, Dept. of Psychology, University of London Goldsmiths' College, Nelo Cross, London SE14 6NW, U.K.

PART SIX

PSYCHOLOGY AND THE CLAIMS OF PSI

27

EVALUATION OF THE MILITARY'S TWENTY-YEAR PROGRAM ON PSYCHIC SPYING

Ray Hyman

In 1995 the Central Intelligence Agency contracted for an outside evaluation of 20 years of until-then classified government-funded programs in "remote viewing" (ESP). Psychologist Ray Hyman was one of two experts asked to review the research. Beginning on this page Hyman presents an overview of the program and of the evaluation. In a second article (chapter 28), Hyman criticizes statements by some paranormal proponents about the evaluation and provides critical perspective not just on this government-sponsored "Stargate" program, but also on the latest claims about parapsychology overall.—EDITOR

n the early 1970s the Central Intelligence Agency supported a program to see if a form of extrasensory perception (ESP) called "remote viewing" could assist with intelligence gathering. The program consisted of laboratory studies conducted at Stanford Research Institute (SRI) under the direction of Harold Puthoff and Russel Targ. In addition to the laboratory research, psychics were employed to provide information on targets of interest to the intelligence community.

The CIA abandoned this program in the late 1970s because it showed no promise. The Defense Intelligence Agency (DIA) took over the program and continued supporting it until it was suspended in the spring of 1995. Under the DIA the program was named *Stargate* and consisted of three components. One component kept track of what foreign countries were doing in the area of psychic warfare and intelligence gathering. A second component, called the "Operations Program," involved six, and later three, psychics on the government payroll who were available to any government agency that wanted to use their services. The third component was the laboratory research on psychic phenomena first carried out at SRI and later transferred to Science Applications International Corporation (SAIC) in Palo Alto, California.

This article first appeared in *Skeptical Inquirer* 20, no. 2 (March/April 1996): 21–23.

This program was secret until it was declassified in early 1995. The declassification was done to enable an outside evaluation of the program. Because of some controversies within the program, a Senate committee decided to transfer the program from the DIA back to the CIA. The CIA, before deciding the fate of the program, contracted with the American Institutes for Research (A.I.R.), Washington, D.C., to conduct the evaluation. The A.I.R. hired Jessica Utts, a statistician at the University of California at Davis, and me, a psychologist at the University of Oregon, as the evaluation panel.

The idea was to have a balanced evaluation by hiring an expert who was known to support the reality of psychic phenomena and one who was skeptical about the existence of psi. Utts, in addition to being a highly regarded statistician, has written and argued for the existence of psychic phenomena and has been a consultant to the SRI and SAIC remote-viewing experiments.

Most recently, I served on the National Research Council committee that issued a report stating that the case for psychic phenomena had no scientific justification (*Skeptical Inquirer,* Fall 1988). In the January 1995 issue of *Psychological Bulletin* I supplied a skeptical commentary on the article by Daryl Bem and Charles Honorton that argued that the recent ganzfeld studies provided evidence for replicable experiments on ESP (See *Skeptical Inquirer,* Fall 1985).

At the beginning of last summer, Utts and I were each supplied with copies of all the reports that had been generated by the remote-viewing program during the 20 years of its existence. This consisted of three large cartons of documents.

We met with Edwin May, the principal investigator who took over this remote-viewing research project (after Puthoff and Targ left SRI in the 1980s); representatives of the CIA; and representatives of A.I.R. The purpose of the meeting was to coordinate our efforts as well as to focus our efforts on those remote-viewing studies that offered the most promise of being scientifically respectable. May helped identify the ten best studies for Utts and me to evaluate.

While Utts and I focused on the best laboratory studies, the two psychologists from A.I.R. conducted an evaluation of the recent operational uses of the three remote viewers (psychics) then on the government payroll. We all agreed that any scientifically meaningful evaluation of these operational psychic intelligence uses was impossible. The operational program had been kept separate from the laboratory research, and the work of the remote viewers was conducted in ways that precluded meaningful evaluation. Nevertheless, we all cooperated in developing a structured interview that the A.I.R. staff could use on the program officer, the three psychics, and the individuals or agencies that had used the services of these remote viewers.

The users said, through the interviews, that the remote viewers did not supply information that was useful in intelligence or other contexts.

The remote-viewing experiments that Utts and I evaluated had, for the most part, been conducted since 1986 and presumably had been designed to meet the objections that the National Research Council and other critics had aimed at the remote-viewing experiments conducted before 1986. These experiments varied in a number of ways but the typical experiment had these components:

1. The remote viewers were always selected from a small pool of previously "successful" viewers. May emphasized that, in his opinion, this ability is possessed by approximately one in every 100 persons. Therefore, they used the same set of "gifted" viewers in each experiment.

2. The remote viewer would be isolated with an experimenter in a secure location. At another location, a sender would look at a target that had been randomly chosen from a pool of targets. The targets were usually pictures taken from the *National Geographic.* During the sending period the viewer would describe and draw whatever impressions came to mind. After the session, the viewer's description and a set of five pictures (one of them being the actual target picture) would be given to a judge. The judge would then decide which picture was closest to the viewer's description. If the actual target was judged closest to the description, this was scored as a "hit."

In this simplified example I have presented, we would expect one hit by chance 20 percent of the time. If a viewer consistently scored more hits than chance, this was taken as evidence for psychic functioning. This description captures the spirit of the experimental evidence although I have simplified matters for convenience of exposition. In fact, the judging was somewhat more complex and involved rank ordering each potential target against the description.

A hit rate better than the chance baseline of 20 percent can be considered evidence for remote viewing, of course, only if all other nonpsychic possibilities have been eliminated. Obvious nonpsychic possibilities would be inadequacies of the statistical model, inadequacies of the randomization procedure in selecting targets or arranging them for judging, sensory leakage from target to viewer *or* from target to judge, and a variety of other sources of bias.

The elimination of these sources of above-chance hitting is no easy task. The history of psychical research and parapsychology presents example after example of experiments that were advertised as having eliminated all nonpsychic possibilities and that were discovered by subsequent investigators to have had subtle and unsuspected biases. Often it takes years before the difficulties with a new experimental design or program come to light.

Utts and I submitted separate evaluations. We agreed that the newly unclassified experiments seemed to have eliminated the obvious defects of the earlier remote-viewing experiments. We also agreed that these ten best experiments were producing hit rates consistently above the chance baseline. We further agreed that a serious weakness of this set of studies is the fact that only one judge, the principal investigator, was used in all the remote-viewing experiments. We agreed that these results remain problematical until it can be demonstrated that significant hitting will still occur when independent judges are used.

Beyond this we disagreed dramatically. Utts concluded that these results, when taken in the context of other contemporary parapsychological experiments —especially the ganzfeld experiments—prove the existence of psychic functioning. I find it bizarre to jump from these cases of statistically significant hitting to the conclusion that a paranormal phenomenon has been proven. As I pointed out, we both agreed that the results of the new remote-viewing experiments have to be independently judged. If independent judges cannot produce the same significant hit rates, this alone would suffice to discard these experiments as evidence of psychic abilities. More to the point, just because these experiments are less than 10 years old and have only recently been opened to public scrutiny, we do not know if they contain hidden and subtle biases or if they can be independently replicated in other laboratories. The history of parapsychology is replete with "successful"experiments that subsequently could not be replicated.

Utts is obviously impressed with consistencies between the new remote-viewing experiments and the current ganzfeld experiments. Where she sees consistencies, I see inconsistencies. The ganzfeld experiments all use the subjects as their own judges. The claim is that the results do not show up when independent judges are used. The exact opposite is true of remote-viewing experiments. When subjects are used as their own judges in remote-viewing experiments, the outcome is rarely, if ever, successful. Successful results come about only when the judges are someone other than the remote viewer. The recent ganzfeld experiments get successful results only with dynamic (animated video clips) rather than static targets. The remote-viewing experiments mostly use static targets. I could go on spelling out such inconsistencies, but this would be futile.

Even if the consistent hit rate above chance can be replicated with independent remote-viewing experiments, this would be a far cry from having demonstrated something paranormal. Parapsychologist John Palmer has argued that the successful demonstration of an above-chance statistical anomaly is insufficient to prove a paranormal cause. This is because remote viewing and ESP are currently only defined negatively. ESP is what is left after the experimenter has eliminated all obvious, normal explanations.

Several problems are created by trying to establish the existence of a phenomenon on the basis of a negative definition. For one thing, if ESP is shown by any departure from chance that has no obvious normal explanation, there is no way to show that the observed departures are due to one or several causes. Also, the claim for psi can never be falsified, because any glitch in the data can be used as evidence for psi. What is needed, of course, is a positive theory of psychic functioning that enables us to tell *when psi is present and when it is absent.* As far as I can tell, every other discipline that claims to be a science deals with phenomena whose presence or absence can clearly be decided.

The evidence for N-rays, mitogenetic radiation, polywater, cold fusion, and a host of other "phenomena" that no longer are considered to exist was much clearer and stronger than the current evidence for psychic functioning. In these cases of alleged phenomena, at least we were given criteria to decide when the reputed phenomena were supposed to be present and when they were not. Nothing like this exists in parapsychology. Yet the claim is being made that a phenomenon has been clearly demonstrated.

Fortunately, we do not have to squabble over whether the current remote-viewing experiments do or do not prove the existence of an anomalous phenomenon. We can follow the normal and accepted scientific process of (1) waiting to see if independent laboratories can replicate the above chance hitting conditions using appropriate controls; (2) seeing whether the researchers can devise positive tests to enable us to decide when psi is present and when it is absent; (3) seeing whether they can specify conditions under which we can reliably observe the phenomenon; (4) showing that the phenomenon varies in lawful ways with specifiable variables. Every science—except parapsychology—has met this accepted procedure. So far, parapsychology has not even come close to meeting any of these criteria. It is premature to draw any conclusions. We will simply have to wait and see. If history is a guide, then this will be a long wait, indeed.

28

THE EVIDENCE FOR PSYCHIC FUNCTIONING: CLAIMS VS. REALITY

Ray Hyman

(T) he recent media frenzy over the Stargate report violated the truth. Sober scientific assessment has little hope of winning in the public forum when pitted against unsubstantiated and unchallenged claims of "psychics" and psychic researchers—especially when the claimants shamelessly indulge in hyperbole. While this situation may be depressing, it is not unexpected. The proponents of the paranormal have seized an opportunity to achieve by propaganda what they have failed to achieve through science.

Most of these purveyors of psychic myths should not be taken seriously. However, when one of the persons making extreme claims is Jessica Utts, who is a professor of statistics at the University of California at Davis, this is another matter. Utts has impressive credentials and she marshals the evidence for her case in an effective way. So it is important to look at the basis for what I believe are extreme claims, even for a parapsychologist. Here is what Utts writes in her report on the Stargate program:

> Using the standards applied to any other area of science, it is concluded that psychic functioning has been well established. The statistical results of the studies examined are far beyond what is expected by chance. Arguments that these results could be due to methodological flaws in the experiments are soundly refuted. Effects of similar magnitude to those found in government-sponsored research at SRI [Stanford Research Institute] and SAIC [Science Applications International Corporation] have been replicated at a number of laboratories across the world. Such consistency cannot be readily explained by claims of flaws or fraud.... [Psychic functioning] is reliable enough to be replicated in properly conducted experiments, with sufficient trials to achieve the long-run statistical results needed for replicability.... Precognition, in which the answer is known to no one until a future time, appears to work quite well.... There is little benefit

This article first appeared in *Skeptical Inquirer* 20, no. 2 (March/April 1996): 24–26.

to continuing experiments designed to offer proof, since there is little more to be offered to anyone who does not accept the current collection of data.

For what it is worth, I happen to be one of those "who does not accept the current collection of data" as proving psychic functioning. Indeed, I do not believe that "the current collection of data" justifies that an anomaly of any sort has been demonstrated, let alone a paranormal anomaly. Although Utts and I—in our capacities as coevaluators of the Stargate project—evaluated the same set of data, we came to very different conclusions. If Utts's conclusion is correct, then the fundamental principles that have so successfully guided the progress of science from the days of Galileo and Newton to the present must be drastically revised. Neither relativity theory nor quantum mechanics in their present versions can cope with a world that harbors the psychic phenomena so boldly proclaimed by Utts and her parapsychological colleagues.

So, it is worth looking at the evidence that Utts uses to buttress her case. Unfortunately, many of the issues that this evidence raises are technical or require long and tedious refutations. This is not the place to develop this lengthy rebuttal. Instead, I will briefly list the sources of Utts's evidence and try to provide at least one or two simple reasons why they do not, either singly or taken together, justify her conclusions. As I understand it, Utts supports her conclusion with the following sources of evidence:

1. META–ANALYSES OF PREVIOUS PARAPSYCHOLOGICAL EXPERIMENTS

In a meta-analysis, an investigator uses statistical tools to pool the data from a series of similar experiments published over a period of time that may involve several different investigators and laboratories. Although some or many of the individual experiments might have yielded weak or nonsignificant results, the pooled data can be highly significant from a statistical viewpoint. In addition to getting an overall measure of significance, the meta-analyses typically also grade each study for quality on one or more dimensions. The idea is to see if the successful outcomes are correlated with poor quality. If so, this counts against the evidence for paranormal functioning. If not, then this is proclaimed as evidence that the successful outcomes were not due to flaws.

In the four major meta-analyses of previous parapsychological research, the pooled data sets produced astronomically significant results while the correlation between successful outcome and rated quality of the experiments was essentially zero.

Much can be written at this point. The major point I would make, however, is that drawing conclusions from meta-analytic studies is like having your cake and eating it too. The same data are being used to generate and test a hypothesis. The proper use of meta-analysis is to generate hypotheses, which then must be independently tested on new data. As far as I know, this has yet to be done. The correlation between quality and outcome also must be suspect because the ratings are not done blindly.

As far as I can tell, I was the first person to do a meta-analysis on parapsychological data. I did a meta-analysis of the original ganzfeld experiments as part of my critique of those experiments. My analysis demonstrated that certain flaws, especially quality of randomization, did correlate with outcome. Successful outcomes correlated with inadequate methodology. In his reply to my critique, Charles Honorton did his own meta-analysis of the same data. He too scored for flaws, but he devised scoring schemes different from mine. In his analysis, his quality ratings did not correlate with outcome. This came about because, in part, Honorton found more flaws in unsuccessful experiments than I did. On the other I found more flaws in successful experiments than Honorton did. Presumably, both Honorton and I believed we were rating quality in an objective and unbiased way. Yet, both of us ended up with results that matched our preconceptions.

So far, other than my meta-analysis, all the meta-analyses evaluating quality and outcome have been carried out by parapsychologists. We might reasonably expect that the findings will differ with skeptics as raters.

These are just two, but very crucial, reasons why the meta-analyses conducted so far on parapsychological data cannot be used as evidence for psi.

2. THE ORIGINAL GANZFELD EXPERIMENTS

These consisted of 42 experiments (by Honorton's count) of which 55 percent had been claimed as producing significant results in favor of ESP. My meta-analysis and evaluation of these experiments showed that this database did not justify concluding that ESP was demonstrated. Honorton's meta-analysis and rebuttal suggests otherwise. Utts naturally relies on Honorton's meta-analysis and ignores mine. In our joint paper, both Honorton and I agreed that there were sufficient problems with this original database that nothing could be concluded until further replications, conducted according to specified criteria, appeared.

3. THE AUTOGANZFELD EXPERIMENTS

This series of experiments, conducted over a period of six years, is so named because the collection of data was partially automated. When this set of experiments was first published in the *Journal of Parapsychology* in 1990, it was presented as a successful replication of the original ganzfeld experiments. Moreover, these experiments were said to have been conducted according to the criteria set out by Honorton and me. This indeed seemed to be the case with the strange exception of the procedure for randomizing targets at presentation and judging. Even in writing our joint paper, Honorton argued with me that careful randomization was not necessary in the ganzfeld experiments because each subject appears only once. I disagreed with Honorton, but even by his own reasoning, randomization is not as important if you believe that the subject is the sole source of the final judgment. But this was blatantly not the case in the autoganzfeld experiments. The experimenter, who was not so well shielded from the sender as the subject, interacted with the subject during the judging process. Indeed, during half of the trials the experimenter deliberately prompted the subject during the judging procedure. This means that the judgments from trial to trial were not strictly independent.

However, from the original published report, I had little reason to question the methodology of these experiments. What I did question was the claim that they were consistent with the original ganzfeld experiments. I pointed out a number of ways that the two outcomes were inconsistent. Not until I was asked to write a response to a new presentation of these experiments in the January 1994 issue of the *Psychological Bulletin* did I get an opportunity to scrutinize the raw data. Unfortunately, I did not get all of the data, especially the portion that I needed to make direct tests of the randomizing procedures. But my analyses of what I did get uncovered some peculiar and strong patterns in the data. All of the significant hitting was done on the second or later appearance of a target. If we examined the guesses against just the first occurrences of targets, the result is consistent with chance. Moreover, the hit rate rose systematically with each additional occurrence of a target. This suggests to me a possible flaw. Daryl Bem, the coauthor with Honorton of the *Psychological Bulletin* paper, responded that it might reveal another peculiarity of psychic phenomena. The reason why my finding is of concern is that all the targets were on videotape and played on tape players during presentation. At the very least, the peculiar pattern I identified suggests that we need to require that when targets and decoys are presented to the subjects for judging, they all have been run through the machine the exact same number of times. Otherwise there might be nonparanormal reasons why one of the video clips appears different to the subjects.

Subsequent to my response, I have learned about other possible problems with the autoganzfeld experiments. The point of this is to show that it takes time and critical scrutiny to realize that what at first seems like an airtight series of experiments has a variety of possible weaknesses. I concluded, and do so even more strongly now, that the autoganzfeld experiments constitute neither a successful replication of the original ganzfeld experiments nor a sufficient body of data to conclude that ESP has finally been demonstrated. This new set of experiments needs independent replication with tighter controls.

4. APPARENT REPLICATIONS OF THE AUTOGANZFELD EXPERIMENTS

Utts points to some apparent replications of the ganzfeld experiments that have been reported at parapsychological meetings. The major one is a direct attempt to replicate the autoganzfeld experiments with better controls, done at the University of Edinburgh. The reported results were apparently significant but were due to just one of the three experimenters. The two experienced experimenters produced only chance hitting. There are some inconsistencies in these unpublished reports. Utts points to three different replications that were apparently successful. I have heard of at least two large-scale replications that were unsuccessful. None of these replications, however, has been reported in a refereed journal and none has had the opportunity to be critically scrutinized. So we cannot count these one way or the other at this time until we know the details.

5. THE SAIC EXPERIMENTS

Utts and I were hired as the evaluation panel to assess the results of 20 years of previously classified research on remote viewing and related ESP phenomena. In the time available to us, it was impossible to scrutinize carefully all the of documents generated by this program. Instead, we focused our efforts on evaluating the ten studies done at Science Applications International Corporation (SAIC) during the early 1990s. These were selected, in consultation with the principal investigator, as representing the best experiments in the set. These ten experiments included two that examined physiological correlates of ESP. The results were negative. Another study found a correlation between when a subject was being observed (via remote camera) and galvanic skin reactions. The remaining studies, in one way or another, dealt with various target and other factors that might influence remote viewing ability. In these studies the same set of viewers produced descriptions that were successfully matched against the correct target consistently better than chance (with some striking exceptions).

Neither Utts nor I had the time or resources to fully scrutinize the laboratory procedures or data from these experiments. Instead, we relied on what we could glean from reading the technical reports. Two of the experiments had recently been published in the *Journal of Parapsychology*. The difficulty here is that these newly declassified experiments have not been in the public arena for a sufficient time to have been carefully and critically scrutinized. As with the original ganzfeld data base and the autoganzfeld experiments, it takes careful scrutiny and a period of a few years to find the problems of newly published or revealed parapsychological experiments. One obvious problem with the SAIC experiments is that the remote viewing results were all judged by one person—the director of the program. I believe that Utts agrees with me that we have to withhold judgments on these experiments until it can be shown that independent judges can produce the same results. Beyond this, we would require, as with any other set of newly designed experiments, replication by independent laboratories before we decide that the reported outcomes can be trusted.

6. PRIMA FACIE EVIDENCE

Utts and other parapsychologists also talk about *prima facie* evidence in connection with the operational stories of the psychics (or remote viewers) employed by the government. Everyone agrees there is no way to evaluate the accounts of these attempts to use input from remote viewers in intelligence activities. This is because the data were collected in haphazard and nonsystematic ways. No consistent records are available; no attempt was made to interrogate the viewers in non-suggestive ways; no contemporary systematic attempts to evaluate the results are there, etc.

The attempts to evaluate these operational uses after the fact are included in the American Institutes for Research (A.I.R.) report and they do not justify concluding anything about the effectiveness or reality of remote viewing. Some stories, especially those involving cases that occurred long ago and/or that are beyond actual verification, have been put forth as evidence of apparently striking hits. The claim is that these remote viewers are right on—are actually getting true psychic signals—about 20 percent of the time.

Call it *prima facie* or whatever, none of this should be considered as evidence for anything. In situations where we do have some control comparisons, we find the same degree of hitting for wrong targets (when the judge does not realize it is the wrong target) as for the correct targets. A sobering example of this with respect to remote viewing can be found in David Marks and Richard Kammann's book *The Psychology of the Psychic* (Prometheus Books, Amherst, New York, 1980).

Psychologists, such as myself, who study subjective validation find nothing striking or surprising in the reported matching of reports against targets in the Stargate data. The overwhelming amount of data generated by the viewers is vague, general, and way off target. The few apparent hits are just what we would expect if nothing other than reasonable guessing and subjective validation are operating.

7. CONSISTENCY AMONG THE DIFFERENT SOURCES

Utts points to consistencies in effect sizes across the studies. More important, she points out several patterns such as bigger effect sizes with experienced subjects, etc. I do not have time or space to detail all the problems with these apparent consistencies. Many of them happen to relate to the fact that the average effect sizes in these cases are arbitrary combinations of heterogeneous sources. Moreover, where Utts detects consistencies, I find inconsistencies. I have documented some of these elsewhere; I will do so again in the near future.

CONCLUSIONS

When we examine the basis of Utts's strong claim for the existence of psi, we find that it relies on a handful of experiments that have been shown to have serious weaknesses after undergoing careful scrutiny, and another handful of experiments that have yet to undergo scrutiny or be successfully replicated. What seems clear is that the scientific community is not going to abandon its fundamental ideas about causality, time, and other principles on the basis of a handful of experiments whose findings have yet to be shown to be replicable and lawful.

Utts does assert that the findings from parapsychological experiments can be replicated with well-controlled experiments given adequate resources. But this is a hope or promise. Before we abandon relativity and quantum mechanics in their current formulations, we will require more than a promissory note. We will want, as is the case in other areas of science, solid evidence that these findings can, indeed, be produced under specified conditions.

Again, I do not have time to develop another part of this story. Because even if Utts and her colleagues are correct and we were to find that we could reproduce the findings under specified conditions, this would still be a far cry from concluding that psychic functioning has been demonstrated. This is because the current claim is based entirely upon a negative outcome—the sole basis for arguing for ESP is that extra-chance results can be obtained that apparently cannot be explained by normal means. But an infinite variety of normal possibilities exist and it is not clear than one can control for all of them in a single experiment. You need a positive theory to guide you as to what needs to be controlled, and what can be ignored. Parapsychologists have not come close to this as yet.

29

PSI IN PSYCHOLOGY
Susan Blackmore

(**M**) ost academic psychologists do not yet accept the existence of psi." So begins the abstract of an important new paper on parapsychology.

The claim is undoubtedly true, and many readers of *Skeptical Inquirer* will think it ought to stay that way. Yet this quotation implies that the psychologists in question might soon be changing their minds.

It comes, not from a tirade against the skepticism of academia but from an article published in one of psychology's most prestigious academic journals, *Psychological Bulletin.*

Titled "Does Psi Exist? Replicable Evidence for an Anomalous Process of Information Transfer," the article is by parapsychologist Charles Honorton (who died in 1992—see *Skeptical Inquirer* 17: 306–308, Spring 1993) and Cornell psychologist Daryl Bem. Bem's high profile and the respect he is accorded by psychologists will ensure that this article is taken seriously and that perhaps some of them will begin to wonder about psi.

So how good is this "replicable evidence"? Is it sufficient, as the authors claim, to suggest the existence of "anomalous processes of information or energy transfer (like telepathy or other forms of extrasensory perception) that are currently unexplained in terms of known physical or biological mechanisms"?

The evidence in question is an overview of research on the ganzfeld. This technique of partial sensory deprivation of subjects in ESP experiments, pioneered by Honorton in the mid-1970s, produced the great Ganzfeld Debate in the mid-1980s (Hyman 1985; Honorton 1985; see also *Skeptical Inquirer* 10: 2–7, Fall 1985), and finally culminated in fully automated testing procedures (Honorton et al. 1990). The paper reviews the various meta-analyses of ganzfeld findings and, as Honorton has done before, argues that obvious problems like multiple analysis,

---

This article first appeared in *Skeptical Inquirer* 18, no. 4 (Summer 1994): 351–55.

selective reporting, and methodological flaws cannot be responsible for the high replication rates. It then presents data from "11 new ganzfeld studies."

In 1988 the National Research Council produced a report on enhancing human performance that included a highly negative conclusion on parapsychology (Druckman and Swets 1988; see *Skeptical Inquirer* 13: 34–45, Fall 1988). Bem and Honorton have some new light to cast on this report that might seem out of place in an academic article. The NRC apparently solicited a background paper from Harris and Rosenthal, and this paper noted the impressive ganzfeld results. According to Bem and Honorton, the chair of the NRC committee telephoned Rosenthal and asked him to delete this section of the paper. Rosenthal refused to do so, but the section in question did not appear in the final NRC report.

This is troubling indeed if it means that prejudice is coming before genuine scientific inquiry. It serves more than anything to highlight how difficult it is to stick to purely scientific issues in this controversial area.

Bem and Honorton also accuse the NRC report of not being an independent examination of the ganzfeld because it was heavily based on Ray Hyman's original critique. As for their own paper, I would say it could also be misleading. The strength of some of Hyman's arguments is not at all apparent and the impression is given that many laboratories have found replicable effects.

This claim receives some support from Honorton's analysis, which shows that the overall effect does not depend on just one or two labs. Nevertheless, by far the greatest number of studies were contributed by just two researchers. Out of 28 studies, Honorton contributed 5 and Carl Sargent, at Cambridge, 9. So Sargent's are by far the largest number of studies. However, I have shown (Blackmore 1987) that there were very serious problems with nearly all of Sargent's ganzfeld studies. By failing to mention this, Bem and Honorton imply that these nine studies are reliable research.

The story of my visit to Sargent's laboratory is no secret. I went there to try to find out why Sargent was successful with the ganzfeld while I was not. I observed 13 sessions, of which 6 were direct hits. I then considered whether the hits might be due to sensory leakage, experimental error, fraud, or psi, and made predictions based on all these hypotheses. I was satisfied that sensory leakage was not a problem but found serious errors in the randomization procedure. This cumbersome procedure required sets of sealed, unmarked envelopes containing letters specifying the target selection (Sargent 1980). On the basis of my various hypotheses I predicted specific errors in the placement and contents of some of these envelopes and, when I was able to open all the envelopes, I found several such errors. I published the findings and Sargent and his colleagues responded, claiming that they were due to minor experimental error (Harley and Matthews 1987; Sargent 1987).

Whatever one's favored interpretation, it is my opinion that these problems are serious and we should not consider Sargent's studies a valid part of the ganzfeld database. It may be for this reason that parapsychologists now rarely cite them as evidence for psi. So I was disappointed to find that Bem and Honorton did not mention this at all.

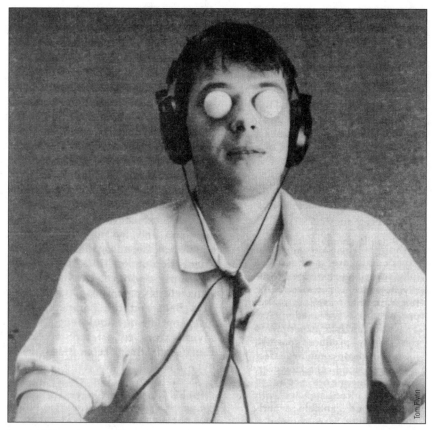

In a ganzfeld experiment, a subject relaxes in a chair in a soundproof room with half a Ping-Pong ball taped over each eye. The only sound is white noise coming from the headphones. At the same time, in a series of trials, a "sender" in another room is given a set of four pictures and asked to concentrate on one. No one else knows which pictures are chosen until the experiment is completed and the subject has reported any impressions he "received."

They then go on to provide data from "11 new ganzfeld studies." This will perhaps give readers the impression that the data are being presented for the first time. In fact they were previously published, in almost the same form, in the *Journal of Parapsychology* in 1990 (Honorton et al. 1990).

Admittedly they are impressive and deserve presentation to a wider audience than the readers of parapsychology journals. They are the series of experiments using the autoganzfeld—or fully automated ganzfeld system. This was designed to meet the "stringent standards" set by Hyman and Honorton (1986) in their joint communique—which ended the great Ganzfeld Debate.

In this new paper Bem and Honorton claim that these "stringent standards" have been met—a view with which I widely concur. As the experiments are presented here, and in the previous 1990 paper, there are no obvious methodological flaws. The results are highly significant and the effect size is comparable to that

RAY HYMAN RESPONDS TO BEM AND HONORTON

Bem and Honorton's *Psychological Bulletin* paper is followed in the same issue by a six-page critique from Ray Hyman, Department of Psychology, University of Oregon ("Anomaly or Artifact? Comments on Bem and Honorton," pp. 19–24).

States Hyman:

> Although the autoganzfeld experiments are methodologically superior to previous parapsychological experiments, the tests of their randomization procedures were inadequate. The autoganzfeld experiments consistently produced positive hit rates, whose combined effect was highly significant. However, these experiments produced important inconsistencies with the previous ganzfeld experiments. They also showed a unique pattern in the data that may reflect a systematic artifact. Because of these unique features, we have to wait for independent replications of these experiments before we can conclude that a replicable anomaly or psi has been demonstrated.

Hyman concludes: "The autoganzfeld experiments are a praiseworthy improvement in methodological sophistication and experimental rigor over the previous ganzfeld experiments. Despite these improvements, the experiments fall disappointingly short in the critical area of justifying the randomization procedures. Even though all but one of the individual studies produced a positive effect size and the overall effect was significant, the autoganzfeld experiments do not constitute a successful replication of the original ganzfeld experiments." Hyman says Bem and the parapsychologists' optimism should encourage investigators to attempt replications, and "these replications will eventually decide the issue."

Bem follows Hyman's critique with his "Response to Hyman" (pp. 25–27), in which he says: "I argue that our claims about the consistency of the autoganzfeld results with the earlier database are quite modest and challenge his [Hymen's] counterclaim that the results are inconsistent with it. In response to his methodological point, I present new analyses that should allay apprehensions about the adequacy of the randomization procedures."

found in previous ganzfeld studies. There is also confirmation of findings that emerged from earlier meta-analyses. For example, selected subjects who are artistic and believe in psi do better than unselected subjects. They do better when friends act as senders and dynamic targets are better than static ones.

I think one's response to this should be optimistic. The work so far looks promising—let's see whether the results can be replicated by other researchers.

Honorton is no longer with us, but fortunately he was in Edinburgh long enough to help set up the autoganzfeld there for others to continue. Just a few months ago the first results were reported at the Society for Psychical Research annual conference in Glasgow (Morris et al. 1993).

Two pilot experiments were reported. In the first, 16 unselected subjects, with friends as senders, tried to detect both dynamic and static targets. Results were at chance. In a second study 32 subjects were selected for artistic or musical ability and a positive attitude toward psi. They were tested in pairs using dynamic targets. This time the hit rate was over 40 percent (25 percent is expected by chance), which is comparable to Honorton's previous results.

What can we now conclude? These were only pilot studies carried out by students and they used a ganzfeld setup that was not quite completed. Will these results continue when the promised refinements are in place? Will the final setup really rule out methodological flaws and the chance of experimenter or subject fraud? I await the next set of results with more than bated breath.

In my opinion Bem and Honorton have overstated the strength of the earlier ganzfeld evidence. They have completely ignored the problems with Sargent's work and understated some of Hyman's objections. However, I agree with them that the autoganzfeld procedure appears sound and the results are important. It is too soon for those "academic psychologists" to change their minds about psi, but it is no bad thing that they have been given the chance to consider it.

REFERENCES

Bem, V. J., and C. Honorton. 1994. Does psi exist? Replicable evidence for an anomalous process of information transfer. *Psychological Bulletin* 115(1): 4–18.

Blackmore, S. J. 1987. A report of a visit to Carl Sargent's laboratory. *Journal of the Society for Psychical Research* 54: 186–98.

Druckman, D., and J. A. Swets. eds. 1988. *Enhancing Human Performance: Issues, Theories and Techniques.* Washington, D.C. National Academy Press.

Harley, T., and G. Matthews. 1987. Cheating, psi, and the appliance of science: A reply to Blackmore. *Journal of the Society for Psychical Research* 54: 199–207.

Honorton, C. 1985. Meta-analysis of psi ganzfeld research: A response to Hyman. *Journal of Parapsychology* 49: 51–86.

Honorton, C., R. E. Berger, M. P. Varvoglis, M. Quant, P. Derr, E. I. Schechter, and D. C. Ferrari. 1990. Psi communication in the ganzfeld. *Journal of Parapsychology* 54: 99–139.

Hyman, R. 1985. The ganzfeld psi experiment: A critical appraisal. *Journal of Parapsychology* 49: 3–49.

Hyman, R., and C. Honorton. 1986. A joint communique: The psi ganzfeld controversy. *Journal of Parapsychology* 50: 351–64.

Morris, R. L., S. Cunningham, S. McAlpine, and R. Taylor. 1993. Toward replication and extension of autoganzfeld results. Paper presented at the 17th International Conference of the Society for Psychical Research, Glasgow, September 1993.

Sargent, C. L. 1980. Exploring psi in the ganzfeld. *Parapsychological Monographs* 17. New York: Parapsychological Foundation.

———. 1987. Sceptical fairytales from Bristol. *Journal of the Society for Psychical Research* 54: 208–18.

30

MYTHS OF SUBLIMINAL PERSUASION: THE CARGO–CULT SCIENCE OF SUBLIMINAL PERSUASION

Anthony R. Pratkanis

I magine that it is the late 1950s—a time just after the Korean War, when terms like *brainwashing* and *mind control* were on the public's mind and films like *The Manchurian Candidate* depicted the irresistible influence of hypnotic trances. You and your friend are off to see *Picnic*, one of the more popular films of the day. However, the movie theater, located in Fort Lee, New Jersey, is unlike any you have been in before. Unbeknownst to you, the projectors have been equipped with a special device capable of flashing short phrases onto the movie screen at such a rapid speed that you are unaware that any messages have been presented. During the film, you lean over to your companion and whisper, "Gee, I'd love a tub of buttered popcorn and a Coke right now." To which he replies, "You're always hungry and thirsty at movies, shhhhh." But after a few moments he says, "You know, some Coke and popcorn might not be a bad idea."

A short time later you hear that you and your friend weren't the only ones desiring popcorn and Coke at the theater that day. According to reports in newspapers and magazines, James Vicary, an advertising expert, had secretly flashed, at a third of a millisecond, the words "Eat Popcorn" and "Drink Coke" onto the movie screen. His studies, lasting six weeks, involved thousands of moviegoing subjects who received a subliminal message every five seconds during the film. Vicary claimed an increase in Coke sales of 18 percent and a rise in popcorn sales of almost 58 percent. Upon reading their newspapers, most people were outraged and frightened by a technique so devilish that it could bypass their conscious intellect and beam *subliminal* commands directly to their subconscious. (See Moore, chapter 31 in this book, for a definition of "subliminal.")

In an article titled "Smudging the Subconscious," Norman Cousins (1957) captured similar feelings as he pondered the true meaning of such a device. As he put it, "If the device is successful for putting over popcorn, why not politicians or

This article first appeared in *Skeptical Inquirer* 16, no. 3 (Spring 1992): 260–72.

anything else?" He wondered about the character of people who would dream up a machine to "break into the deepest and most private parts of the human mind and leave all sorts of scratchmarks." Cousins concluded that the best course of action would be "to take this invention and everything connected to it and attach it to the center of the next nuclear explosive scheduled for testing."

Cousins's warnings were taken to heart. The Federal Communications Commission immediately investigated the Vicary study and ruled that the use of subliminal messages could result in the loss of a broadcast license. The National Association of Broadcasters prohibited the use of subliminal advertising by its members. Australia and Britain banned subliminal advertising. A Nevada judge ruled that subliminal communications are not protected as free speech.

The Vicary study also left an enduring smudge on Americans' consciousness—if not their *sub*conscious. As a teacher of social psychology and a persuasion researcher, one of the questions I am most frequently asked is, "Do you know about the 'Eat Popcorn/Drink Coke' study that *they* did?" At cocktail parties, I am often pulled aside and, in hushed tones, told about the "Eat Popcorn/Drink Coke" study. Indeed, my original interest in subliminal persuasion was motivated by an attempt to know how to respond to such questions.

Three public-opinion polls indicate that the American public shares my students' fascination with subliminal influence (Haber 1959; Synodinos 1988; Zanot, Pincus, and Lamp 1983). By 1958, just nine months after the Vicary subliminal story first broke, 41 percent of survey respondents had heard of subliminal advertising. This figure climbed to 81 percent in the early 1980s, with more than 68 percent of those aware of the term believing that it was effective in selling products. Most striking, the surveys also revealed that many people learn about subliminal influence through the mass media and through courses in high school and college.

But there is a seamier side to the "Eat Popcorn/Drink Coke" study—one that is rarely brought to public attention. In a 1962 interview with *Advertising Age,* James Vicary announced that the original study was a fabrication intended to increase customers for his failing marketing business. The circumstantial evidence suggests that this time Vicary was telling the truth. Let me explain by recounting the story of the "Eat Popcorn/Drink Coke" study as best I can, based on various accounts published in academic journals and trade magazines (see Advertising Research Foundation 1958; "ARF Checks" 1958; Danzig 1962; McConnell, Cutler, and McNeil 1958; "Subliminal Ad" 1958; "Subliminal Has" 1958; Weir 1984).

Advertisers, the FCC, and research psychologists doubted Vicary's claims from the beginning and demanded proof. To meet these demands, Vicary set up demonstrations of his machine. Sometimes there were technical difficulties in getting the machine to work. When the machine did work, the audience felt little compulsion to comply with subliminal commands, prompting an FCC commissioner to state, "I refuse to get excited about it—I don't think it works" ("Subliminal Has" 1958).

In 1958, the Advertising Research Foundation pressed Vicary to release his data and a detailed description of his procedures. They argued that it had been more than a year since the results were made public and yet there had been no formal write-up of the experiment, which was necessary to evaluate the claims. To this day, there has been no primary published account of the study, and scientists interested in replicating the results must rely on accounts published in such magazines as the *Senior Scholastic* ("Invisible Advertising" 1957), which, although intended for junior-high students, presents one of the most detailed accounts of the original study.

Pressures for a replication accumulated. Henry Link, president of Psychological Corporation, challenged Vicary to a test under controlled conditions and supervised by an independent research firm. No change occurred in the purchase of either Coke or popcorn (Weir 1984). In one of the more interesting attempted replications, the Canadian Broadcast Corporation, in 1958, subliminally flashed the message "Phone Now" 352 times during a popular Sundaynight television show called *Close-up* ("Phone Now" 1958). Telephone usage did not go up during that period. Nobody called the station. When asked to guess the message, viewers sent close to five hundred letters, but not one contained the correct answer. However, almost half of the respondents claimed to be hungry or thirsty during the show. Apparently, they guessed (incorrectly) that the message was aimed at getting them to eat or drink.

Finally, in 1962 James Vicary lamented that he had handled the subliminal affair poorly. As he stated, "Worse than the timing, though, was the fact that we hadn't done any research, except what was needed for filing for a patent. I had only a minor interest in the company and a small amount of data—too small to be meaningful. And what we had shouldn't have been used promotionally" (Danzig 1962). This is not exactly an affirmation of a study that supposedly ran for six weeks and involved thousands of subjects.

My point in presenting the details of the Vicary study is twofold. First, the "Eat Popcorn/Drink Coke" affair is not an isolated incident. The topic of subliminal persuasion has attracted the interest of Americans on at least four separate occasions: at the turn of the century, in the 1950s, in the 1970s, and now in the late 1980s and early 1990s. Each of these four flourishings of subliminal persuasion show a similar course of events. First, someone claims to find an effect; next, others attempt to replicate that effect and fail; the original finding is then criticized on methodological grounds; nevertheless the original claim is publicized and gains acceptance in lay audiences and the popular imagination. Today we have reached a point where one false effect from a previous era is used to validate a false claim from another. For example, I recently had the occasion to ask a manufacturer of subliminal self-help audiotapes for evidence of his claim that his tapes had therapeutic value. His reply: "You are a psychologist. Don't you know about the study they did where they flashed 'Eat Popcorn and Drink Coke' on the movie screen?"

During the past few years, I have been collecting published articles on subliminal processes—research that goes back over a hundred years (Suslowa 1863) and includes more than a hundred articles from the mass media and more than two hundred academic papers on the topic (Pratkanis and Greenwald 1988). In none of these papers is there clear evidence in support of the proposition that subliminal messages influence behavior. Many of the studies fail to find an effect,

"YOU KNOW A COKE AND A POPCORN MIGHT NOT BE A BAD IDEA RIGHT NOW."

and those that do either cannot be reproduced or are fatally flawed on one or more methodological grounds, including: the failure to control for subject expectancy and experimenter bias effects, selective reporting of positive over negative findings, lack of appropriate control treatments, internally inconsistent results, unreliable dependent measures, presentation of stimuli in a manner that is not truly subliminal, and multiple experimental confounds specific to each study. As Moore (chapter 31) points out, there is considerable evidence for subliminal perception or the detection of information outside of self-reports of awareness. However, subliminal perception should *not* be confused with subliminal persuasion or influence—motivating or changing behavior—for which there is little good evidence (see McConnell, Cutler, and McNeil 1958; Moore 1982 and 1988).

My second reason for describing the Vicary study in detail is that it seems to me that our fascination with subliminal persuasion is yet another example of what Richard Feynman (1985) called "cargo-cult science." For Feynman, a cargo-cult science is one that has all the trappings of science—the illusion of objectivity, the appearance of careful study, and the motions of an experiment—but lacks one important ingredient: skepticism, or a leaning over backward to see if one might be mistaken. The essence of science is to doubt your own interpretations and theories so that you may improve upon them. This skepticism is often missing in the interpretation of studies claiming to find subliminal influence. Our theories and wishes for what we would like to think the human mind is capable of doing interferes with our ability to see what it actually does.

The cargo-cult nature of subliminal research can be seen in some of the first studies on the topic done at the turn of the century. In 1900, Dunlap reported a subliminal Müller-Lyer illusion—a well-known illusion in which a line is made to appear shorter or longer depending on the direction of angles placed at its ends. Dunlap flashed an "imperceptible shadow" or line to subliminally create this illusion. He claimed that his subjects' judgment of length was influenced by the imperceptible shadows. However, Dunlap's results could not be immediately replicated either by Titchener and Pyle (1907) or by Manro and Washburn (1908). Nevertheless, this inconsistency of findings did not stop Hollingworth (1913) from discussing the subliminal Müller-Lyer illusion in his advertising textbook or from drawing the conclusion that subliminal influence is a powerful tool available to the advertiser.

I contend that it was no accident that subliminal influence was first investigated in America at the turn of the century. The goal of demonstrating the power of the subliminal mind became an important one for many people at that time. It was a time of great religious interest, as illustrated by academic books on the topic, religious fervor among the populace, and the further development of a uniquely American phenomenon—the spiritual self-help group. One such movement, popular in intellectual circles, was called "New Thought," which counted William James among its followers. The doctrine of New Thought stated that the mind possesses an unlimited but hidden power that could be tapped—if one knew how—to bring about a wonderful, happy life and to exact physical cures. Given the rise of industrialization and the anonymity of newly formed city life, one can see how a doctrine of the hidden power of the individual in the face of realistic powerlessness would be well received in some circles.

The historian Robert Fuller (1982; 1986) traces the origins of New Thought and similar movements to early American interest in the teachings of Franz Anton Mesmer. Fuller's point is that the powerful unconscious became a replacement for religion's "soul." Mesmer's doctrines contended that each person possessed a hidden, though strong, physical force, which he termed *animal magnetism*. This force could be controlled by the careful alignment of magnets to effect personality changes and physical cures. On one level, mesmerism can be viewed as a secularization of the metaphor of spiritual humans that underlies witchcraft. Animal magnetism replaced the soul, and good and bad magnets replaced angels and devils that could invade the body and affect their will. Mesmerism was introduced to America at the beginning of the nineteenth century and, characteristic of Yankee ingenuity, self-help movements soon sprang up with the goal of improving on Mesmer's original magnet therapy; they did so by developing the techniques of hypnotism, séances, the healing practices of Christian Science, positive thinking, and the speaking cure.

With the distance of a century, we overlook the fact that many journals of the nineteenth century were devoted to archiving the progress of mesmerism and with documenting the influence of the unconscious on the conscious. As Dunlap (1900) said in the introduction to his article on the subliminal Müller-Lyer illusion, "If such an effect is produced, then we have evidence for the belief that under certain conditions things of which we are not and can not become conscious have their immediate effects upon consciousness." In other words, we would have one of the first scientific demonstrations that the unconscious can powerfully influence the conscious. A simple step perhaps, but who knows what wonderful powers of the human mind wait to be unleashed.

As a postscript to the subliminal Müller-Lyer affair, I should point out that 30 years later Joseph Bressler—a student of Hollingworth—was able to reconcile the empirical differences between Dunlap and his opponents. Bressler (1931) found that as the subliminal angles increased in intensity—that is, as they approached the threshold of awareness—the illusion was more likely to be seen. This finding, along with many others, served as the basis for concluding that there is no absolute threshold of awareness—it can vary as a function of individual and situational factors—and led to the hypothesis that, on some trials, subjects could see enough of the stimulus to improve their guessing at what might be there. (See also Holender 1986 and Cheesman and Merikle's 1985 distinction between objective and subjective thresholds.)

Other manifestations of "subliminal-mania" illustrate additional aspects of a cargo-cult science. In the early 1970s, during the third wave of popular interest in subliminal persuasion, the best-selling author Wilson Bryan Key (1973, 1976, 1980, 1989) advanced the cargo-cult science of subliminal seduction in two ways. (See also Creed 1987.) First, Key argued that subliminal techniques were not just limited to television and movies. Cleverly hidden messages aimed at inducing sexual arousal are claimed to be embedded in the photographs of print advertisements. Key found the word sex printed on everything from Ritz crackers to the ice cubes in a Gilbey Gin ad. Second, Key was successful in linking the concept of subliminal persuasion to the issues of his day. The 1970s were a period of distrust by Americans of their government, businesses, and institutions. Key claimed that big advertisers and big government are in a conspiracy to control our minds using subliminal implants.

The legacy of Key's cargo-cult science is yet with us. I often ask my students at the University of California, Santa Cruz, if they have heard of the term *subliminal persuasion* and, if so, where. Almost all have heard of the term and about half report finding out about it in high school. Many received an assignment from their teachers to go to the library and look through magazine ads for subliminal implants.

These teachers miss an opportunity to teach science instead of cargo-cult science. Key (1973) reports a study where more than a thousand subjects were shown the Gilbey Gin ad that supposedly contained the word sex embedded in ice cubes. Sixty-two percent of the subjects reported feeling "aroused," "romantic," "sensuous." Instead of assuming that Key was right and sending students out to find subliminals, a science educator would encourage a student to ask, "But where is the control group in the Gilbey Gin ad study? Perhaps an even higher percentage would report feeling sexy if the subliminal "sex" was removed—perhaps the same, perhaps less. One just doesn't know.

Now in the late 1980s and early 1990s, we see a fourth wave of interest in subliminal influence. Entrepreneurs have created a $50-million-plus industry offering subliminal self-help audio- and video-tapes designed to improve everything from self-esteem to memory, to employee and customer relations, to sexual responsiveness, and—perhaps most controversial—to overcoming the effects of family and sexual abuse (Natale 1988). The tapes work, according to one manufacturer, because "subliminal messages bypass the conscious mind, and imprint directly on the subconscious mind, where they create the basis for the kind of life you want." Part of the popularity of such tapes no doubt springs from the tenets of New Age. Like its predecessor New Thought, New Age also postulates a powerful hidden force in the human personality that can be controlled for the good, not by magnets, but by crystals, and can be redirected with subliminal commands.

Accusations concerning the sinister use of subliminal persuasion continue as well. In the summer of 1990, the rock band Judas Priest was placed on trial for allegedly recording, in one of their songs, the subliminal implant "Do it." This message supposedly caused the suicide deaths of Ray Belknap and James Vance.

What is the evidence that subliminal influence, despite not working in the 1900s, 1950s, and 1970s, is now effective in the 1990s? Tape company representatives are likely to provide you with a rather lengthy list of "studies" demonstrating their claims. Don't be fooled. The studies on these lists fall into two camps—those done by the tape companies and for which full write-ups are often not available, and those that have titles that sound as if they apply to subliminal influence, but really don't. For example, one company lists many subliminal-perception studies to support its claims. It is a leap of faith to see how a lexical priming study provides evidence that a subliminal self-help tape will cure insomnia or help overcome the trauma of being raped. Sadly, the trick of claiming that something that has nothing to do with subliminal influence really does prove the effectiveness of subliminal influence goes back to the turn of the century. In the first footnote to their article describing a failure to replicate Dunlap's subliminal Müller-Lyer effect, Titchener and Pyle (1907) state: "Dunlap finds a parallel to his own results in the experiments of Pierce and Jastrow on small difference of sensations. There is, however, no resemblance whatever between the two investigations." In a cargo-cult science, any evidence—even irrelevant facts—is of use and considered valuable.

Recently, there have been a number of studies that directly tested the effectiveness of subliminal self-help tapes. I conducted one such study in Santa Cruz with my colleagues Jay Eskenazi and Anthony Greenwald (Pratkanis, Eskenazi, and Greenwald 1990). We used mass-marketed audiotapes with subliminal messages designed to improve either self-esteem or memory abilities. Both types of tapes contained the same supraliminal content—various pieces of classical music. However, they differed in their subliminal content. According to the manufacturer, the self-esteem tapes contained subliminal messages like "I have high self-worth and high self-esteem." The memory tape contained subliminal messages like "My ability to remember and recall is increasing daily."

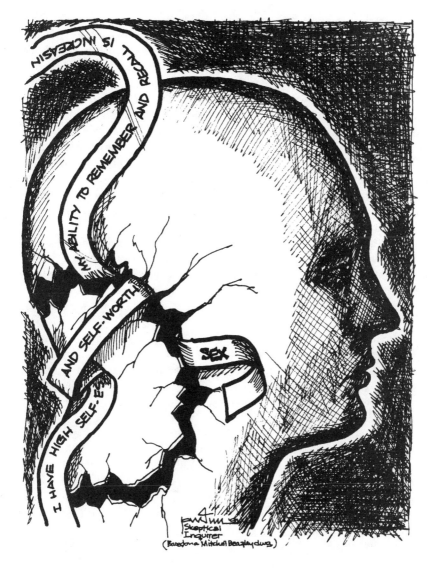

Using public posters and ads placed in local newspapers, we recruited volunteers who appeared most interested in the value and potential of subliminal self-help therapies (and who were probably similar to those likely to buy such tapes). On the first day of the study, we asked our volunteers to complete three different self-esteem and three different memory measures. Next they randomly received their subliminal tape, but with an interesting twist. Half of the tapes were mislabeled so that some of the subjects received a memory tape, but thought it was intended to improve self-esteem, whereas others received a self-esteem tape that had been mislabeled as memory improvement. (Of course half the subjects received correctly labeled tapes.)

The volunteers took their tapes home and listened to them every day for five weeks (the period suggested by the manufacturer for maximum effectiveness). During the listening phase, we attempted to contact each subject about once a week to encourage their daily listening. Only a handful of subjects were unable to complete the study, suggesting a high level of motivation and interest in subliminal therapy. After five weeks of daily listening, they returned to the laboratory and once again completed self-esteem and memory tests and were also asked to indicate if they believed the tapes to be effective.

The results: the subliminal tapes produced *no* effect (improvement or decrement) on either self-esteem or memory. But our volunteers did not believe this to be the case. Subjects who thought they had listened to a self-esteem tape (regardless of whether they actually did or not) were more likely to be convinced that their self-esteem had improved, and those who thought they had listened to a memory tape were more likely to believe that their memory had improved as a result of listening to the tape. We called this an illusory placebo effect—placebo, because it was based on expectations; illusory, because it wasn't real. In sum, the subliminal tapes did nothing to improve self-esteem or memory abilities but, to some of our subjects, they appeared to have an effect. As we put it in the title of our report of this study, "What you expect is what you believe, but not necessarily what you get."

Our results are not a fluke. We have since repeated our original study twice using different tapes and have yet to find an effect of subliminal messages upon behavior as claimed by the manufacturer (Greenwald, Spangenberg, Pratkanis, and Eskenazi 1991). By combining our data from all three studies, we gain the statistical power to detect quite small effects. Still, there is no evidence of a subliminal effect consistent with the manufacturers' claims.

Other researchers are also finding that subliminal self-help tapes are of no benefit to the user. In a series of three experiments, Auday, Mellett, and Williams (1991) tested the effectiveness of bogus and real subliminal tapes designed either to improve memory, reduce stress and anxiety, or increase self-confidence. The subliminal tapes proved ineffective on all three fronts. Russell, Rowe, and Smouse (1991) tested subliminal tapes designed to improve academic achievement and found the tapes improved neither grade point average nor final examination scores. Lenz (1989) had 270 Los Angeles police recruits listen for 24 weeks to music with and without subliminal implants designed to improve either knowledge of the law or marksmanship. The tapes did not improve either. In a recent test, Merikle and Skanes (1991) found that overweight subjects who listened to subliminal weight-loss tapes for five weeks showed no more weight loss than did control subjects. In sum, independent researchers have conducted 9 studies to evaluate the effectiveness of subliminal self-help tapes. All 9 studies failed to find an effect consistent with the manufacturers' claims. (See also Eich and Hyman 1991.)

It appears that, despite the claims in books and newspapers and on the backs of subliminal self-help tapes, subliminal-influence tactics have not been demonstrated to be effective. Of course, as with anything scientific, it may be that someday, somehow, someone will develop a subliminal technique that may work, just as someday a chemist may find a way to transmute lead to gold. l am personally not purchasing lead futures on this hope however.

The history of the subliminal controversy teaches us much about persuasion—but not the subliminal kind. If there is so little scientific evidence of the effectiveness of subliminal influence, why then do so many Americans believe it works? In a nutshell, I must conclude, with Feynman, that despite enjoying the fruits of science, we are not a scientific culture, but one of ill-directed faith as defined in Hebrews 11:1 (KJV): "Now faith is the substance of things hoped for, the evidence of things not seen."

We can see the workings of this faulty faith, not science, in the more than a hundred popular press articles on the topic of subliminal persuasion. Many of the articles (36 percent) deal with ethical and regulatory concerns of subliminal practices—assuming them to be effective. Only 18 percent of the articles declare flatly that subliminal influence is ineffective, with the remaining either claiming that it works or suggesting a big "maybe" to prompt readers' concern. In general, popular press articles fail to rely on scientific evidence and method to critically evaluate subliminal findings. Positive findings are emphasized and null results rarely reported. Problems with positive subliminal findings, such as lack of control groups, expectancy effects setting subliminal thresholds, and so on, are rarely mentioned. If negative information is given, it is often presented at the end of the article, giving the reader the impression that, at best, the claims for subliminal effectiveness are somewhat controversial. Recent coverage of subliminal self-help tapes, however, has been less supportive of subliminal claims—but this may reflect more of an attack on big business than an embrace of science.

Instead of the scientific method, those accused of subliminal persuasion (mostly advertisers) are subjected to what can be termed the "witch test." During the Middle Ages, one common test of witchcraft was to tie and bind the accused and throw her into a pond. If she floats, she is a witch. If she drowns, then her innocence is affirmed. Protestations by the accused were taken as further signs of guilt.

How do we know that subliminals work and that advertisers use them? As Key notes, advertisers spend a considerable amount of money on communications that contain subliminal messages. Why would they spend such vast sums if subliminal persuasion is ineffective? The fact that these subliminal messages cannot be readily identified or seen and that the advertisers deny their use further demonstrates the craftiness of the advertiser. After all, witches are a wily lot, carefully covering their tracks. It appears that the only way that advertisers can prove their innocence, by the logic of the witch test, is to go out of business at the bottom of the pond, thereby showing that they do not possess the arts of subliminal sorcery. In contrast, just as the motives of the Inquisition for power and fortune went unquestioned, so too the motives of the proponents of subliminal seduction, who frequently profit by the sale of more newspapers, books, or audiotapes, are rarely (or have only recently been) questioned.

The proponents of subliminal persuasion make use of our most sacred expectations, hopes, and fears. Each manifestation of interest in subliminal influence has been linked to the important philosophies and thinking of the day—New

Thought in the 1900s, brainwashing in the 1950s, the corruption of big governments in the 1970s, and New Age philosophy today.

But the belief in subliminal persuasion provides much more for the individual. We live in an age of propaganda; the average American will see approximately seven million advertisements in a lifetime. We provide our citizens with very little education concerning the nature of these persuasive processes. The result is that many may feel confused and bewildered by basic social processes (see Pratkanis and Aronson 1992). The negative side of subliminal persuasion is presented as an irrational force outside the control of the message recipient. As such, it takes on a supernatural "devil made me do it" quality capable of justifying and explaining why Americans are often persuaded and can seemingly engage in irrational behavior. Why then did I buy this worthless product at such a high price? Subliminal sorcery. On the positive side, a belief in subliminal persuasion imbues the human spirit at least with the possibility of overcoming the limitations of being human and of living a mundane existence. We can be like the gods—healing ourselves, finding enjoyment in everything we do, working for the benefit of humankind by tapping our own self potentials. Perhaps our theories of what should be or what we would like to be have caused us to be a little less critical of the claims for the power of subliminal influence.

But belief in subliminal persuasion is not without its cost. Perhaps the saddest aspect of the subliminal affair is that it distracts our attention from more substantive issues. By looking for subliminal influences, we may ignore more powerful, blatant influence tactics employed by advertisers and sales agents. We may ignore other, more successful ways—such as science—for reaching our human potentials.

Consider the tragic suicide deaths of teenagers Ray Belknap and James Vance that were brought to light in the recent trial of Judas Priest. They lived troubled lives—lives of drug and alcohol abuse, run-ins with the law, learning disabilities, family violence, and chronic unemployment. What issues did the trial and the subsequent mass-media coverage emphasize? Certainly not the need for drug treatment centers; there was no evaluation of the pros and cons of America's juvenile justice system, no investigation of the schools, no inquiry into how to prevent family violence, no discussion of the effects of unemployment on a family. Instead, our attention was mesmerized by an attempt to count the number of subliminal demons that can dance on the end of a record needle.

In this trial, Judge Jerry Carr Whitehead (*Vance & Belknap* v. *Judas Priest & CBS Records* 1990) ruled in favor of Judas Priest, stating: "The scientific research presented does not establish that subliminal stimuli, even if perceived, may precipitate conduct of this magnitude. There exist other factors which explain the conduct of the deceased independent of the subliminal stimuli." Perhaps now is the time to lay the myth of subliminal sorcery to rest and direct our attention to other, more scientifically documented ways of understanding the causes of human behavior and improving our condition.

ACKNOWLEDGMENTS

I thank Elliot Aronson, Timothy E. Moore, Marlene E. Turner, and Rick Stoltz for helpful comments. Portions of this paper were presented at the American Psychological Association meetings on August 12, 1990, in Boston, Massachusetts, and the CSICOP Conference on May 3, 1991, in Berkeley/Oakland, California.

REFERENCES

Advertising Research Foundation. 1958. *The Application of Subliminal Perception in Advertising.* New York, N.Y.

ARF checks data on subliminal ads; verdict "Insufficient." 1958. *Advertising Age,* September 15.

Auday, B. C., J. L. Mellett, and P. M. Williams. 1991. Self-improvement Using Subliminal Self-help Audiotapes: Consumer Benefit or Consumer Fraud?" Paper presented at the meeting of the Western Psychological Association, San Francisco Calif., April.

Bressler, J. 1931. Illusion in the case of subliminal visual stimulation. *Journal of General Psychology* 5: 244–51.

Cheesman, J., and P. M. Merikle. 1985. Word recognition and consciousness. In *Reading Research Advances in Theory and Practice,* vol. 5, edited by D. Besner, T. G. Waller, and G. E. MacKinnon, 311–52. New York: Academic Press.

Cousins, N. 1957. Smudging the subconscious. *Saturday Review,* October 5, p. 20.

Creed, T. T. 1987. Subliminal deception: Pseudoscience on the college lecture circuit. *Skeptical Inquirer* 11: 358–66.

Danzig, F. 1962. Subliminal advertising—Today it's just historic flashback for researcher Vicary. *Advertising Age,* September 17.

Dunlap, K. 1900. The effect of imperceptible shadows on the judgment of distance. *Psychological Review* 7: 435–53.

Eich, E., and R. Hyman. 1991. "Subliminal Self-help." In *In the Mind's Eye: Enhancing Human Performance,* edited by D. Druckman and R. A. Bjork, 107–19. Washington D.C.: National Academy Press.

Feynman, R. P. 1985. *Surely You're Joking, Mr. Feynman!* New York: Bantam Books.

Fuller, R. C. 1982. *Mesmerism and the American Cure of Souls.* Philadelphia: University of Pennsylvania Press.

———. 1986. *Americans and the Unconscious.* New York: Oxford University Press.

Greenwald, A. G., E. R. Spangenberg, A. R. Pratkanis, and J. Eskenazi. 1991. Double-blind tests of subliminal self-help audiotapes. *Psychological Science* 2: 119–22.

Haber, R. N. 1959. Public attitudes regarding subliminal advertising. *Public Opinion Quarterly* 23: 291–93.

Holender, D. 1986. Semantic activation without conscious identification in dichotic listening, parafoveal vision, and visual masking: A survey and appraisal. *Behavior and Brain Sciences* 9: 1–66.

Hollingworth, H. L. 1913. *Advertising and Selling.* New York: D. Appleton

Invisible Advertising. 1957. *Senior Scholastic,* October 4.

Key, W. B. 1973. *Subliminal Seduction.* Englewood Cliffs, N.J.: Signet.

———. 1976. *Media Sexploitation.* Englewood Cliffs, N.J.: Signet.

———. 1980. *The Clam-plate Orgy.* Englewood Cliffs, N.J.: Signet.

———. 1989. *The Age of Manipulation.* New York: Holt.

Lenz, S. 1989. "The Effect of Subliminal Auditory Stimuli on Academic Learning and Motor Skills Performance Among Police Recruits." Unpublished doctoral dissertation, California School of Professional Psychology, Los Angeles, Calif.

Manro, H. M., and M. F. Washburn. 1908. The effect of imperceptible line on the judgment of distance. *American Journal of Psychology* 19: 242.

McConnell, J. V., R. I. Cutler, and E. B. McNeil. 1958. Subliminal stimulation: An overview. *American Psychologist* 13: 229–42.

Merikle, P., and H. E. Skanes. 1991. "Subliminal Self-help Audiotapes: A Search for Placebo Effects." Unpublished manuscript, University of Waterloo, London, Ontario.

Moore, T. E. 1982. Subliminal advertising: What you see is what you get. *Journal of Marketing* 46: 38–47.

———. 1988. The case against subliminal manipulation. *Psychology & Marketing* 5: 297–316.

Natale, J. A. 1988. Are you open to suggestion? *Psychology Today,* September, pp. 28–30.

"Phone now," said CBC subliminally—but nobody did. 1958. *Advertising Age,* February 10, p. 8.

Pratkanis, A. R., and E. Aronson. 1992. *Age of Propaganda: The Everyday Use and Abuse of Persuasion.* New York: W. H. Freeman.

Pratkanis, A. R., and A. G. Greenwald. 1988. Recent perspectives on unconscious processing: Still no marketing applications. *Psychology & Marketing* 5: 339–55.

Pratkanis. A. R., J. Eskenazi, and A. G. Greenwald. 1990. "What You Expect Is What You Believe (But Not Necessarily What You Get): On the Effectiveness of Subliminal Self-help Audiotapes." Paper presented at the meeting of the Western Psychological Association, Los Angeles, Calif., April.

Russell, T. G., W. Rowe, and A. D. Smouse. 1991. Subliminal self-help tapes and academic achievement: An evaluation. *Journal of Counseling & Development* 69: 359–62.

Subliminal ad is transmitted in test but scores no popcorn sales. 1958. *Advertising Age,* January 20.

Subliminal has a test; can't see if it works. 1958. *Printer's Ink,* January 17.

Suslowa, M. 1863. Veranderungen der hautgefule unter dem einflusse electrischer reizung. *Zeitschrift fur Rationelle Medicin* 18: 155–60.

Synodinos, N. E. 1988. Subliminal stimulation: What does the public think about it? *Current Issues & Research in Advertising* 11: 157–87.

Titchener, E. B., and W. H. Pyle. 1907. The effect of imperceptible shadows on the judgment of distance. *Proceedings of the American Philosophical Society* 46: 94–109.

Vance and Belknap v. *Judas Priest and CBS Records.* 86-5844186-3939. Second District Court of Nevada. August 24, 1990.

Weir, W. 1984. Another look at subliminal "facts." *Advertising Age,* October 15, p. 46.

Zanot, E. J., J. D. Pincus, and E. J. Lamp. 1983. Public perceptions of subliminal advertising. *Journal of Advertising* 12: 37–45.

31

SUBLIMINAL PERCEPTION: FACTS AND FALLACIES

Timothy E. Moore

$\left(\mathbf{C}\right)$ an the meaning of a stimulus affect the behavior of observers in some way in the absence of their awareness of the stimulus? In a word, *yes*. While there is some controversy, there is also respectable scientific evidence that observers' responses can be shown to be affected by stimuli they claim not to have seen. To a cognitive psychologist this is not particularly earthshaking, but the media and the public have often responded to the notion of subliminal perception with trepidation.

What is subliminal perception? Should we be worried (or perhaps enthused) about covert manipulation of thoughts, attitudes, and behaviors? My reviews (Moore 1982; 1988) have dealt primarily with the validity of the more dramatic claims made on behalf of subliminal techniques and devices. Such an appraisal requires a working definition of "subliminal perception." Then we need to determine whether the conditions under which it occurs and the means by which it is achieved are reflected in the products on the market.

How should "awareness" be defined? One way is simply to ask observers whether or not they are "aware" of a stimulus. If the observer denies any awareness, then the stimulus is, by definition, below an awareness threshold. Using this approach, unconscious perception consists of demonstrating that observers can be affected by stimuli *whose presence they do not report*. Another way to define "awareness" involves requiring observers to distinguish between two or more stimuli that are presented successively. With fast exposure durations, observers may be unable to distinguish between stimuli, or between a stimulus's presence or absence. This method was advocated by Eriksen (1960) and defines consciousness as the observer's ability to discriminate between two or more alternative stimuli in a forced-choice task. In this context, unconscious perception consists of a demonstration that observers are affected by stimuli *whose presence they cannot detect*. The

---

This article first appeared in *Skeptical Inquirer* 16, no. 3 (Spring 1992): 273–81.

approaches are different and involve different sorts of evidence. In the former case the stimuli are not reported; in the latter instance the stimuli cannot be detected.

These two methods of defining consciousness have been referred to as "subjective" and "objective," respectively, by Merikle and his coworkers (Cheesman and Merikle 1986; Merikle and Cheesman 1986). Higher levels of visibility are typically associated with subjective thresholds. The disadvantage of a subjective definition is that a failure to report a stimulus's presence may result from response bias (i.e., the observer is ambivalent about the stimulus's presence and elects to report its absence). As Merikle (1984) has argued, the use of subjective thresholds implies that each participant provides his or her own idiosyncratic definition of "awareness." Consequently, awareness thresholds could (and would) vary greatly from subject to subject.

Some recent studies (e.g., Cheesman and Merikle 1986) have looked at performance when both subjective *and* objective thresholds have been assessed. Such studies indicate that subliminal perception is most appropriately viewed as perception in the absence of concurrent phenomenal experience. We sometimes receive information when subjectively we feel that nothing useful has been "seen." Investigators can establish that perception has occurred in the face of disavowals from participants by forcing them to guess. Respondents may object that they have no basis for making a decision, but by using a forced-choice task we can see that their guesses are more accurate than they would be if they were guessing at random. Clearly, some information is being utilized.

When respondents' guesses are at chance in a detection task, there is no well-established evidence for perception. Thus, subliminal perception is not perception in the absence of a detectable signal. Rather, it occurs under conditions where subjects can detect a signal on at least some proportion of trials. Subjects may claim to be guessing without realizing that their guesses are better than chance. According to Merikle, the dissociation between these two indicators of perception (signal detection vs. introspective reports) defines the necessary empirical conditions for demonstrating subliminal perception. There is an inconsistency between what observers know and what "they know they know."

Recent reviews of research findings in subliminal perception have provided very little evidence that stimuli below observers' subjective thresholds influence motives, attitudes, beliefs, or choices (Moore 1988, 1991b; Pratkanis and Greenwald 1988; Greenwald 1992). In most studies, the stimuli do not consist of directives, commands, or imperatives, and there is no reliable evidence that subliminal stimuli have any pragmatic impact or effects on intentions. Studies that do purport to find such effects are either unreplicated or methodologically flawed in one or more ways (Pratkanis, chapter 30 in this book). There is very little evidence for any perceptual processing *at all* (let alone any pragmatic consequences) when perceptual awareness is equated with an objective threshold.

How do the dramatic claims regarding undetectable stimuli stack up against the preceding review? What are these claims and what is their status? I shall confine my comments to claims involving advertising applications and self-help auditory tapes.

ADVERTISING

Many people believe that most advertisements contain hidden sexual images or words that affect our susceptibility to the ads. This belief is widespread even though there is no evidence for such practices, let alone evidence for such effects. "Embedded" stimuli are difficult to characterize in terms of signal-detection theory or threshold-determination procedures because most of them remain unidentifiable even when focal attention is directed to them. Nevertheless, the use of the term *subliminal* is a fait accompli, and belief in such an influence is primarily the consequence of the writings and lectures of just one person—Wilson Bryan Key (1973, 1976, 1980, 1990). Key offers no scientific evidence to support the existence of subliminal images; nor does he provide any empirical documentation of their imputed effects (Creed 1987; see also Vokey and Read 1985).

In a review of Key's most recent book, John O'Toole, president of the American Association of Advertising Agencies wondered: "Why is there a market for yet another rerun of this troubled man's paranoid nightmares?" (O'Toole 1989: 26). Part of the reason that Key's books sell so well may be that they are not what they appear to be. The information is not presented as the subjective fantasies of one person. Instead, it is presented as scientific, empirical fact. Science is respectable. Consequently, if claims are cloaked in scientific jargon, and if propositions are asserted to be scientifically valid, people can be fooled. Key knows this and uses it to his advantage. His intent is to persuade; and if he can do so by misrepresenting scientific data and findings, he is apparently prepared to do so.

Key provided pretrial testimony at the Judas Priest trial in Reno, Nevada, in the summer of 1990. Two teenagers had committed suicide. Their parents sued Judas Priest and CBS Records Inc., alleging that subliminal messages in Judas Priest's music contributed to the suicides. Key was testifying on behalf of the plaintiffs, and at the trial he responded to a question about scientific methodology by saying: "Science is pretty much what you can get away [with] at any particular point in history and you can get away with a great deal" (*Vance/Roberson v. CBS/Judas Priest,* 1990, 60). This unabashed disdain for anything approaching scientific integrity has not endeared him to the scientific community.

Attempting to apply scientific criteria to propositions for which there is no pretense at scientific foundation is a relatively futile exercise. Key's only interest in science seems to be in the persuasive power of adopting a scientific posture or style. The use of scientific jargon does not necessarily reflect scientific attitudes or methods. Under these circumstances, even to apply the term *pseudoscience* seems unwarranted.

Extravagant claims notwithstanding, advertising may affect us in subtle and indirect ways. While there is no scientific evidence for the existence of "embedded" figures or words, let alone effects from them, the images and themes contained in advertisements may well influence viewers' attitudes and values without their awareness. In other words, the viewer may be well aware of the stimulus, but not necessarily aware of the connection between the stimulus and responses or reactions to it. For example, there was a television commercial a few years ago for

---

"Neither Theoretical Foundation Nor Experimental Evidence"

The committee's review of the available research literature leads to our conclusion that, at this time, there is neither theoretical foundation nor experimental evidence to support claims that subliminal self-help tapes enhance human performance....

Several sociopsychological phenomena, including effort justification and expectancy or placebo effects, may contribute to an erroneous judgment that self-help products are effective, even in the absence of any actual improvements in emotion, appearance, attitude, or any other physical or psychological quality.

—*In the Mind's Eye,* Committee on Techniques for the Enhancement of Human Performance National Research Council (National Academy Press, Washington, 1991, pp. 15–16).

---

skin cream in which a mother and daughter were portrayed. The viewer was challenged to distinguish mother from daughter. According to Postman (1988), the unstated message is that in our culture it is desirable that a mother not look older than her daughter. A number of social scientists believe that advertising may play a role in the development of personal identity and social values (Leiss, Klein, and Jhally 1986; Schudson 1984; Wachtel 1983). It is difficult, however, to isolate advertising's role from the many other social forces at work. Moreover, most research on advertising effects consists of content analyses of the ads themselves. Such studies leave many unanswered questions about the impact of that content on the viewing public.

SUBLIMINAL AUDITORY SELF-HELP TAPES

When claims about covert advertising were raised in September 1957, the *New Yorker* lamented that "minds had been broken and entered" (Moore 1982). More than three decades later claims of covert subliminal manipulation persist. Television commercials, magazine ads, and bookstores promote subliminal tapes that promise to induce dramatic improvements in mental and psychological health. These devices are ostensibly capable of producing many desirable effects, including weight loss, breast enlargement, improvement of sexual function, and relief from constipation.

Subliminal tapes represent a change in modality from visual to auditory, and now subliminal stimulation is supposedly being harnessed for a more noble purpose—psychotherapy, clearly a less crass objective than that of covert advertising. However, the scientific grounds for substantiating the utility of today's self-help tapes is as poor as was the documentation for advertising effects 30 years ago. Proponents seem to have assumed that for obtaining subliminal effects one modality is as good as another. Claims about the utility of subliminal tapes are essentially claims about the subliminal perception of speech—a phenomenon for which there is very little evidence (Moore 1988). The basic problem is that the few studies that purport to have demonstrated effects of subliminal speech used such crude methods for defining subliminality that the findings are quite uninteresting (e.g., Henley 1975; Borgeat et al. 1985).

It is not obvious what the analogue to visual masking is for a speech signal. Masking, in the visual domain, is procedurally defined with relative precision. The mask does not mutilate or change the target stimulus—it simply limits the time available for perceiving the preceding target. In the absence of the mask, the target is easily perceived. In the auditory domain, the target signal is reduced in volume and further attenuated by the superimposition of other supraliminal material. Often the subliminal "message" is accelerated or compressed to such a degree that the message is unintelligible, even when supraliminal. It is an extraordinary claim that an undetectable speech signal engages our nervous system and is perceived— consciously or not. Signal detection is an implicit sine qua non of most theories of speech perception (Massaro 1987). To assert that "subliminal speech" is unconsciously perceived appears to call into question some very fundamental principles of sensory physiology. What is the nature of the signal that arrives at the basilar membrane? If the critical signal is washed out or drowned out by other sounds, then on what basis are we to suppose that the weaker of the two signals becomes disentangled, and comprehensible?

The tapes also have a dubious conceptual rationale in their assumed therapeutic impact. Even if the message could achieve semantic representation, how or why should it affect motivation? Answering the question "How?" is important, because it provides the theoretical justification for the practice.

> There are subliminally embedded messages at work. You won't be able to hear them consciously. But your subconscious will. And it will obey. [Zygon]

> To gain control, it is necessary to speak to the subconscious mind in a language that it comprehends— we have to speak to it subliminally. [Mind Communications Inc.]

Is there a pipeline to the id? Can we sneak directives into the unconscious through the back door? There may be a fundamental misconception at work here, consisting of equating unconscious perceptual processes with the psychodynamic unconscious (Eagle 1987; Marcel 1988). Cognitive psychologists use the term *unconscious* to refer to perceptual processes and effects of which we have no phenomenal awareness. Induced movement is an example of an unconscious perceptual process. Tacit knowledge of and conformity to grammatical rules is another example of unconscious proceasing. No one would want to argue, however, that either of these domains of activity has anything to do with the psychodynamic unconscious. Psychodynamic theorists use the term *unconscious* as a noun with a capital U, to refer to, for lack of a better term, the id—"a cauldron full of seething excitations," as Freud expressed it. Because semantic activation without conscious awareness can be demonstrated, some observers have jumped to the conclusion that subliminal stimulation provides relatively direct access to the id. This assumption has neither theoretical nor empirical support.

While tape distributors often claim that their products have been scientifically validated, there is no evidence of therapeutic effectiveness (e.g., Auday et al. 1991; Greenwald et al. 1991; Merikle and Skanes 1992; Russell et al. 1991). In addition, both Merikle (1988) and Moore (1991a) have conducted studies that showed that many tapes do not appear to contain the sort of signal that could, in principle, allow subliminal perception to occur.

Quite apart from the lack of empirical support, there is little or no theoretical motivation for expecting therapeutic effects from such stimuli. The "explanation" consists of attributing to the systemic unconscious whatever mechanisms or processes would be logically necessary in order for the effects to occur. Because there is no independent evidence for such "unconscious" perceptual processes, it is not surprising that there is no evidence for the imputed effects (see Eich and Hyman 1991; Moore 1991b). Furthermore, Greenwald (in press) has recently queried the conventional psychoanalytic conception of a sophisticated unconscious processor, arguing that it is neither theoretically necessary nor empirically substantiated.

The burden of proof of the viability of these materials is on those who are promoting their use. There is no such proof, and therefore the possibility of health fraud could be raised. These tapes sometimes sell for as much as $400 a set. Of even greater concern is the fact that legitimate forms of therapy may go untried in the quest for a fast, cheap "cure."

According to William Jarvis, president of the National Coalition Against Health Fraud, a quack is "anyone who promotes, for financial gain, a remedy known to be false, unsafe, or unproven" (Jarvis 1989, 4). Fraud, on the other hand, implies intentional deception. Consequently, not all quackery is fraud, nor is fraud synonymous with quackery. As Jarvis has pointed out, in some ways quacks may be *worse* than frauds. "The most dangerous quacks are the zealots who will take the poison themselves in their enthusiasm for their nostrums. Sincerity may make quacks more socially tolerable, but it goes far in enhancing their danger to the public" (Jarvis 1989, 4).

SCIENTISTS, THE MEDIA, AND THE POPULARIZATION OF SCIENCE

The popularity and interest in the topic of subliminal influences—both inside and outside academic circles—can be attributed, in part, to media coverage (cf. Pratkanis, chapter 30 in this book). Conspiracy theories make good copy, and in subliminal advertising we have a large-scale technological conspiracy to control people's minds with invisible stimuli. With subliminal tapes you can allegedly change your behavior and your personality in profound and important ways—effortlessly and painlessly. The quick fix of psychotherapy is an intriguing notion. It is therefore small wonder that it continues to be a popular topic for writers.

Carl Sagan (1987) has suggested that pseudoscience flourishes because the scientific community does a poor job of communicating its findings. To propose that we can be influenced in dramatic ways by undetectable stimuli is a remarkable claim with little scientific support, but blaming journalists for promulgating the claim absolves the scientific community from any responsibility in the educational process. Relations between scientists and the press could be improved if scientists communicated more clearly. Researchers take such great pains to avoid making absolute pronouncements that they often err in the opposite direction. We sometimes speak with a tentativeness that belies the facts, understating our confidence that some propositions are true and that others are false (Rothman 1989). When we talk to the press, we need to speak plainly. For example, Phil Merikle recently observed that "there's unanimous opinion that subliminal tapes are a complete sham and a fraud" (Rae 1991). Merikle is correct, but such candor is relatively rare. Who will distinguish science from pseudoscience if not the scientists?

Paradoxically, while negative scientific evidence continues to accumulate, the subliminal-tape industry—fueled by aggressive advertising campaigns—thrives. As Burnham (1987) has noted, advertising's authority often derives from the use of scientific regalia. Advertising's purpose is, however, antithetical to that of science: "Advertisers [are] engaged in remystifying the world, not demystifying it" (Burnham 1987, 247). Extraordinary claims, if they are repeated often enough, can perpetuate extraordinary beliefs. When nonsense masquerades as science and magic is disguised as therapy, the result is not always laughable. Consider the self-help tape for survivors of sexual abuse; the user is informed that lasting relief from the trauma of abuse is contingent upon the victim's acknowledgment of their own role in causing the abuse in the first place (Moore 1991b).

CONCLUSION

Subliminal advertising and psychotherapeutic effects from subliminal tapes are ideas whose scientific status appears to be on a par with wearing copper bracelets to cure arthritis. Not even the most liberal speculations regarding the use of subliminal techniques for "practical" purposes impute any potential utility to these practices (Bornstein 1989). The interesting question to ask is not "Do subliminal advertising techniques or subliminal auditory tapes work?" but, rather, "How did these implausible ideas ever acquire such an undeserved mantle of scientific respectability?" The answer involves a complex interplay of public attitudes toward science, how social science is popularized in the mass media, and how the scientific community communicates to those outside the scientific community. Carl Sagan may be right—pseudoscience *will* flourish if scientists don't take more responsibility for the accurate dissemination of scientific information.

According to Burnham (1987), superstition has triumphed over rationalism and skepticism partly because scientists no longer engage in the popularization of science—summarizing, simplifying, and translating scientific findings for lay audiences. The function of popularizing science and health is now carried out by journalists and educators. Consequently, many topics, including this one, receive coverage that is, at best, deficient in background information and meaningful context, and at worst, fragmented and misleading. Further confusion is caused by the tendency among journalists to manufacture controversies where none exists by juxtaposing the pronouncements of "authorities" who contradict one another. If all authorities (including those with financial stakes in their positions) are equally admissible, controversies abound.

NOTE

I am grateful to Phil Merikle and Anthony Pratkanis for comments on an earlier draft of this paper, portions of which were presented to the American Speech and Hearing Association's annual convention in St. Louis, November 18, 1989, to the American Psychological Association's annual convention in Boston, August 12, 1990, and to the annual convention of the Committee for the Scientific Investigation of Claims of the Paranormal, Berkeley, Calif., May 2, 1991. Inquiries should be directed to Timothy E. Moore, Department of Psychology, Glendon College, York University, 2275 Bayview Ave., Toronto, Ont. M4N 3M6. E-mail: GL2500020@YUVENUS.

REFERENCES

Auday, B. C., J. L. Mellett, and P. M. Williams. 1991. "Self-improvement Using Subliminal Self-help Audiotapes: Consumer Benefit or Consumer Fraud?" Paper presented at the meeting of the Western Psychological Association, San Francisco April.

Borgeat, F., R. Elie, L. Chaloult and R. Chabot. 1985. Psychophysiological responses to masked auditory stimuli. *Canadian Journal of Psychiatry* 30: 22–27.

Bornstein, R. F. 1989. Subliminal techniques as propaganda tools: Review and critique. *Journal of Mind and Behavior* 10: 231–62.

Burnham, J. C. 1987. *How Superstition Won and Science Lost: Popularizing Science and Health in the United States.* New Brunswick, N.J.: Rutgers University Press.

Cheesman, J., and P. M. Merikle. 1986. Distinguishing conscious from unconscious perceptual processes. *Canadian Journal of Psychology* 40: 343–67.

Creed, T. L. 1987. Subliminal deception: Pseudoscience on the college lecture circuit. *Skeptical Inquirer* 11: 358–66.

Eagle, M. 1987. "The Psychoanalytic and the Cognitive Unconscious." In *Theories of the Unconscious and Theories of the Self,* edited by R. Stern. Hillsdale, N.J.: Analytic Press.

Eich, E., and R. Hyman. 1991. Subliminal self-help. In *In the Mind's Eye: Enhancing Human Performance,* edited by D. Druckman and R. Bjork. Washington, D.C.: National Academy Press.

Eriksen, C. W. 1960. Discrimination and learning without awareness: A methodological survey and evaluation. *Psychological Review* 67: 279–300.

Greenwald, A. G. 1992. New look 3: A paradigm shift reclaims the unconscious. *American Psychologist* 47: 766–79.

Greenwald, A. G., E. R. Spangenberg, and J. Eskenazi. 1991. Double-blind tests of subliminal self-help audiotapes. *Psychological Science* 2: 119–22.

Henley, S. 1975. Cross-modal effects of subliminal verbal stimuli. *Scandinavian Journal of Psychology* 16: 30–36.

Jarvis, W. 1989. What constitutes quackery? *NCAHF Newsletter* 12(4): 4–5.

Key, W. B. 1973. *Subliminal Seduction.* Englewood Cliffs, N.J.: Signet.

———. 1976. *Media Sexploitation.* Englewood Cliffs, N.J.: Prentice-Hall.

———. 1980. *The Clam-Plate Orgy.* Englewood Cliffs, N.J.: Prentice-Hall.

———. 1990. *The Age of Manipulation: The Con in Confidence, the Sin in Sincere.* Englewood Cliffs, N.J.: Prentice-Hall.

Leiss, W., S. Kline, and S. Jhally. 1986. *Social Communication in Advertising.* Toronto: Methuen.

Marcel, A. J. 1988. "Electrophysiology and Meaning in Cognitive Science and Dynamic Psychology—Comments on 'Unconscious Conflict: A Convergent Psychodynamic and Electrophysical Approach.'" In *Psychodynamics and Cognition,* edited by M. J. Horowitz. Chicago: University of Chicago Press.

Massaro, D. W. 1987. *Speech Perception by Ear and Eye: A Paradigm for Psychological Inquiry.* Hillsdale, N.J.: Erlbaum.

Merikle, P., and H. E. Skanes. 1992. Subliminal self-help audiotapes: A search for placebo effects. *Journal of Applied Psychology* 77: 772–76.

Merikle, P. M. 1984. Toward a definition of awareness. *Bulletin of the Psychonomic Society* 22: 449–50.

———. 1988. Subliminal auditory tapes: An evaluation. *Psychology & Marketing,* 46: 355–72.

Moore, T. E. 1982. Subliminal advertising: What you see is what you get. *Journal of Marketing* 46: 38–47.

———. 1988. The case against subliminal manipulation. *Psychology & Marketing* 46: 297–316.

———. 1991a. "Evaluating Subliminal Auditory Tapes: Is There Any Evidence for Subliminal Perception?" Unpublished manuscript, Glendon College, York University, Toronto.

———. 1991b. "Subliminal Auditory Self-help Tapes." In *Self-Care: A Symposium on Self-Help Therapies.* Symposium conducted at the 99th convention of the American Psychological Association, San Francisco, August.

O'Toole, P. 1989. Those sexy ice cubes are back. *Advertising Age,* October 2, p. 26.

Postman, N. 1988. *Conscientious Objections.* New York: Knopf.

Pratkanis, A. R., and A. G. Greenwald. 1988. Recent perspectives on unconscious processing: Still no marketing applications. *Psychology & Marketing* 5: 337–53.

Rae, S. 1991. Brain waves: Subliminal Self-help messages while you sleep? *Wile,* April, pp. 118–19.

Rothman, M. 1989. Myths about science . . . and belief in the paranormal. *Skeptical Inquirer* 14: 25–34.

Russell T. G., W. Rowe, and A. Smouse. 1991. Subliminal self-help tapes and academic achievement: An evaluation. *Journal of Counselling & Development* 69: 359–62.

Sagan, C. 1987. The burden of skepticism. *Skeptical Inquirer* 12: 38–46.

Schudson, M. 1984. *Advertising, the Uneasy Persuasion.* New York: Basic Books.

Vance/Roberson v. *CBS Inc./Judas Priest.* 1990. No. 86-5844 and 86-3939 (Washoe County, 2nd Judicial District Court of Nevada, Motion for Summary Judgment, June 3, 1989).

Vokey, J. R., and J. D. Read. 1985. Subliminal messages: Between the devil and the media. *American Psychologist* 40: 1231–39

Wachtel, R. 1983. *The Poverty of Affluence.* New York: Macmillan.

AN UPDATE ON SUBLIMINAL INFLUENCE

Timothy E. Moore and Anthony R. Pratkanis provide this collaborative update of the topics in their two preceding articles on subliminal influence:

What has happened in the area of subliminal influence in the six years since we wrote our reviews for *Skeptical Inquirer?* In a nutshell, a little bit more science, a little less hysteria, and some still long-term, unresolved issues.

In the "little bit more" category, recent scientific evidence continues to support our original appraisal that actions, motives, and beliefs are not susceptible to manipulation through the use of briefly (i.e., subliminally) presented messages or directives. If anything the case against subliminal manipulation is stronger now than ever as a result of some recent research designed to address Anthony Greenwald's (1992) "two-word challenge"—to create an experimental demonstration that multiple words presented subliminally could be understood as a unit and more than the sum of the parts.

Numerous studies had previously demonstrated semantic activation of single words under conditions in which subjects had no phenomenal awareness of the stimulus, as we noted in our reviews. However, no priming study had shown that multiple words, presented subliminally, were capable of semantic activation. Such a demonstration would be essential for validating claims that phrases such as "Eat Popcorn/Drink Coke" and "Mommy and I are one" could affect human motivation and behavior. (The latter phrase comes from the work of Lloyd Silverman and the 'subliminal psychodynamic activation' paradigm. Silverman had claimed therapeutic effects for his technique; see Silverman and Weinberger 1985; Weinberger and Hardaway 1990). If phrases such as "Mommy and I are One," subliminally presented, produce important consequences for the viewer, the entire sentence, including all semantic relations inherent in the sentence's syntactic structure, would have to achieve internal representation. A recent study by S. C. Draine (1997) has cast considerable doubt on the proposition that multiple words presented subliminally can be comprehended. In his work, Draine established that priming effects of word pairs are a function of individual word meanings, rather than their combined meaning. For example, the pair of words "Not Dirty" was perceived to be evaluatively negative. The impact of the prime was uninfluenced by its negation. Draine concluded that two-word grammatical combinations are beyond the analytic powers of unconscious cognition. (See also Greenwald and Liu 1985.)

In the "little less of" category, we are pleased to report that much of the furor over subliminal influence has died down. There have been no new rock bands labeled as "subliminal criminals." In 1991 a suit brought against Ozzy Osbourne in Georgia was dismissed because the judge found no evidence of subliminal material on the record in question. The news media, including ABC, CBS News, and CNN Headline News, have made numerous accurate presentations of the scientific data showing the ineffectiveness of subliminal influence in general and subliminal self-help tapes in particular. The National Academy of Sciences and the

British Psychological Association both issued statements concerning the lack of efficacy of subliminal tapes. The level of promotion of subliminal self-help tapes seems to have declined. The message seems to have gotten out that when it comes to subliminal tapes, "Buyer Beware," although sales of such tapes continue. We wish we could end this update on a happy note and state that the hysterical claims for the power of subliminal influence have finally been laid to rest. However, we noted in our reviews that interest in subliminal influence is often cyclical—first appearing before the turn of century, then again in the 1950s, 1970s, and today. And while the most recent manifestations of interest in things subliminal appear to have died down, the underlying reasons for this interest remain. These include a general lack of scientific literacy, unclear standards for qualifying experts in court (Moore 1996), confusing unconscious perceptual processes with the psychodynamic unconscious, a mass media interested in ratings and the sensational, and a desire for quick solutions to difficult problems and quick scapegoats when the quick solutions don't work. There is also a feature of subliminal stimulation that is somewhat unique compared to other urban myths—namely that evidence that could disconfirm the presence of subliminal stimuli is not readily available to the viewer (or listener). A subliminal stimulus is, by definition, outside of conscious awareness. Consequently, NOT seeing (or hearing) subliminal messages when one suspects their presence, confirms their presence in the minds of those who have been encouraged to believe in subliminal persuasion. For this reason, subliminal conspiracies will no doubt continue to crop up. Time will tell if, 5 or 10 or 20 years from now, some new researchers will need to write articles similar to the ones we wrote 6 years ago to alert the public to the fact that no scientific evidence is available showing that subliminal stimulation can significantly influence human motivation and behavior.

REFERENCES

Draine, S. C. 1997. Analytic limitations of unconscious language processing. Unpublished doctoral dissertation. Department of Psychology, University of Washington, Seattle.

Greenwald, A. G. 1992. New look 3: Unconscious cognition reclaimed. *American Psychologist* 47: 766–79.

Greenwald, A. G., and T. J. Liu. 1985, November. "Limited Unconscious Processing of Meaning." Paper presented at meetings of the Psychonomic Society, Boston.

Moore, T. E. 1996. Scientific consensus and expert testimony: Lessons from the Judas Priest trial. *Skeptical Inquirer* 20(6): 32–38.

Silverman, L. H., and J. Weinberger. 1985. Mommy and I are one: Implications for psychotherapy. *American Psychologist* 40(12): 1296–1308.

Weinberger, J., and R. Hardaway. 1990. Separating science from myth in subliminal psychodynamic activation. *Clinical Psychology Review* 10: 727–56.

32

PSYCHIC CRIME DETECTIVES: A NEW TEST FOR MEASURING THEIR SUCCESSES AND FAILURES

Richard Wiseman, Donald West, and Roy Stemman

(M) any psychics claim to be able to help the police solve serious crime. Recent surveys suggest that approximately 35 percent of urban United States police departments and 19 percent of rural departments (Sweat and Durm 1993) admit to having used a psychic at least once in their investigations. In addition, Lyons and Truzzi (1991) report the widespread use of psychic detectives in several other countries including Britain, Holland, Germany, and France.

Most of these psychics' claims are supported only by anecdotal evidence. This is unfortunate because it is often extremely difficult to rule out nonpsychic explanations. For example, Hoebens (1985) described how some psychics have made several (often conflicting) predictions relating to an unsolved crime. Once the crime was solved, the incorrect predictions were forgotten while the correct ones were exhibited as evidence of paranormal ability. Rowe (1993) cites examples of psychics making vague and ambiguous predictions that later were interpreted to fit the facts of the crime. Lyons and Truzzi (1991) noted that it is often difficult to obtain "baseline" information for many of these predictions. For example, a psychic may state that a murder weapon will be discovered "near, or in, a large body of water." Although this may later prove to be accurate, it is difficult to know how many criminals dump incriminating objects in areas that could be seen as "large bodies of water" (e.g., streams, lakes, rivers, the ocean, etc.) and therefore establish a statistical baseline for the prediction.

Some investigators have overcome these problems by carrying out controlled tests of psychic detection abilities. One of the earliest controlled studies was conducted by a Dutch police officer, Filippus Brink. Brink carried out a one-year study using four psychics. These psychics were shown various photographs and objects and asked to describe the crimes that had taken place. Some of the photographs and objects were connected with actual crimes; others were not. In a

This article first appeared in *Skeptical Inquirer* 20, no. 1 (January/February 1996): 38–40, 58.

report to INTERPOL, Brink (1960) noted that the psychics had failed to provide any information that would have been of any use to an investigating officer. However, this report is brief and, as noted by Lyons and Truzzi (1991, 51): "Because Brink gives us few details of his method and analysis in this report, the strength, if not the value, of his conclusions cannot really be evaluated."

Studies have been carried out by Martin Reiser of the Los Angeles Police Department. An initial study by Reiser, Ludwig, Saxe, and Wagner (1979) involved twelve psychics. Each psychic was presented with several sealed envelopes containing physical evidence from four crimes (two solved, two unsolved). The psychics were asked to describe the crimes that had taken place. They were then allowed to open the envelopes and describe any additional impressions they received from the object. The study was double-blind, as neither the psychics nor the experimenters had any prior knowledge of the details of the crimes.

The psychics' statements were then coded into several categories (e.g., crime committed, victim, suspect, etc.) and compared with the information known about the crime. For each of the psychics' predictions that matched the actual information, they were awarded one point. The psychics' performances were less than impressive. For example, the experimenters knew that 21 key facts were true of the first crime. The psychics identified an average of only 4. Similarly, of the 33 known facts concerning the second crime, the psychics correctly identified an average of only 1.8. This data caused Reiser et al. to conclude: "The research data does not support the contention that psychics can provide significant additional information leading to the solution of major crime" (pp. 21–22).

Reiser and Klyver (1982) also carried out a follow-up study that used three groups of participants: psychic detectives, students, and police homicide detectives. Four crimes were used (two solved and two unsolved) and again physical evidence from each crime was presented to participants in sealed envelopes. Reiser and Klyver report that the data produced by the three groups was quite different in quantity and character. The psychic detectives produced descriptions that were, on average, six times the length of the student descriptions. In addition, the psychic detectives' statements sounded more confident and dramatic than those produced by either the students or the homicide detectives. Parts of the descriptions were separated into several categories (e.g., sex of criminal, age, height, etc.) and, if correct, assigned one point. A comparison between the three groups showed that although the psychics produced the greatest number of predictions, they were not any more accurate than either the students or the homicide detectives.

In August 1994 the authors of this article were contacted by a British television company involved in making a major documentary series on the paranormal (Arthur C. Clarke's "Mysterious Universe"). One of their programs was to be devoted to psychic detectives, and the producers were eager to film a well-controlled test of three British psychics. The company approached the authors and asked if we would design and carry out these tests. We agreed.

This was the first test of its type in Britain and one of only a handful carried out anywhere in the world. In addition, the methods used during previous studies have been the subject of some criticism (see Lyons and Truzzi 1991) so the authors thought it worthwhile to devise a new method for testing the claims of psychic detection.

This test compared the performance of two groups of participants: psychic detectives and a "control" group of college students. Two of the psychics were professional while the third (who will be referred to as "Psychic 1") was not, but had recently received a great deal of attention from the British media. The psychic's local police force (Hertfordshire Police Force) described him as follows:

> When [psychic's name] comes to the police with his dreams, he is taken seriously and the information that he passes on to his established contact, Sgt. Richard MacGregor, is acted upon immediately (*Psychic News*, November 26, 1994, p. 1)[1]

None of the students claimed to be psychic or had any special interest in criminology.

Each participant was shown three items that had been involved in one of three crimes: a bullet, a scarf, and a shoe. They were asked to handle each of the objects and speak aloud any ideas, images, or thoughts that might be related to these crimes. Participants were told that they were free to take as long as they wished and to say as little or as much as they thought necessary. During the test they were left alone in the room, but everything they said and did was filmed.

After they had finished commenting on all three objects, the participants were given three response sheets (one for each object), each containing 18 statements. Six of each of the 18 statements were true of each crime. The participants were then asked to mark the 6 statements that they believed were true about the crime in question.

Table 1 presents the individual scores for each of the six participants. None of the scores of any of the individuals was statistically significant or impressive.

It could be argued that the above method of testing might *underestimate* participants' psychic ability. For example, a participant may have made several accurate comments describing the crime in question but, nevertheless, obtained a low score if this information was not included on the list of 18 statements. For this reason, a judge not involved in the test transcribed and separated all of the comments made by the participants as the participants handled the objects. The order of these statements was then randomized within each crime and presented to two

	Individual scores (min=0, max=6)	Group scores (min=0, max=6)	Z-score	P-value (2 tailed)
Psychic 1	2.3 (1.15)	2.09 (0.68)	.24	.8
Psychic 2	2.66 (0.57)		.73	.46
Psychic 3	1.33 (0.57)		-.73	.46
Student 1	2 (0.5)	2.33 (0.57)	0	1
Student 2	2 (0.5)		0	1
Student 3	3 (1.73)		1.21	.22

Table 1: Individual/group means, standard deviations (in brackets), z-scores, and p-values.

	Accuracy rating (min=0, max=7)	Number of statements	Group scores (min=0, max=7)
Psychic 1	3.87 (2.57)	15	
Psychic 2	3.65 (1.83)	16	3.83 (0.17)
Psychic 3	4.00 (1.96)	8	
Student 1	6.37 (0.64)	8	
Student 2	4.14 (2.62)	7	5.63 (1.28)
Student 3	5.10 (2.13)	5	

Table 2: Individual/group accuracy means, standard deviations (in brackets), and number of statements.

additional judges. These judges were asked to read about each crime and rate the accuracy of each statement from 1 (very inaccurate) to 7 (very accurate). Table 2 contains the average of the two judges' ratings (inter-rater reliability = .77).

Overall, the psychics made a total of 39 statements while the students made 20 statements. A paired t-test showed no significant differences for the accuracy ratings of students and psychics (t = 2.38, df = 4, p[2 tailed] = .074). This supports Reiser and Klyver's finding that even though psychics tend to make more predictions than students, they are no more accurate.

After their predictions had been recorded, the participants were told about the crimes associated with each of the target objects. This debriefing was filmed, and it is interesting to review the way in which the participants reacted to finding out the truth about each crime:

Crime 1. The Moat Farm murder, 1889–1903. In 1889 an army sergeant major named Samuel Herbert Dougal wished to have an affair with his maid but first needed to dispose of his wife. On May 16, 1889, he and his wife went out for a horse-and-trap ride into the town. During the trip Dougal shot his wife in the head and buried her in a ditch. The body remained buried for four years before the police eventually discovered it. The shoes worn by the corpse were identified by a cobbler as belonging to the dead woman, and Dougal was hung for the murder in 1903.

Crime 2. The murder of Constable Gutteridge, 1927. In 1927 a police officer (Constable George William Gutteridge from the Essex Police Force) stopped a stolen car. The driver suddenly pulled out a gun and fired two shots— one into each of Constable Gutteridge's eyes. The car was later found abandoned in Brixton, London. A six-month-long investigation resulted in two men having been caught and hanged. An important part of the incriminating evidence was the bullet removed from the scene of the crime.

Crime 3. The killing of Margery Pattison, 1962. Margery Pattison, a 71-year-old widow, returned to her flat and disturbed her milkman who had entered through an unlocked door and had started to look for money. An argument ensued and the man grabbed the scarf around her neck, pulled it tight, and strangled her. The man was later caught and charged with murder.

All three psychics thought that they had been successful. On hearing that Crime 2 involved the killing of a police officer, Psychic 1 noted that one of his precognitive dreams involved Police Constable Keith Blakelock (who had been killed on duty in London a few years earlier). This participant noted that he thought at the time the dream was related to Blakelock's murder, but that he now believed it related to the killing of Constable Gutteridge. The same participant remarked that he felt he had given a successful description of Crime 1, as he had said it involved a woman having been raped and murdered and that "that is the fundamental theme of the crime." Psychic 1 failed to recall that he had also said the woman was murdered by a black man and that it happened on Tottenham Court Road. Both of these statements were incorrect. This lends support to the notion that some psychic detection may appear to work, in part, because inaccurate predictions may be forgotten about later, whereas successful ones are recalled and elaborated on.

Psychic 2 remarked that he believed that the experiment showed a "good conclusion all round" and that "my colleagues and I have put the jigsaw puzzle together." He emphasized that all three psychics believed that the scarf was involved in a suffocation, had had trouble with Crime 2, but had predicted that the shoe-related crime involved some form of burial.

Psychic 3 also thought that there had been a consensus on the scarf and shoe. Remarking on the lack of information forthcoming on Crime 2, the psychic noted that "sometimes access to information is not appropriate at certain times." Despite this, he said that he was "relatively pleased with the outcome."

In short, this study provided no evidence to support the claims of psychic detection and, as such, the results are in accordance with other controlled studies. The study utilized a novel method of evaluating psychic detection. The way in which the participants responded to being told the true nature of the crimes gives some insight into some of the mechanisms that might cause individuals to believe erroneously that they are able to solve crimes by psychic means.

NOTES

This research was carried out with support from the Committee for the Scientific Investigation of Claims of the Paranormal.

The authors would like to thank Granite Television, London, Melvin Harris, and Sergeant Fred Feather for helping to set up our study described in this paper. Thanks also to Matthew Smith for helping to run the experiment, and Carol Hurst for carrying out the qualitative analysis of the data. Finally, our thanks to the psychics and students who kindly gave up their time to act as subjects. Correspondence regarding this article should be addressed to Richard Wiseman.

1. Richard Wiseman contacted Sgt. Richard MacGregor of the Hertfordshire Police Force concerning this matter and received confirmation that the above statement was correct (personal communication, December 19, 1994).

REFERENCES

Brink, F. 1960. Parapsychology and criminal investigations. *International Criminal Police Review* 134: 3–9.

Hoebens, P. H. 1985. Reflections on psychic sleuths. Edited by Marcello Truzzi in *A Skeptic's Handbook of Parapsychology,* edited by P. Kurtz, part 6, pp. 631–43. Amherst, N.Y.: Prometheus Books.

Lyons, A. and M. Truzzi. 1991. *The Blue Sense.* New York: Warner Books.

Herts police admit to using psychic help. 1994. *Psychic News,* November 26, 3259: 1.

Reiser, M., L. Ludwig, S. Saxe, and C. Wagner. 1979. An evaluation of the use of psychics in the investigation of major crimes. *Journal of Police Science and Administration,* 7(1): 1825. (Reprinted in Nickel, J. [ed.], *Psychic Sleuths,* Amherst, N.Y.: Prometheus Books, 1994.)

Reiser, M., and N. Klyver. 1982. A comparison of psychics, detectives, and students in the investigation of major crimes. In *Police Psychology: Collected Papers* by M. Reiser, Los Angeles, Calif.: Lehi.

Rowe, W. F. 1993. Psychic detectives: A critical examination. *Skeptical Inquirer* 17(2): 159–65.

Sweat, J. A., and M. W. Durm. 1993. Psychics: Do police departments really use them? *Skeptical Inquirer* 17(2): 148–58.

PART SEVEN

PSYCHOLOGY AND THE ANOMALOUS EXPERIENCE

33

NEAR-DEATH EXPERIENCES
Susan Blackmore

What is it like to die? Although most of us fear death to a greater or lesser extent, there are now more and more people who have "come back" from states close to death and have told stories of usually very pleasant and even joyful experiences at death's door.

For many experiencers, their adventures seem unquestionably to provide evidence for life after death, and the profound effects the experience can have on them is just added confirmation. By contrast, for many scientists these experiences are just hallucinations produced by the dying brain and of no more interest than an especially vivid dream.

So which is right? Are near-death experiences (NDEs) the prelude to our life after death or the very last experience we have before oblivion? I shall argue that neither is quite right: NDEs provide no evidence for life after death, and we can best understand them by looking at neurochemistry, physiology, and psychology; but they are much more interesting than any dream. They seem completely real and can transform people's lives. Any satisfactory theory has to understand that too—and that leads us to questions about minds, selves, and the nature of consciousness.

DEATHBED EXPERIENCES

Toward the end of the last century the physical sciences and the new theory of evolution were making great progress, but many people felt that science was forcing out the traditional ideas of the spirit and soul. Spiritualism began to flourish, and people flocked to mediums to get in contact with their dead friends and relatives "on the other side." Spiritualists claimed, and indeed still claim, to have found proof of survival.

This article first appeared in *Skeptical Inquirer* 16, no. 1 (Fall 1991): 34–45.

In 1882, the Society for Psychical Research was founded, and serious research on the phenomena began; but convincing evidence for survival is still lacking over one hundred years later (Blackmore 1988). In 1926, a psychical researcher and Fellow of the Royal Society, Sir William Barrett (1926), published a little book on deathbed visions. The dying apparently saw other worlds before they died and even saw and spoke to the dead. There were cases of music heard at the time of death and reports of attendants actually seeing the spirit leave the body.

With modern medical techniques, deathbed visions like these have become far less common. In those days people died at home with little or no medication and surrounded by their family and friends. Today most people die in the hospital and all too often alone. Paradoxically it is also improved medicine that has led to an increase in quite a different kind of report—that of the near-death experience.

CLOSE BRUSHES WITH DEATH

Resuscitation from ever more serious heart failure has provided accounts of extra-ordinary experiences (although this is not the only cause of NDEs). These remained largely ignored until about 15 years ago, when Raymond Moody (1975), an American physician, published his best-selling *Life After Life*. He had talked with many people who had "come back from death," and he put together an account of a typical NDE. In this idealized experience a person hears himself pronounced dead. Then comes a loud buzzing or ringing noise and a long, dark tunnel. He can see his own body from a distance and watch what is happening. Soon he meets others and a "being of light" who shows him a playback of events from his life and helps him to evaluate it. At some point he gets to a barrier and knows that he has to go back. Even though he feels joy, love, and peace there, he returns to his body and life. Later he tries to tell others; but they don't understand, and he soon gives up. Nevertheless the experience deeply affects him, especially his views about life and death.

Many scientists reacted with disbelief. They assumed Moody was at least exaggerating, but he claimed that no one had noticed the experiences before because the patients were too frightened to talk about them. The matter was soon settled by further research. One cardiologist had talked to more than 2,000 people over a period of nearly 20 years and claimed that more than half reported Moody-type experiences (Schoonmaker 1979). In 1982, a Gallup poll found that about 1 in 7 adult Americans had been close to death and about 1 in 20 had had an NDE. It appeared that Moody, at least in outline, was right. In my own research I have come across numerous reports like this one, sent to me by a woman from Cyprus:

> An emergency gastrectomy was performed. On the 4th day following that operation I went into shock and became unconscious for several hours.... Although thought to be unconscious, I remembered, for years afterwards, the entire, detailed conversation that passed between the surgeon and anesthetist present.... I was lying above my own body, totally free of pain, and looking down at my own self with compassion for the agony I could see on the face; I was floating peace-

fully. Then ... I was going elsewhere, floating towards a dark, but not frightening, curtain-like area.... Then I felt total peace....

Suddenly it all changed—I was slammed back into my body again, very much aware of the agony again.

Within a few years some of the basic questions were being answered. Kenneth Ring (1980), at the University of Connecticut, surveyed 102 people who had come close to death and found almost 50 percent had had what he called a "core experience." He broke this into five stages: peace, body separation, entering the darkness (which is like the tunnel), seeing the light, and entering the light. He found that the later stages were reached by fewer people, which seems to imply that there is an ordered set of experiences waiting to unfold.

One interesting question is whether NDEs are culture specific. What little research there is suggests that in other cultures NDEs have basically the same structure, although religious background seems to influence the way it is interpreted. A few NDEs have even been recorded in children. It is interesting to note that nowadays children are more likely to see living friends than those who have died, presumably because their playmates only rarely die of diseases like scarlet fever or smallpox (Morse et al. 1986).

Perhaps more important is whether you have to be nearly dead to have an NDE. The answer is clearly no (e.g., Morse et al. 1989). Many very similar experiences are recorded of people who have taken certain drugs, were extremely tired, or, occasionally, were just carrying on their ordinary activities.

I must emphasize that these experiences seem completely real—even more real (whatever that may mean) than everyday life. The tunnel experience is not like just imagining going along a tunnel. The view from out of the body seems completely realistic, not like a dream, but as though you really are up there and looking down. Few people experience such profound emotions and insight again during their lifetimes. They do not say, "I've been hallucinating," "I imagined I went to heaven," or "Can I tell you about my lovely dream?" They are more likely to say, "I have been out of my body" or "I saw Grandma in heaven."

Since not everyone who comes close to death has an NDE, it is interesting to ask what sort of people are more likely to have them. Certainly you don't need to be mentally unstable. NDEers do not differ from others in terms of their psychological health or background. Moreover, the NDE does seem to produce profound and positive personality changes (Ring 1984). After this extraordinary experience people claim that they are no longer so motivated by greed and material achievement but are more concerned about other people and their needs. Any theory of the NDE needs to account for this effect.

EXPLANATIONS OF THE NDE

Astral Projection and the Next World: Could we have another body that is the vehicle of consciousness and leaves the physical body at death to go on to another world?

This, essentially, is the doctrine of astral projection. In various forms it is very popular and appears in a great deal of New Age and occult literature.

One reason may be that out-of-body experiences (OBEs) are quite common, quite apart from their role in NDEs. Surveys have shown that anywhere from 8 percent (in Iceland) to as much as 50 percent (in special groups, such as marijuana users) have had OBEs at some time during their lives. In my own survey of residents of Bristol I found 12 percent. Typically these people had been resting or lying down and suddenly felt they had left their bodies, usually for no more than a minute or two (Blackmore 1984).

A survey of more than 50 different cultures showed that almost all of them believe in a spirit or soul that could leave the body (Shells 1978). So both the OBE and the belief in another body are common, but what does this mean? Is it just that we cannot bring ourselves to believe that we are nothing more than a mortal body and that death is the end? Or is there really another body?

You might think that such a theory has no place in science and ought to be ignored. I disagree. The only ideas that science can do nothing with are the purely metaphysical ones—ideas that have no measurable consequences and no testable predictions. But if a theory makes predictions, however bizarre, then it can be tested.

The theory of astral projection is, at least in some forms, testable. In the earliest experiments mediums claimed they were able to project their astral bodies to distant rooms and see what was happening. They claimed not to taste bitter aloes on their real tongues, but immediately screwed up their faces in disgust when the substance was placed on their (invisible) astral tongues. Unfortunately these experiments were not properly controlled (Blackmore 1982).

In other experiments, dying people were weighed to try to detect the astral body as it left. Early this century a weight of about one ounce was claimed, but as the apparatus became more sensitive the weight dropped, implying that it was not a real effect. More recent experiments have used sophisticated detectors of ultraviolet and infrared, magnetic flux or field strength, temperature, or weight to try to capture the astral body of someone having an out-of-body experience. They have even used animals and human "detectors," but no one has yet succeeded in detecting anything reliably (Morris et al. 1978).

If something really leaves the body in OBEs, then you might expect it to be able to see at a distance, in other words to have extrasensory perception (ESP). There have been several experiments with concealed targets. One success was Tart's subject, who lay on a bed with a five-digit number on a shelf above it (Tart 1968). During the night she had an OBE and correctly reported the number, but critics argued that she could have climbed out of the bed to look. Apart from this one, the experiments tend, like so many in parapsychology, to provide equivocal results and no clear signs of any ESP.

So, this theory has been tested but seems to have failed its tests. If there really were astral bodies I would have expected us to to have found something out about them by now—other than how hard it is to track them down!

In addition there are major theoretical objections to the idea of astral bodies. If you imagine that the person has gone to another world, perhaps along some "real" tunnel, then you have to ask what relationship there is between this world and the other one. If the other world is an extension of the physical, then it ought to be observable and measurable. The astral body, astral world, and tunnel ought to be detectable in some way, and we ought to be able to say where exactly the tunnel is going. The fact that we can't, leads many people to say the astral world is "on another plane," at a "higher level of vibration," and the like. But unless you can specify just what these mean the ideas are completely empty, even though they may sound appealing. Of course we can never prove that astral bodies don't exist, but my guess is that they probably don't and that this theory is not a useful way to understand OBEs.

Birth and the NDE: Another popular theory makes dying analogous with being born: that the out-of-body experience is literally just that— reliving the moment when you emerged from your mother's body. The tunnel is the birth canal and the white light is the light of the world into which you were born. Even the being of light can be "explained" as an attendant at the birth.

This theory was proposed by Stanislav Grof and Joan Halifax (1977) and popularized by the astronomer Carl Sagan (1979), but it is pitifully inadequate to explain the NDE. For a start the newborn infant would not see anything like a tunnel as it was being born. The birth canal is stretched and compressed and the baby usually forced through it with the top of its head, not with its eyes (which are closed anyway) pointing forward. Also it does not have the mental skills to recognize the people around, and these capacities change so much during growing up that adults cannot reconstruct what it was like to be an infant.

"Hypnotic regression to past lives" is another popular claim. In fact much research shows that people who have been hypnotically regressed give the appearance of acting like a baby or a child, but it is no more than acting. For example, they don't make drawings like a real five-year-old would do but like an adult imagines children do. Their vocabulary is too large and in general they overestimate the abilities of children at any given age. There is no evidence (even if the idea made sense) of their "really" going back in time.

Of course the most important question is whether this theory could be tested, and to some extent it can. For example, it predicts that people born by Caesarean section should not have the same tunnel experiences and OBEs. I conducted a survey of people born normally and those born by Caesarean (190 and 36 people, respectively). Almost exactly equal percentages of both groups had had tunnel experiences (36 percent) and OBEs (29 percent). I have not compared the type of birth of people coming close to death, but this would provide further evidence (Blackmore 1982b).

In response to these findings some people have argued that it is not one's own birth that is relived but the idea of birth in general. However, this just reduces the theory to complete vacuousness.

Just Hallucinations: Perhaps we should give up and conclude that all the experiences are "just imagination" or "nothing but hallucinations." However, this is the weakest theory of all. The experiences must, in some sense, be hallucinations, but this is not, on its own, any explanation. We have to ask why are they these kinds of hallucinations? Why tunnels?

Some say the tunnel is a symbolic representation of the gateway to another world. But then why always a tunnel and not, say, a gate, doorway, or even the great River Styx? Why the light at the end of the tunnel? And why always above the body, not below it? I have no objection to the theory that the experiences are hallucinations. I only object to the idea that you can explain them by saying, "They are just hallucinations." This explains nothing. A viable theory would answer these questions without dismissing the experiences. That, even if only in tentative form, is what I shall try to provide.

The Physiology of the Tunnel: Tunnels do not only occur near death. They are also experienced in epilepsy and migraine, when falling asleep, meditating, or just relaxing, with pressure on both eyeballs, and with certain drugs, such as LSD, psilocybin, and mescaline. I have experienced them many times myself. It is as though the whole world becomes a rushing, roaring tunnel and you are flying along it toward a bright light at the end. No doubt many readers have also been there, for surveys show that about a third of people have— like this terrified man of 28 who had just had the anesthetic for a circumcision.

> I seemed to be hauled at "lightning speed" in a direct line tunnel into outer space; (not a floating sensation . . .) but like a rocket at a terrific speed. I appeared to have left my body.

In the 1930s, Heinrich Kluver, at the University of Chicago, noted four form constants in hallucinations: the tunnel, the spiral, the lattice or grating, and the cobweb. Their origin probably lies in the structure of the visual cortex, the part of the brain that processes visual information. Imagine that the outside world is mapped onto the back of the eye (on the retina), and then again in the cortex. The mathematics of this mapping (at least to a reasonable approximation) is well known.

Jack Cowan, a neurobiologist at the University of Chicago, has used this mapping to account for the tunnel (Cowan 1982). Brain activity is normally kept stable by some cells inhibiting others. Disinhibition (the reduction of this inhibitory activity) produces too much activity in the brain. This can occur near death (because of lack of oxygen) or with drugs like LSD, which interfere with inhibition. Cowan uses an analogy with fluid mechanics to argue that disinhibition will induce stripes of activity that move across the cortex. Using the mapping it can easily be shown that stripes in the cortex would appear like concentric rings or spirals in the visual world. In other words if you have stripes in the cortex you will seem to see a tunnel-like pattern of spirals or rings.

This theory is important in showing how the structure of the brain could produce the same hallucination for everyone. However, I was dubious about the idea of these moving stripes, and also Cowan's theory doesn't readily explain the bright light at the center. So Tom Troscianko and I, at the University of Bristol, tried to develop a simpler theory (Blackmore and Troscianko 1989). The most obvious thing about the representation in the cortex is that there are lots of cells representing the center of the visual field but very few for the edges. This means that you can see small things very clearly in the center, but if they are out at the edges you cannot. We took just this simple fact as a starting point and used a computer to simulate what would happen when you have gradually increasing electrical noise in the visual cortex.

The computer program starts with thinly spread dots of light, mapped in the same way as the cortex, with more toward the middle and very few at the edges. Gradually the number of dots increases, mimicking the increasing noise. Now the center begins to look like a white blob and the outer edges gradually get more and more dots. And so it expands until eventually the whole screen is filled with light. The appearance is just like a dark speckly tunnel with a white light at the end, and the light grows bigger and bigger (or nearer and nearer) until it fills the whole screen. (See Figure 1.)

If it seems odd that such a simple picture can give the impression that you are moving, consider two points. First, it is known that random movements in the periphery of the visual field are more likely to be interpreted by the brain as outward than inward movements (Georgeson and Harris 1978). Second, the brain infers our own movement to a great extent from what we see. Therefore, presented with an apparently growing patch of flickering white light your brain will easily interpret it as yourself moving forward into a tunnel.

The theory also makes a prediction about NDEs in the blind. If they are blind because of problems in the eye but have a normal cortex, then they too should see tunnels. But if their blindness stems from a faulty or damaged cortex, they should not. These predictions have yet to be tested.

According to this kind of theory there is, of course, no real tunnel. Nevertheless there is a real physical cause of the tunnel experience. It is noise in the visual cortex. This way we can explain the origin of the tunnel without just dismissing the experiences and without needing to invent other bodies or other worlds.

Out of the Body Experiences: Like tunnels, OBEs are not confined to near death. They too can occur when just relaxing and falling asleep, with meditation, and in epilepsy and migraine. They can also, at least by a few people, be induced at will. I have been interested in OBEs since I had a long and dramatic experience myself (Blackmore 1982a).

It is important to remember that these experiences seem quite real. People don't describe them as dreams or fantasies but as events that actually happened. This is, I presume, why they seek explanations in terms of other bodies or other worlds.

However, we have seen how poorly the astral projection and birth theories cope with OBEs. What we need is a theory that involves no unmeasurable entities or untestable other worlds but explains why the experiences happen and why they seem so real.

I would start by asking why anything seems real. You might think this is obvious—after all, the things we see out there are real aren't they? Well no, in a sense they aren't. As perceiving creatures all we know is what our senses tell us. And our senses tell us what is "out there" by constructing models of the world with ourselves in it. The whole of the world "out there" and our own bodies are really constructions of our minds. Yet we are sure, all the time, that this construction—if you like, this "model of reality"—is "real" while the other fleeting thoughts we have are

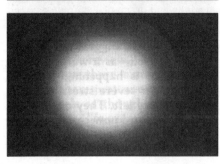

Figure 1: A computer simulation of the "tunnel" some see near death.

unreal. We call the rest of them daydreams, imagination, fantasies, and so on. Our brains have no trouble distinguishing "reality" from "imagination." But this distinction is not given. It is one the brain has to make for itself by deciding which of its own models represents the world "out there." I suggest it does this by comparing all the models it has at any time and choosing the most stable one as "reality."

This will normally work very well. The model created by the senses is the best and most stable the system has. It is obviously "reality," while that image I have of the bar I'm going to go to later is unstable and brief. The choice is easy. By comparison, when you are almost asleep, very frightened, or nearly dying, the model from the senses will be confused and unstable. If you are under terrible stress or suffering oxygen deprivation, then the choice won't be so easy. All the models will be unstable.

So what will happen now? Possibly the tunnel being created by noise in the visual cortex will be the most stable model and so, according to my supposition, this will seem real. Fantasies and imagery might become more stable than the sensory model, and so seem real. The system will have lost input control.

What then should a sensible biological system do to get back to normal? I would suggest that it could try to ask itself—as it were—"Where am I? What is happening?" Even a person under severe stress will have some memory left. They might recall the accident, or know that they were in hospital for an operation, or remember the pain of the heart attack. So they will try to reconstruct, from what little they can remember, what is happening.

Now we know something very interesting about memory models. Often they are constructed in a bird's-eye view. That is, the events or scenes are seen as though from above. If you find this strange, try to remember the last time you went to a pub or the last time you walked along the seashore. Where are "you" looking from in this recalled scene? If you are looking from above you will see what I mean.

So my explanation of the OBE becomes clear. A memory model in bird's-eye view has taken over from the sensory model. It seems perfectly real because it is the best model the system has got at the time. Indeed, it seems real for just the same reason anything ever seems real.

This theory of the OBE leads to many testable predictions, for example, that people who habitually use bird's-eye views should be more likely to have OBEs. Both Harvey Irwin (1986), an Australian psychologist, and myself (Blackmore 1987) have found that people who dream as though they were spectators have more OBEs, although there seems to be no difference for the waking use of different viewpoints. I have also found that people who can more easily switch viewpoints in their imagination are also more likely to report OBEs.

Of course this theory says that the OBE world is only a memory model. It should only match the real world when the person has already known about something or can deduce it from available information. This presents a big challenge for research on near death. Some researchers claim that people near death can actually see things that they couldn't possibly have known about. For example, the American cardiologist Michael Sabom (1982) claims that patients reported the exact behavior of needles on monitoring apparatus when they had their eyes closed and appeared to be unconscious. Further, he compared these descriptions with those of people imagining they were being resuscitated and found that the real patients gave far more accurate and detailed descriptions.

There are problems with this comparison. Most important, the people really being resuscitated could probably feel some of the manipulations being done on them and hear what was going on. Hearing is the last sense to be lost and, as you will realize if you ever listen to radio plays or news, you can imagine a very clear visual image when you can only hear something. So the dying person could build up a fairly accurate picture this way. Of course hearing doesn't allow you to see the behavior of needles, and so if Sabom is right I am wrong. We can only await further research to find out.

The Life Review: The experience of seeing excerpts from your life flash before you is not really as mysterious as it first seems. It has long been known that stimulation of cells in the temporal lobe of the brain can produce instant experiences that seem like the reliving of memories. Also, temporal-lobe epilepsy can produce similar experiences, and such seizures can involve other limbic structures in the brain, such as the amygdala and hippocampus, which are also associated with memory.

Imagine that the noise in the dying brain stimulates cells like this. The memories will be aroused and, according to my hypothesis, if they are the most stable model the system has at that time they will seem real. For the dying person they may well be more stable than the confused and noisy sensory model.

The link between temporal-lobe epilepsy and the NDE has formed the basis of a thorough neurobiological model of the NDE (Saavedra-Aguilar and Gomez-Jeria 1989). They suggest that the brain stress consequent on the near-death episode leads to the release of neuropeptides and neurotransmitters (in particular the endogenous endorphins). These then stimulate the limbic system and other connected areas. In addition, the effect of the endorphins could account for the blissful and other positive emotional states so often associated with the NDE.

Morse provided evidence that some children deprived of oxygen and treated with opiates did not have NDE-like hallucinations, and he and his colleagues (Morse et al. 1986) have developed a theory based on the role of the neurotransmitter serotonin, rather than the endorphins. Research on the neurochemistry of the NDE is just beginning and should provide us with much more detailed understanding of the life review.

Of course there is more to the life review than just memories. The person feels as though she or he is judging these life events, being shown their significance and meaning. But this too, I suggest, is not so very strange. When the normal world of the senses is gone and memories seem real, our perspective on our life changes. We can no longer be so attached to our plans, hopes, ambitions, and fears, which fade away and become unimportant, while the past comes to life again. We can only accept it as it is, and there is no one to judge it but ourselves. This is, I think, why so many NDEers say they faced their past life with acceptance and equanimity.

Other Worlds: Now we come to what might seem the most extraordinary parts of the NDE; the worlds beyond the tunnel and OBE. But I think you can now see that they are not so extraordinary at all. In this state the outside world is no longer real, and inner worlds are. Whatever we can imagine clearly enough will seem real. And what will we imagine when we know we are dying? I am sure for many people it is the world they expect or hope to see. Their minds may turn to people they have known who have died before them or to the world they hope to enter next. Like the other images we have been considering, these will seem perfectly real.

Finally, there are those aspects of the NDE that are ineffable—they cannot be put into words. I suspect that this is because some people take yet another step, a step into nonbeing. I shall try to explain this by asking another question. What is consciousness? If you say it is a thing, another body, a substance, you will only get into the kinds of difficulty we got into with OBEs. I prefer to say that consciousness is just what it is like being a mental model. In other words, all the mental models in any person's mind are all conscious, but only one is a model of "me." This is the one that I think of as myself and to which I relate everything else. It gives a core to my life. It allows me to think that I am a person, something that lives on all the time. It allows me to ignore the fact that "I" change from moment to moment and even disappear every night in sleep.

Now when the brain comes close to death, this model of self may simply fall apart. Now there is no self. It is a strange and dramatic experience. For there is no longer an experiencer—yet there is experience.

This state is obviously hard to describe, for the "you" who is trying to describe it cannot imagine not being. Yet this profound experience leaves its mark. The self never seems quite the same again.

The After Effects: I think we can now see why an essentially physiological event can change people's lives so profoundly. The experience has jolted their usual (and erroneous) view of the relationship between themselves and the world. We all too easily assume that we are some kind of persistent entity inhabiting a perishable body. But, as the Buddha taught, we have to see through that illusion. The world is only a construction of an information-processing system, and the self is too. I believe that the NDE gives people a glimpse into the nature of their own minds that is hard to get any other way. Drugs can produce it temporarily, mystical experiences can do it for rare people, and long years of practice in meditation or mindfulness can do it. But the NDE can out of the blue strike anyone and show them what they never knew before, that their body is only that— a lump of flesh—that they are not so very important after all. And that is a very freeing and enlightening experience.

And Afterwards? If my analysis of the NDE is correct, we can extrapolate to the next stage. Lack of oxygen first produces increased activity through disinhibition, but eventually it all stops. Since it is this activity that produces the mental models that give rise to consciousness, then all this will cease. There will be no more experience, no more self, and so that, as far as my constructed self is concerned, is the end.

So, are NDEs in or out of the body? I should say neither, for neither experiences nor selves have any location. It is finally death that dissolves the illusion that we are a solid self inside a body.

NOTE

In November 1990 I visited the Netherlands to give two lectures. The first, on parapsychology, was part of a series organized by the Stadium Generale of the University of Utrecht and titled "Science Confronts the Paranormal." The second was at the Skepsis Conference. Skepsis refers to the very active Dutch skeptics organization called Stichting Skepsis, which means "skeptical foundation." Cornelis de Jager, professor emeritus in astronomy, is the Chair. Skepsis was established in 1987 and publishes the journal *Skepter.* Stichting Skepsis also publishes conference proceedings and monographs on subjects like reincarnation, spiritism, and homeopathy. As its purpose is to educate the public, Skepsis received a starting grant from the government but is now self-supporting, thanks to many generous donations. This is the lecture I presented at the organization's 1990 conference, on "Belief in the Paranormal."

REFERENCES

Barrett, W. 1926. *Death-bed Visions.* London: Methuen.

Blackmore, S. J. 1982a. *Beyond the Body.* London: Heinemann.

———. 1982b. Birth and the OBE: An unhelpful analogy. *Journal of the American Society for Psychical Research* 77: 229–38.

———. 1984. A postal survey of OBEs and other experiences. *Journal of the Society for Psychical Research* 52: 225–44.

Blackmore, Susan. 1987. Where am I? Perspectives in imagery and the out-of-body experience. *Journal of Mental Imagery* 11: 53–66.

———. 1988. Do we need a new psychical research? *Journal of the Society for Psychical Research* 55: 49–59.

Blackmore, S. J., and T. S. Troscianko. 1989. The physiology of the tunnel. *Journal of Near-Death Studies* 8: 15–28.

Cowan, J. D. 1982. Spontaneous symmetry breaking in large-scale nervous activity. *International Journal of Quantum Chemistry* 22: 1059–82.

Georgeson, M. A., and M. A. Harris. 1978. Apparent foveo-fugal drift of counterphase gratings. *Perception* 7: 527–36.

Grof, S., and J. Halifax. 1977. *The Human Encounter with Death.* London: Souvenir Press.

Irwin, H. J. 1986. Perceptual perspectives of visual imagery in OBEs, dreams and reminiscence. *Journal of the Society for Psychical Research* 53: 210–17.

Moody, R. 1975. *Life After Life.* Covinda, Ga.: Mockingbird.

Morris, R. L., S. B. Harary, J. Janis, J. Hartwell, and W. G. Roll. 1978. Studies of communication during out-of-body experiences. *Journal of the Society for Psychical Research* 72: 1–22.

Morse, J., P. Castillo, D. Venecia, J. Milstein, and D. C. Tyler. 1986. Childhood near-death experiences. *American Journal of Diseases of Children* 140: 1110–14.

Morse, J., D. Venecia, and J. Milstein. 1989. Near-death experiences: A neurophysiological explanatory model. *Journal of Near-Death Studies* 8: 45–53.

Ring, K. 1980. *Life at Death.* New York: Coward, McCann & Geoghegan.

———. 1986. *Heading Toward Omega.* New York: Morrow.

Saavedra-Aguilar, J. C., and J. S. Gomez-Jeria. 1989. A new biological model for near-death studies. *Journal of Near-Death Studies* 7: 205–22.

Sabom, M. 1982. *Recollections of Death.* New York: Harper & Row.

Sagan, C. 1979. *Broca's Brain.* New York: Random House.

Schoonmaker, F. 1979. Denver cardiologist discloses findings after 18 years of near-death research. *Anabiosis* 1: 1–2.

Sheils, D. 1978. A cross-cultural study of beliefs in out-of-the-body experiences. *Journal of the Society for Psychical Research* 49: 697–741.

Tart, C. T. 1978. A psychophysiological study of out-of-the-body experiences in a selected subject. *Journal of the Society for Psychical Research* 62: 3–27.

34

LUCID DREAMS
Susan Blackmore

(W) hat could it mean to be conscious in your dreams? For most of us, dreaming is something quite separate from normal life. When we wake up from being chased by a ferocious tiger, or seduced by a devastatingly good-looking Nobel Prize winner we realize with relief or disappointment that "it was only a dream."

Yet there are some dreams that are not like that. Lucid dreams are dreams in which you know *at the time* that you are dreaming. That they are different from ordinary dreams is obvious as soon as you have one. The experience is something like waking up in your dreams. It is as though you "come to" and find you are dreaming.

Lucid dreams used to be a topic within psychical research and parapsychology. Perhaps their incomprehensibility made them good candidates for being thought paranormal. More recently, however, they have begun to appear in psychology journals and have dropped out of parapsychology—a good example of how the field of parapsychology shrinks when any of its subject matter is actually explained.

Lucidity has also become something of a New Age fad. There are machines and gadgets you can buy and special clubs you can join to learn how to induce lucid dreams. But this commercialization should not let us lose sight of the very real fascination of lucid dreaming. It forces us to ask questions about the nature of consciousness, deliberate control over our actions, and the nature of imaginary worlds.

A REAL DREAM OR NOT?

The term *lucid dreaming* was coined by the Dutch psychiatrist Frederik van Eeden in 1913. It is something of a misnomer since it means something quite different from just clear or vivid dreaming. Nevertheless we are certainly stuck with it. Van Eeden explained that in this sort of dream "the re-integration of the psychic func-

This article first appeared in *Skeptical Inquirer* 15, no. 4 (Summer 1991): 362–70.

tions is so complete that the sleeper reaches a state of perfect awareness and is able to direct his attention, and to attempt different acts of free volition. Yet the sleep, as I am able confidently to state, is undisturbed, deep, and refreshing."

This implied that there could be consciousness during sleep, a claim many psychologists denied for more than 50 years. Orthodox sleep researchers argued that lucid dreams could not possibly be real dreams. If the accounts were valid, then the experiences must have occurred during brief moments of wakefulness or in the transition between waking and sleeping, not in the kind of deep sleep in which rapid eye movements (REMs) and ordinary dreams usually occur. In other words, they could not really be dreams at all.

This presented a challenge to lucid dreamers who wanted to convince people that they really were awake in their dreams. But of course when you are deep asleep and dreaming you cannot shout, "Hey! Listen to me. I'm dreaming right now." All the muscles of the body are paralyzed.

It was Keith Hearne (1978), of the University of Hull, who first exploited the fact that not all the muscles are paralyzed. In REM sleep the eyes move. So perhaps a lucid dreamer could signal by moving the eyes in a predetermined pattern. Just over ten years ago, lucid dreamer Alan Worsley first managed this in Hearne's laboratory. He decided to move his eyes left and right eight times in succession whenever he became lucid. Using a polygraph, Hearne could watch the eye movements for signs of the special signal. He found it in the midst of REM sleep. So lucid dreams are real dreams and do occur during REM sleep.

Further research showed that Worsley's lucid dreams most often occurred in the early morning, around 6:30 A.M., nearly half an hour into a REM period and toward the end of a burst of rapid eye movements. They usually lasted for two to five minutes. Later research showed that they occur at times of particularly high arousal during REM sleep (Hearne 1978).

It is sometimes said that discoveries in science happen when the time is right for them. It was one of those odd things that at just the same time, but unbeknown to Hearne, Stephen LaBerge, at Stanford University in California, was trying the same experiment. He too succeeded, but resistance to the idea was very strong. In 1980, both *Science* and *Nature* rejected his first paper on the discovery (LaBerge 1985). It was only later that it became clear what an important step this had been.

AN IDENTIFIABLE STATE?

It would be especially interesting if lucid dreams were associated with a unique physiological state. In fact this has not been found, although this is not very surprising since the same is true of other altered states, such as out-of-body experiences and trances of various kinds. However, lucid dreams do tend to occur in periods of higher cortical arousal. Perhaps a certain threshold of arousal has to be reached before awareness can be sustained.

The beginning of lucidity (marked by eye signals, of course) is associated with pauses in breathing, brief changes in heart rate, and skin response changes, but there is no unique combination that allows the lucidity to be identified by an observer.

In terms of the dream itself, there are several features that seem to provoke lucidity. Sometimes heightened anxiety or stress precedes it. More often there is a kind of intellectual recognition that something "dreamlike" or incongruous is going on (Fox 1962; Green 1968; LaBerge 1985).

It is common to wake from an ordinary dream and wonder, "How on earth could I have been fooled into thinking that I was really doing pushups on a blue beach?" A little more awareness is shown when we realize this in the dream. If you ask yourself, "Could this be a dream?" and answer "No" (or don't answer at all), this is called a pre-lucid dream. Finally, if you answer "Yes," it becomes a fully lucid dream.

It could be that once there is sufficient cortical arousal it is possible to apply a bit of critical thought; to remember enough about how the world ought to be to recognize the dream world as ridiculous, or perhaps to remember enough about oneself to know that these events can't be continuous with normal waking life. However, tempting as it is to conclude that the critical insight produces the lucidity, we have only an apparent correlation and cannot deduce cause and effect from it.

BECOMING A LUCID DREAMER

Surveys have shown that about 50 percent of people (and in some cases more) have had at least one lucid dream in their lives. (See, for example, Blackmore 1982; Gackenbach and LaBerge 1988; Green 1968.) Of course surveys are unreliable in that many people may not understand the question. In particular, if you have never had a lucid dream, it is easy to misunderstand what is meant by the term. So overestimates might be expected. Beyond this, it does not seem that surveys can find out much. There are no very consistent differences between lucid dreamers and others in terms of age, sex, education, and so on (Green 1968; Gackenbach and LaBerge 1988).

For many people, having lucid dreams is fun, and they want to learn how to have more or to induce them at will. One finding from early experimental work was that high levels of physical (and emotional) activity during the day tend to precede lucidity at night. Waking during the night and carrying out some kind of activity before falling asleep again can also encourage a lucid dream during the next REM period and is the basis of some induction techniques.

Many methods have been developed (Gackenbach and Bosveld 1989; Tart 1988; Price and Cohen 1988). They roughly fall into three categories.

One of the best known is LaBerge's MILD (Mnemonic Induction of Lucid Dreaming). This is done on waking in the early morning from a dream. You should wake up fully, engage in some activity like reading or walking about, and then lie down to go to sleep again. Then you must imagine yourself asleep and dreaming, rehearse the dream from which you woke, and remind yourself, "Next time I dream this I want to remember I'm dreaming."

A second approach involves constantly reminding yourself to become lucid throughout the day rather than the night. This is based on the idea that we spend most of our time in a kind of waking daze. If we could be more lucid in waking life, perhaps we could be more lucid while dreaming. German psychologist Paul Tholey suggests asking yourself many times every day, "Am I dreaming or not?" This sounds easy but is not. It takes a lot of determination and persistence not to forget all about it. For those who do forget, French researcher Clerc suggests writing a large "C" on your hand (for "conscious") to remind you (Tholey 1983; Gackenbach and Bosveld 1989).

This kind of method is similar to the age-old technique for increasing awareness by meditation and mindfulness. Advanced practitioners of meditation claim to maintain awareness through a large proportion of their sleep. TM is often claimed to lead to sleep awareness. So perhaps it is not surprising that some recent research finds associations between meditation and increased lucidity (Gackenbach and Bosveld 1989).

The third and final approach requires a variety of gadgets. The idea is to use some sort of external signal to remind people, while they are actually in REM sleep, that they are dreaming. Hearne first tried spraying water onto sleepers' faces or hands but found it too unreliable. This sometimes caused them to incorporate water imagery into their dreams, but they rarely became lucid. He eventually decided to use a mild electric shock to the wrist. His "dream machine" detects changes in breathing rate (which accompany the onset of REM) and then automatically delivers a shock to the wrist (Hearne 1990).

Meanwhile, in California, LaBerge was rejecting taped voices and vibrations and working instead with flashing lights. The original version was laboratory based and used a personal computer to detect the eye movements of REM sleep and to turn on flashing lights whenever the REMs reached a certain level. Eventually, however, all the circuitry was incorporated into a pair of goggles. The idea is to put the goggles on at night, and the lights will flash only when you are asleep and dreaming. The user can even control the level of eye movements at which the lights begin to flash.

The newest version has a chip incorporated into the goggles. This will not only control the lights but will store data on eye-movement density during the night and when and for how long the lights were flashing, making fine tuning possible. At the moment, the first users have to join in workshops at LaBerge's Lucidity Institute and learn how to adjust the settings, but within a few months he hopes the whole process will be fully automated. (See LaBerge's magazine, *DreamLight*.)

LaBerge tested the effectiveness of the Dream Light on 44 subjects who came into the laboratory, most for just one night. Fifty-five percent had at least one lucid dream and two had their first-ever lucid dream this way. The results suggested that this method is about as successful as MILD, but using the two together is the most effective (LaBerge 1985).

LUCID DREAMS AS AN EXPERIMENTAL TOOL

There are a few people who can have lucid dreams at will. And the increase in induction techniques has provided many more subjects who have them frequently. This has opened the way to using lucid dreams to answer some of the most interesting questions about sleep and dreaming.

How long do dreams take? In the last century, Alfred Maury had a long and complicated dream that led to his being beheaded by a guillotine. He woke up terrified, and found that the headboard of his bed had fallen on his neck. From this, the story goes, he concluded that the whole dream had been created in the moment of awakening.

This idea seems to have got into popular folklore but was very hard to test. Researchers woke dreamers at various stages of their REM period and found that those who had been longer in REM claimed longer dreams. However, accurate timing became possible only when lucid dreamers could send "markers" from the dream state.

LaBerge asked his subjects to signal when they became lucid and then count a ten-second period and signal again. Their average interval was 13 seconds, the same as they gave when awake. Lucid dreamers, like Alan Worsley, have also been able to give accurate estimates of the length of whole dreams or dream segments (Schatzman, Worsley, and Fenwick 1988).

DREAM ACTIONS

As we watch sleeping animals it is often tempting to conclude that they are moving their eyes in response to watching a dream, or twitching their legs as they dream of chasing prey. But do physical movements actually relate to the dream events?

Early sleep researchers occasionally reported examples like a long series of left-right eye movements when a dreamer had been dreaming of watching a ping-pong game, but they could do no more than wait until the right sort of dream came along.

Lucid dreaming made proper experimentation possible, for the subjects could be asked to perform a whole range of tasks in their dreams. In one experiment with researchers Morton Schatzman and Peter Fenwick, in London, Worsley planned to draw large triangles and to signal with flicks of his eyes every time he did so. While he dreamed, the electromyogram, recording small muscle movements, showed not only the eye signals but spikes of electrical activity in the right forearm just afterward. This showed that the preplanned actions in the dream produced corresponding muscle movements (Schatzman, Worsley, and Fenwick 1988).

Further experiments, with Worsley kicking dream objects, writing with umbrellas, and snapping his fingers, all confirmed that the muscles of the body show small movements corresponding to the body's actions in the dream. The question about eye movements was also answered. The eyes do track dream objects. Worsley could even produce slow scanning movements, which are very difficult to produce in the absence of a "real" stimulus (Schatzman, Worsley, and Fenwick 1988).

LaBerge was especially interested in breathing during dreams. This stemmed from his experiences at age five when he had dreamed of being an undersea pirate who could stay under water for very long periods without drowning. Thirty years later he wanted to find out whether dreamers holding their breath in dreams do so physically as well. The answer was yes. He and other lucid dreamers were able to signal from the dream and then hold their breath. They could also breathe rapidly in their dreams, as revealed on the monitors. Studying breathing during dreamed speech, he found that the person begins to breathe out at the start of an utterance just as in real speech (LaBerge and Dement 1982a).

HEMISPHERIC DIFFERENCES

It is known that the left and right hemispheres are activated differently during different kinds of tasks. For example, singing uses the right hemisphere more, while counting and other, more analytical tasks use the left hemisphere more. By using lucid dreams, LaBerge was able to find out whether the same is true in dreaming.

In one dream he found himself flying over a field. (Flying is commonly associated with lucid dreaming.) He signaled with his eyes and began to sing "Row, row, row your boat...." He then made another signal and counted slowly to ten before signaling again. The brainwave records showed just the same patterns of activation that you would expect if he had done these tasks while awake (LaBerge and Dement 1982b).

DREAM SEX

Although it is not often asked experimentally, I am sure plenty of people have wondered what is happening in their bodies while they have their most erotic dreams.

LaBerge tested a woman who could dream lucidly at will and could direct her dreams to create the sexual experiences she wanted. (What a skill!) Using appropriate physiological recording, he was able to show that her dream orgasms were matched by true orgasms (LaBerge, Greenleaf, and Kedzierski 1983).

Experiments like these show that there is a close correspondence between actions of the dreamer and, if not real movements, at least electrical responses. This puts lucid dreaming somewhere between real actions, in which the muscles work to move the body, and waking imagery, in which they are rarely involved at all. So what exactly is the status of the dream world?

THE NATURE OF THE DREAM WORLD

It is tempting to think that the real world and the world of dreams are totally separate. Some of the experiments already mentioned show that there is no absolute dividing line. There are also plenty of stories that show the penetrability of the boundary.

Alan Worsley describes one experiment in which his task was to give himself a prearranged number of small electric shocks by means of a machine measuring his eye movements. He went to sleep and began dreaming that it was raining and he was in a sleeping bag by a fence with a gate in it. He began to wonder whether he was dreaming and thought it would be cheating to activate the shocks if he was awake. Then, while making the signals, he worried about the machine, for it was out there with him in the rain and might get wet (Schatzman, Worsley, and Fenwick 1988).

This kind of interference is amusing, but there are dreams of confusion that are not. The most common and distinct are called false awakenings. You dream of waking up but in fact, of course, are still asleep. Van Eeden (1913) called these "wrong waking up" and described them as "demoniacal, uncanny, and very vivid and bright, with...a strong diabolical light." The French zoologist Yves Delage, writing in 1919, described how he had heard a knock at his door and a friend calling for his help. He jumped out of bed, went to wash quickly with cold water, and when that woke him up he realized he had been dreaming. The sequence repeated four times before he finally actually woke up—still in bed.

A student of mine described her infuriating recurrent dream of getting up, cleaning her teeth, getting dressed, and then cycling all the way to the medical school at the top of a long hill, where she finally would realize that she had dreamed it all, was late for lectures, and would have to do it all over again for real.

The one positive benefit of false awakenings is that they can sometimes be used to induce out-of-body experiences (OBEs). Indeed, Oliver Fox (1962) recommends this as a method for achieving the OBE. For many people OBEs and lucid dreams are practically indistinguishable. If you dream of leaving your body, the experience is much the same. Also recent research suggests that the same people tend to have both lucid dreams and OBEs (Blackmore 1988; Irwin 1988).

All of these experiences have something in common. In all of them the "real" world has been replaced by some kind of imaginary replica. Celia Green, of the Institute of Psychophysical Research at Oxford, refers to all such states as "metachoric experiences."

Jayne Gackenbach, a psychologist from the University of Alberta, Canada, relates these experiences to UFO-abduction stories and near-death experiences (NDEs). The UFO abductions are the most bizarre but are similar in that they too involve the replacement of the perceived world by a hallucinatory replica.

There is an important difference between lucid dreams and these other states. In the lucid dream one has insight into the state (in fact that defines it). In false awakening, one does not (again by definition). In typical OBEs, people think they have *really* left their bodies. In UFO "abductions" they believe the little green men are "really there", and in NDEs, they are convinced they are rushing down a real tunnel toward a real light and into the next world. It is only in the lucid dream that one realizes it is a dream.

I have often wondered whether insight into these other experiences is possible and what the consequences might be. So far I don't have any answers.

WAKING UP

The oddest thing about lucid dreams—and, to many people who have them, the most compelling—is how it feels when you wake up. Upon waking up from a normal dream, you usually think, "Oh, that was only a dream." Waking up from a lucid dream is more continuous. It feels more real, it feels as though you were conscious in the dream. Why is this? I think the reason can be found by looking at the mental models the brain constructs in waking, in ordinary dreaming, and in lucid dreams.

I have previously argued that what seems real is the most stable mental model in the system at any time. In waking life, this is almost always the input-driven model, the one that is built up from the sensory input. It is firmly linked to the body image to make a stable model of "me, here, now." It is easy to decide that this represents "reality" while all the other models being used at the same time are "just imagination" (Blackmore 1988).

Now consider an ordinary dream. In that case there are lots of models being built but no input-driven model. In addition there is no adequate self-model or body image. There is just not enough access to memory to construct it. This means, if my hypothesis is right, that whatever model is most stable at any time will seem real. But there is no recognizable self to whom it seems real. There will just be a series of competing models coming and going. Is this what dreaming feels like?

Finally, we know from research that in the lucid dream there is higher arousal. Perhaps this is sufficient to construct a better model of self. It is one that includes such important facts as that you have gone to sleep, that you intended to signal with your eyes, and so on. It is also more similar to the normal waking self than those fleeting constructions of the ordinary dream. This, I suggest, is what makes the dream seem more real on waking up. Because the *you* who remembers the dream is more similar to the *you* in the dream. Indeed, because there was a better model of *you*, *you* were more conscious.

If this is right, it means that lucid dreams are potentially even more interesting than we thought. As well as providing insight into the nature of sleep and dreams, they may give clues to the nature of consciousness itself.

REFERENCES

Blackmore, S. J. 1982. *Beyond the Body.* London: Heinemann.

————. 1988. A theory of lucid dreams and OBEs. In *Conscious Mind, Sleeping Brain,* edited by J. Gackenbach and S. LaBerge, 373–87. New York: Plenum.

Delage, Y. 1919. *Le Reve.* Paris: Les Presses Universitaires de France.

Fox, O. 1962. *Astral Projection.* New York: University Books.

Gackenbach, J., and J. Bosveld. 1989. *Control Your Dreams.* New York: Harper & Row.

Gackenbach, J., and S. LaBerge, eds. 1988. *Conscious Mind, Sleeping Brain.* New York: Plenum.

Green, C. E. 1968. *Lucid Dreams.* London: Hamish Hamilton.

Hearne, K. 1978. *Lucid Dreams: An Electrophysiological and Psychological Study.* Unpublished Ph.D. thesis, University of Hull.

Hearne, K. 1990. *The Dream Machine.* Northants: Aquarian.

Irwin, H. J. 1988. Out-of-body experiences and dream lucidity: Empirical perspectives. In *Conscious Mind, Sleeping Brain,* edited by J. Gackenbach and S. LaBerge, 353–71. New York: Plenum.

LaBerge, S. 1985. *Lucid Dreaming.* Los Angeles: Tarcher.

LaBerge, S., and W. Dement. 1982a. Voluntary control of respiration during REM sleep. *Sleep Research* 11: 107.

———. 1982b. Lateralization of alpha activity for dreamed singing and counting during REM sleep. *Psychophysiology* 19: 331–32.

LaBerge, S., W. Greenleaf, and B. Kedzierski. 1983. Physiological responses to dreamed sexual activity during lucid REM sleep. *Psychophysiology* 20: 454–55.

Price, R. F., and D. B. Cohen. 1988. Lucid dream induction: An empirical evaluation. In *Conscious Mind, Sleeping Brain,* edited by J. Gackenbach and S. LaBerge, 105–34. New York: Plenum.

Schatzman, M., A. Worsley, and P. Fenwick. 1988. Correspondence during lucid dreams between dreamed and actual events. In *Conscious Mind, Sleeping Brain,* edited by J. Gackenbach and S. LaBerge, 155–79. New York: Plenum.

Tart, C. 1988. From spontaneous event to lucidity: A review of attempts to consciously control nocturnal dreaming. In *Conscious Mind, Sleeping Brain,* edited by J. Gackenbach and S. LaBerge, 67–103. New York: Plenum.

Tholey, P. 1983. Techniques for controlling and manipulating lucid dreams. *Perceptual and Motor Skills* 57: 79–90.

Van Eeden, F. 1913. A study of dreams. *Proceedings of the Society for Psychical Research* 26: 431–61.

35

NIGHT TERRORS, SLEEP PARALYSIS, AND DEVIL-STRICKEN TELEPHONE CORDS FROM HELL

Peter Huston

When traveling in foreign countries on a budget, it's ironic that some of my most memorable experiences have been those of just spending time and swapping stories with the people I met.

Such people tend to be very interesting in themselves and, due to the transient nature of the relationship of travelers' crossing paths, much more open than they would be under a more conventional social situation. They talk about things they normally would not mention, and since they do not plan to see the person they are talking to again, they rarely regret it.

It was under such circumstances that I heard one of the most peculiar stories I have ever had reason to believe.

I was sitting in the lounge of a low-cost hostel in Asia and found myself talking to another American, someone of about my age (early twenties, at the time), approximately the same social background, and also in Asia for the first time, intending to find work, to experience something different from the mundane life of growing up in the USA.

At some point, and somehow, the conversation turned to ghosts, spirits, and the supernatural. I said that I thought there might be something to it all and hoped to do some research on it someday.

My new acquaintance said that he did not think it was a good idea to look for ghosts, that it was his experience that "normally they look for you."

He then proceeded to tell me the following story.

Once, back in his "angry, punk rocker" days, in a large American city, he'd gone through a period when he'd frequently been bothered by a ghost, specifically a night hag.

The attacks had started, he said, when he found himself waking up and being unable to move, completely paralyzed and with a feeling of great weight on his

This article first appeared in *Skeptical Inquirer* 17, no. 1 (Fall 1992): 64–69.

chest. Needless to say, he found this disturbing, and his first thought was that it was a sign of some sort of mental illness.

In time things got worse. He would awake in a paralyzed state and find himself forced to watch as an old hag entered the room, floating as she came. She would then alight on his bed and proceed to sit on his chest. He would feel a palpable wave of terror, her weight bearing down and pinning him in place, and sometimes he would experience her putrid breath as she lurched over his face. Finally, after a period of unimaginable terror, she would leave; and he would find himself able to rise and move, but in a terror-stricken state, confused and shaken up.

The "attacks by the night hag" came frequently, and he said he would have been convinced that he needed serious psychiatric help, except that on at least one occasion when he was sleeping with his girlfriend, she awoke during one of his attacks and said that she saw the night hag too. It was then that he abandoned his fears and seized upon a supernatural explanation for his experiences.

I was a bit puzzled by this story and didn't know what to make of it. The obvious explanation to me at the time was that he had just made up the story to see how I would react, but I just didn't think so. It was the way in which he'd told the story, embarrassed and unsure of the truth, beyond the reality that he had experienced, unsure of the way in which I would react. Hesitant.

As time went on, much to our surprise, we found our paths crossing a couple of times more, on one occasion for a very long period of time, as we proceeded to find work, leave countries, and acquire working visas and other such things. I came to know him better and he never told any similar stories or mentioned that one again. His stories about various things were all thoroughly grounded in conventional reality, although sometimes the stranger versions thereof.

Overall, the more time that we spent together, the more he seemed like a relatively honest and normal person, lacking in the motivations for telling an imaginary story about ghost attacks.

On one occasion, when I asked him if he'd had any more experiences with spirits since he came to Asia, he looked very embarrassed, said, no, and quickly changed the subject.

In any event, I did not put his story in the same category as the church-camp ghost stories that I'd been exposed to in junior high. I spent a great deal of time puzzling over it from time to time, questioning the truth of the story itself, and looking for any possible explanations that might have been feasible.

For many years I was unable to find one and even went so far as to have a similar experience myself.

I would awake, find myself unable to move, and feel a great deal of pressure on my chest. I would struggle to move and find myself unable to do anything beyond trembling. It was a terrifying experience and like my friend my first explanation was that my mental health was going. The attacks continued night after night for almost a week. I was at a loss to explain them, fearful, and under a great deal of stress as I tried to understand what was wrong with me.

I was terrified to go to sleep at night because of my fear that I would awaken insane or go into cardiac arrest.

My wife offered the explanation that I was being attacked by the ghost of my grandmother, who had recently passed away on the other side of the world. (I live in Asia, my grandmother died in New Jersey.) I confess that I dismissed that explanation, not on any skeptical, scientific grounds but because it would seem unlike my grandmother to travel around the world to torment me for missing her funeral when many of her nearby grandchildren had missed it as well.

At this time, my wife and I had to make a decision about whether to get a telephone installed in our apartment. (In Taiwan, a telephone line often does not come with an apartment and must be installed by each tenant at a fairly high cost.) I'd been opposed to the idea, but my wife had really wanted one. Eventually, I gave in, even knowing that it would be expensive in terms of both direct costs and the charges for local and long distance phone calls.

Ultimately, I again awoke unable to move and finding that the telephone cord somehow became draped over my body and was in the process of electrocuting me.

In time it was over. I was able to move again, but I was quite confused and upset. I was thoroughly agitated and very angry with my wife for not noticing that I'd been shaking and suffering from this terrible experience.

I was even more confused to discover that the telephone cord was far on the other side of the room, and it was physically impossible for it to have become draped over my body, even if someone had been inclined to pick it up and put it there.

I became convinced that I was suffering the beginnings of some sort of mental breakdown and phoned my parents. My father told me that there had been occasions when he had awakened and been unable to move. He had thought it very strange until he mentioned it to his brother, who said that he'd had similar experiences.

This conversation was a major relief to me, and it was followed by a cessation of such attacks.

I filed these experiences away until I stumbled across a scientific explanation for them.

What my friend and I were experiencing is often referred to as "night terror." My friend's were more like the classic variety, complete with a night hag, while mine was more unusual, involving a "demonic telephone cord from Hell." Such experiences, disturbing and frightening as they may be to the unsuspecting participants, are relatively benign and fall easily within the realm of the "explained" when understood properly.

Such experiences consist of two aspects. The first is sleep paralysis. This is a relatively normal condition characterized by awakening to find oneself unable to move. Although it is one of many indicators of the mental illness narcolepsy, normally, in the absence of other problems, it is a fairly unimportant, relatively common condition.

When one is asleep certain portions of the brain effectively are inhibited and cease to function at their normal level. One of these inhibited functions is movement. With the obvious exception of sleepwalkers (who fall well outside the scope of this article), gross muscle movement is inhibited for sleepers as a normal process. In the case of sleep paralysis, the person awakens before his brain can readjust itself and allow for normal, uninhibited movement. It is seen to be the extension of a sleep phenomenon into the waking period. Reportedly people can be snapped out of sleep paralysis by being touched or hearing their names spoken. In the absence of outside stimulation normal sleep paralysis is of short duration and self-correcting.

A closely related and on occasion concurrent phenomenon is hypnopompic hallucinations. These occur when one awakens but hallucinates and sees imagery. At times these images can be vivid and bizarre and of a frightening quality. My friend's ghost and my electrifying telephone cord are both examples of these. Although both had strange, unbelievable qualities, they were both vividly seen, vividly experienced, and believed by the experiencer at the time, and very confusing to the victim upon the cessation of the experience.

The ability to question subjective reality is another brain function that is inhibited during dreaming. Hence the bizarre but unquestioned nature of conventional dreams. This inhibition of the reality-checking function of the brain apparently also extends into the period of hypnopompic hallucination. In some cases, this can lead to intensified emotion, such as terror, and this is the conventional explanation for such experiences. Although the experience of finding oneself paralyzed and possibly insane is frightening in itself, researchers feel that the terror often felt is due to reasons beyond this.

Hypnopompic hallucinations can involve any or all of the senses, but are most commonly visual or auditory. Such hallucinations are normally not shared. I assume my friend's girlfriend said that she had also seen the "ghost" to appease him. Or possibly she actually thought that she saw something. In any event, never having met her, I have no idea how she would respond when faced with a distraught, confused person who feared for his mental health.

Such hallucinations, and the related phenomenon of hypnogogic hallucinations (which occur just *prior* to sleeping), are not uncommon in children. Thus, when children say they see monsters at night at a time they should be sleeping, they just might really think they are seeing them.

Such experiences have a great deal of relevance to investigators of "unexplained phenomena," not just to those interested in "ghosts" and their supernatural entities. They also provide insight into many cases of "UFO abductions" and similar phenomena. Psychologist Robert A. Baker has suggested hypnopompic hallucinations as the explanation for the experiences described by Whitley Strieber in *Communion*. Strieber claimed to have been abducted and manipulated by little fetuslike-looking men who sneak into his bedroom repeatedly, and ultimately cause him a great deal of distress—and financial gain—by their actions. (This is one of several explanations advanced for this book.)

Budd Hopkins, author of *Missing Time* and one of the foremost proponents of the alien-abduction hypothesis, discussed several such reports in a recent *Omni* magazine article. At least three of these involved children, or adults remembering childhood experiences, seeing aliens or other weird strangers in their bedrooms before going to sleep. Hypnogogic hallucination is an obvious direction for such an investigation and it is one prominently absent from the article.

In *Traditions of Belief,* a study of supernatural beliefs among middle-class women in today's England, by Gillian Bennett, there are many other stories involving strange people in children's bedrooms. These are cited by adults as memories that convinced them that there might just be something to "the mysterious side of life." These experiences might be more easily explained as hypnogogic hallucinations.

In the excellent collection of essays *Phenomenon: Forty Years of Flying Saucers,* Mark Moravec discusses hypnopompic/hypnogogic hallucinations as they may relate to UFO sightings. The author states that such hallucinations may explain many UFO-related phenomena that involve tired drivers and people who awake to find strange things, like lights outside their windows and little creatures in their rooms.

The possible images from hypnopompic hallucinations can be of virtually anything that a person might dream about. Obviously culturally determined UFO phenomena, ghost sightings that frequently involve "night hag"-type terrors, or the less hostile images of recently deceased acquaintances (who have been known to provide information that only "they" would know, hence supposed proof of their supernatural origin), fairies, goblins, weird creatures, bright lights, are all possible, just as I experienced the bizarre demonic telephone cord from Hell.

It is important to understand that such phenomena are not a sign of mental weakness or illness, but are often experienced by people of unquestioned sanity (and me too on one occasion). Anyone seeking to understand paranormal phenomena should be aware of this normal human experience.

BIBLIOGRAPHY

Baker, Robert A. 1987. The aliens among us: Hypnotic regressions revisited. *Skeptical Inquirer,* Winter, 1987–88.

Bennett, Gillian. 1987. *Traditions of Belief.* London: Penguin Books.

Carlson, Neil R. 1986 [1977]. *Physiology of Behavior.* Boston: Allyn and Bacon.

Hobson, J. Allan. 1988. *The Dreaming Brain.* London: Penguin Books.

Hufford, David J. 1982. *The Terror that Comes in the Night.* Philadelphia: University of Pennsylvania Press.

McCarthy, Paul. 1990. True confessions: The unbearable pain, sorrow, and terror of alien abduction. *Omni,* December.

Moravec, Mark. 1988. "Is There a UFO State of Mind?" In *Phenomenon: Forty Years of Flying Saucers,* edited by John Spencer and Hilary Evans. New York: Avon.

Pachulski, Roman. 1990. *Psychiatric Notes.* London: Prentice Hall International.

Reed, Graham. 1988. *The Psychology of Anomalous Experience.* Amherst, N.Y.: Prometheus Books.

36

ANGUISHED SILENCE AND HELPING HANDS: AUTISM AND FACILITATED COMMUNICATION

James A. Mulick, John W. Jacobson, and Frank H. Kobe

(T) he vulnerability of parents of handicapped children to offers of easy or miraculous cures is legendary among health professionals. Like most legends, this one has some truth to it. Parents are, in fact, astute critics of the professionals who work with their children. They react strongly to signs of professional aloofness or apparent disinterest or dislike and become understandably fearful at signs of indecision. They want intensely for their children to overcome the handicap, to grow out of it, to get some swift and effective treatment. They can forgive and forget aloofness or vague reasoning as long as help and hope are forthcoming. This transformation of distrust into trust can happen in the space of a single breath.

This is not abnormal. All parents want the little boy or girl they see before them to have a world full of promise and happiness, to grow into the kind of adult they can so vividly and lovingly imagine. Indeed, they begin nurturing dreams of the person their child will become from the time of their first knowledge of conception. Imagined details change with the passage of time and as experiences unfold, especially for parents of handicapped children as facts about the handicap and resulting limitations become evident; but hope, like the child, is seldom abandoned. It keeps people going.

Parents of handicapped children spend much time visiting and listening to a bewildering variety of professional people who sometimes present conflicting information. Despite literature designed to improve parent-professional collaboration (Mulick and Pueschel 1983; Pueschel, Bernier, and Weidenman 1988), misunderstanding is commonplace, and ever-present stress increases miscommunication. Stress affects both parent and professional. Nothing can lessen the emotional shock for the parents who hear that their child has a significant disability likely to result in lifelong limitations. The anguish is profound and tends to be rekindled by everyday events, especially when hoped-for improvements are slow or fail to occur

---

This article first appeared in *Skeptical Inquirer* 17, no. 3 (Spring 1993): 270–80.

(Simons 1987). Professionals empathize easily at such times and may experience similar emotional reactions.

The destruction of what have been termed "highly valued dreams" by dismal facts produces predictable emotional results (Moses 1983). These include guilt, denial, and anger. Of these, only anger, and the aggression or hostility that can occur, has been adequately studied scientifically and understood in terms of its underpinnings in biologically based defensive reactions (e.g., Bandura 1973; Flannelly, Blanchard, and Blanchard 1984). The others, guilt and denial, are essentially cognitive phenomena. While less well understood, they represent highly reliable emotional effects of bad experiences, which also may have adaptive functions. Guilt may motivate problem-solving and independent action. Denial seems to work as a cognitive barrier to the perception of incapacitating or troubling thoughts that might impede an ability to get on with essentials of day-to-day living (Meichenbaum 1985, 74–75); it allows people time to revise their priorities.

Selectively screening out bad news through denial allows people to carry on with plans and relationships that would otherwise have to be abandoned, but which may serve other valued functions. Not believing something that is true also has a darker side. It allows people to go on doing things that could be, in part, bad for someone (including oneself). This is especially problematic when some aspects of continuing a course of action in the face of contradictory information are good for the denier, but bad for someone else. When this is the case, harm can continue for as long as the person doing the denying derives benefit from the thoughts and actions the erroneous beliefs permit.

Denial, a form of avoidant coping, is recognized as a common reaction to the diagnosis of a serious illness or disability in oneself or a loved one. Margalit, Raviv, and Ankonina (1992) demonstrated that parents with disabled children requiring continuous special education differed from demographically similar parents in their increased use of avoidant coping and that they exhibited lower confidence in being able to control and understand their world. Interestingly, a greater tendency to adopt avoidant coping strategies did not appear to prevent them from using active coping strategies (i.e., more direct problem-solving). Rather, families with disabled children used a greater mix of the two problem-solving strategies. Further, families whose disabled children had more socially disruptive behavior seemed to use more avoidant coping than families whose disabled children exhibited fewer socially disruptive acts. These findings also are consistent with the effects of chronic stress on the selection of coping styles.

AUTISM

Autism is a severe developmental disability, fortunately uncommon, but prevalent enough to merit specialized educational and habilitative services for affected individuals in even medium-sized communities. It occurs in 4 to 5 of every 10,000 children (Kiely, Jacobson, Schupf, Zigman, and Silverman 1989; Pueschel et al. 1988). It is normally a lifelong disability, associated with mental retardation in about 60 to 80 percent of cases, defined by seriously delayed and often qualitatively abnormal language and communication, and strongly associated with frequent behavioral abnormalities. Common abnormal behaviors include social withdrawal; ritualistic, self-injurious, or odd repetitive acts; and higher rates of asocial or aggressive types of disruptive behavior. A single neurophysiological cause has not been identified for the syndrome.

An additional tragic fact is that affected children are often otherwise quite normal in appearance, bearing none of the physical deformities associated with many other developmental disabilities. Hence autistic children are sometimes described as "beautiful," probably a poignant reminder that except for their decreased ability to learn many types of things, to communicate, and to interact socially, they are initially as likely to evoke positive emotional commitment from adults as any other children.

There have been a few signs of improved scientific understanding of how to help autistic children learn to behave and communicate more normally (Lovaas 1987), but the necessary techniques are costly and demanding and not always effective. There is an active international research effort. Great strides have been made providing needed educational and medical services to these and all handicapped children in the United States (Pueschel et al. 1988). Still, there is no denying that anyone who works or lives with autistic children, especially when they are young, has the sense and the hope that with just the right nudge—or drug or treatment or environment or *something*—they will snap out of the syndrome and act as appealing and as normal as they look. Parents of autistic children must harbor such thoughts even more often than most of us; unfortunately, most autistic children simply do not snap out of it, or even improve, without much expense, training, and effort.

Parents of autistic children may be expected to employ avoidant coping styles more than most parents (Margalit et al. 1992) and to be highly motivated to obtain services that seem to help their children. For better or worse, these hopes are fueled by the implicit promise of our technological society and its cultural preference for hyping the amazing, the simple, and the ever more dramatic benefits of "discoveries" and "breakthroughs." This sets the stage for increased vulnerability to early adoption of untested and poorly conceived interventions and services.

RESCUERS FROM DOWN UNDER

Enter Professor Douglas Biklen of Syracuse University with an imported miracle cure from Australia for the often utter failure of autistic children to learn to speak normally, a technique referred to as Facilitated Communication (FC). FC is "communication by a person in which the response of that person is expressed through the use of equipment and is dependent on the assistance of another person" (Intellectual Disability Review Panel 1988, iv). FC procedures involve using graduated physical (manual) prompting, in an initial least-to-most-effort hierarchy, with gradual reduction of guidance, to help a person point to or strike the keys of a

---

Defacilitated Communication

Facilitated Communication is not cold fusion or N-rays. Professionals have advocated some form of what is now called FC for 20 years. The principle is unassailable: autistics have deficits in their communication skills and therefore require assistance to communicate—like providing paraplegics with wheelchairs to compensate for their inability to walk.

But the hard-core hysterics miss the point. They see FC not as a means, but as an end. The autistic, they say, will forever be dependent upon the facilitator. So even if the inflated claims for FC are true, what good is it? Unlike a wheelchair, one cannot take a facilitator along to school, to work, or on a family outing. Exchanging dependency upon a parent or caretaker for dependency upon a facilitator is no bargain for the autistic.

My autistic son is five years old. His teacher introduced him to FC at a computer keyboard several months ago, and he was delighted. He needed no facilitator—he was capable of picking out the letters on his own and stringing them into words.

But the charms of the computer paled when his teacher introduced him to an infinitely more powerful and flexible medium: pencil and paper. Now he happily spends hours printing and tracing words. He does not speak, but is learning to write. Once he spontaneously hopped out of his seat, went to the blackboard, and wrote "MARK IS HAPPY." His teacher wept.

FC is merely a stepping stone. The goal is defacilitated communication. The goal is empowerment.

—Mark S. Painter Sr.

Mark S. Painter Sr. is a freelance science-fiction writer in Mont Clare, Pennsylvania, and an attorney.

---

typewriter, a computer keyboard, or a paper facsimile. The intent is to support a person's hand sufficiently to make it more feasible to strike the keys he or she wishes to strike, without affecting the key selection. In practice, the manual guidance is maintained indefinitely, suggesting an opportunity for the facilitator to exert continuing influence on key or picture selection by the person being facilitated.

Biklen has justified and advocated the use of this technique in the United States (Biklen 1991, 1992; Biklen, Morton, Gold, Berrigan, and Swaminathan 1992), an amazingly effective effort. FC has been given wide and relatively uncritical recognition in the popular press (e.g., Spake 1992; Whittemore 1992; in contrast to more skeptical treatment by Shapiro 1992) and scandalously credulous treatment in nationally televised news programs.

This "therapy" is currently used nationally with thousands of handicapped people each day at a likely direct cost of millions of dollars each month. Biklen has bolstered his arguments by developing a novel theory of autism based in reframing its root cause as a form of developmental apraxia (i.e., a disorder of voluntary control of movement). Scientific evidence for developmental apraxia in autism is lacking. Autistic youngsters are often characterized by *better*-developed motor than verbal skills, even real nonverbal problem-solving talent. This presents little conceptual difficulty for Biklen. He also appears to imagine that people with no prior evidence of acquiring letter-recognition skills can quickly begin typing out sentences of fairly complex grammatical structure, albeit with a little help from their facilitator, and indeed have had hidden literacy all along despite previous classification as severely mentally retarded.

A FEEBLE PROFESSIONAL RESPONSE

Speech pathologists are prominent in the clinical promotion of FC, especially via fee-for-service introduction of the basic procedures to parents and paraprofessionals directly involved in day-to-day care of people with handicaps. Two of the present authors (Mulick and Kobe) attended a presentation by a speech pathologist at a recent state autism meeting in Ohio. The speaker (Veale 1992) stressed that facilitators must enter the FC situation with complete trust and belief in the autistic person's communicative competence, with the intention that together you can show others.

Veale did not describe the FC training procedure in any detail. Apparently, you prop the keyboard up on an easel and hold the child's hand near the keys while having a verbal or typed conversation. In fact, procedural detail was specifically characterized as unnecessary for such a "simple" procedure. Additional points made: (a) Clients might not seem to be looking at the keyboard. (b) Interfering behavior problems might need to be opposed with just-right amounts of physical force to prevent movement of the learner's hand away from the keyboard. (It was implied that eventually, through FC, patients will thank you for doing so, as one of Veale's own patients communicated even while apparently trying vigorously to escape her grip.) (c) Because true communication is not composed of factual questions and answers but, rather, of open-ended alternating comments, one shouldn't ask factual questions in FC training; instead, one should just be receptive to what is produced by the handicapped individual.

Veale did not caution the audience about possible facilitator influence. She used testimonial evidence in the form of transcripts of FC conversations and a guest appearance by a grateful parent. Although statements made in conference papers are relatively ephemeral, they can have an inordinate impact on the practices of paraprofessionals and parents who may not be skeptical of obtaining unexpected literacy (especially when they are led to expect it by conference presenters). These features have little resemblance to the valid training in instruction for handicapped people that we have seen (Matson and Mulick 1991).

Novice FC trainees may not be well informed about how to teach or modify behavior. When the advice given is inconsistent, it will surely induce different practices by different trainers. Whereas these vague instructions encourage trainees to believe they need not require clients to look directly at the keyboard, Rosemary Crossley (1992a), the originator of FC in Australia, states that looking at the keyboard is absolutely necessary. Whereas the foregoing descriptions of benefit suggest that substantial progress can be obtained rather quickly, Crossley (1992b), Crossley and Remington-Gurney (1992), and Biklen et al. (1992) state that basic competence (e.g., some understandable content) may require up to six months. DEAL (1992) states that six years may be needed to attain communicative competence.

Unfortunately, no measurable definition of "basic competence" is provided. All sources reviewed by us that favor the use of FC mention some person with autism who demonstrated unexpected literacy upon the first try with FC. Transcripts of apparent conversational content are the only evidence provided (e.g., Biklen, Morton, Saha, Duncan, Gold, Hardardottir, Karna, O'Connor, and Rao 1991). We feel that the perceived unexpected literacy and sudden communicative competence is likely due to facilitator influence in most, if not all, such cases.

SCIENTIFIC EVIDENCE AND FC

There has been no adequate controlled veridical support for any of the crucial claims made by FC proponents. We are aware of no demonstration that complex and meaningful linguistic performance, independent of possible facilitator influence, has been obtained from people who had been diagnosed severely or profoundly mentally retarded using valid methods by qualified diagnosticians. Complex linguistic performance is highly correlated with IQ and level of functioning in people with mental retardation and autism. We are aware of no evidence for rapid emergence of linguistic competence in individuals for whom a clear history of learning cannot be documented. We are aware of little support for a motor-impairment theory of autism. There is scientific evidence relevant to each point.

Because its claimed effects were so marked, so unexpected in the context of existing scientific knowledge, FC provoked academic research interest, first by the government and others in Australia and recently by a few behavioral scientists in the United States. Studies in Australia used message-passing (giving information to a nonverbal person while the "facilitator" was out of the room, and subsequently attempting to verify the nonverbal person's comprehension via the typed content of FC) or question-asking directed to the disabled person with masking of facilitator hearing by white noise. While there were isolated instances of apparent valid communication through FC, a review by Cummins and Prior (1992) concluded that the responses of all people tested in these studies were contaminated by influence over content by the facilitators.

Another message-passing evaluation study in the United States (involving 23 people) reported that the validity of communications could be confirmed for *none* of the participants, despite the fact that all were believed to be conversing with some degree of ability by facilitators and therapists (Szempruch and Jacobson 1992).

A controlled study involving 12 clients (believed by their caregivers to have been routinely communicating via FC for some time) and 9 facilitators, by Wheeler, Jacobson, Paglieri, and Schwartz (1993), was more explicitly revealing. In a procedure where two stimulus pictures were employed, Wheeler et al. (in press) show that clients responded accurately with labels of the pictures only when the same pictures were seen by the facilitators. Sometimes clients were shown the same pictures as the facilitators; sometimes the pictures were different. Clients responded "accurately" with labels of some pictures even when the pictures were seen by the facilitators but *not* by the clients, revealing that the typing was *controlled* by the facilitators.

In yet another study, the recommended training sequence for FC was used over 40 sessions with 21 young people with autism (Eberlin, McConnachie, Ibel, and Volpe 1992); the technique produced no unexpected literacy, and no client was able to better his or her measured expressive language skills using FC over the level that was obtained using controlled testing of these abilities verbally. At present, there are no scientifically controlled studies that unambiguously support benefits in expressive language function from taking part in FC for people with mental retardation or autism.

There have been controlled research studies of very successful communication training in toddlers and preschool-age children with autism using conventional methods (e.g., Lovaas 1987), more limited but still successful reports of speech improvement in older children with autism (Schreibman, Koegel, and Koegel 1989), and growing recognition of effective nonspeech communication approaches. The use of computers with adapted input devices and electronic, printed, or synthesized speech output is well established with motorimpaired people (e.g., Cory, Viall, and Walder 1984; Demasco and Foulds 1982; Mulick, Scott, Gaines, and Campbell 1983). Appropriately configured and adapted computers have long been useful as instructional devices, communication tools, environmental control systems, and game platforms with special populations (Vanderheiden 1982), just as they are for everyone else. Other communication systems successfully used by people with autism include manual sign language and picture or idiogram-based systems. Clearly, the issue is not how communication is mediated, but whether or not it is controlled by someone else, whether or not there are at least two independent participants in a communicative transaction.

MOTOR APRAXIA IN AUTISM

The FC perspective directly challenges neuropsychological and linguistic perspectives, founded in a large body of scientific research, about the language of autistic people. Language is complex and involves elaborate neurological substrata. The speech and language of autistic people differs greatly from that of other people, presumably as a result of neurological impairment. While understanding of the spoken word by autistic people is consistent with their general intelligence, they tend to retain and use the meaning of what is said to them differently, without some of the social information that normally accompanies language and that occurs without much effort for the average person. For example, people with autism who can talk have difficulty sharing nuances of meaning that involve taking another person's perspective. They also have difficulty expressing their emotions and accurately attributing emotions to others. These features greatly affect what autistic people say, how they say things, and in some ways, what they mean by what they say. The language of people with autism who have normal intelligence (and quite a few do) is more concrete, sometimes esoteric, and more oblique compared to the average person. The language of autistic people previously diagnosed with severe mental retardation, but now using FC and cited as examples of success, in contrast, is often rich in interpersonal subtleties well beyond the aspirations of many college English majors (but perhaps not beyond those of philosophically inclined teachers or therapists with master's degrees who promote FC).

How is all this viewed by FC proponents? The motor-impairment perspective on language function set forth in writings on FC is based on the concept of motor apraxia. Apraxia refers to the neurologically determined inability to voluntarily initiate behavior or movement (because one cannot "figure out" how to move). In contrast, the term *aphasia*, which was once the term used for problems in language, now is reserved for more strictly cognitive or conceptual problems in using language. Champions of FC believe that people who have autism are not affected by aphasia but, rather, by apraxia, which involves the neurological substrate of the movement system for organizing or performing speech. At the same time, they argue that autistic people have problems "finding" the words (especially nouns), which should be called dysphasia, and speaking the words, dyspraxia.

Neurological terms have specific meanings. What is described by FC proponents refers to difficulties, rather than inabilities, of people who are considered autistic, in using words in certain ways. The quandary in trying to understand the propositions that proponents offer, in a largely post-hoc fashion, to explain the disorder and why FC is needed is complicated by the fact that they seem to confuse terminological distinctions. The prefix properly related to "difficulty" is *dys*, not *a*. If neuroscientists were to offer the same arguments that are set forth in articles advocating FC they would consistently use the terms *dysphasia* and *dyspraxia*.

There is a great deal of evidence that what autistic people say intrinsically reflects the social character of much of their unusual behavior, how they seem to regard other people, and some features that might accurately be described as specific dysphasia(s) and specific vocal dyspraxia(s). Nevertheless, neurophysiological research has not yet demonstrated even a firm basis for a specific dysphasic disorder. However, aside from using terms that, as they are normally used, do not apply to the effects of autism on how people speak, FC proponents further suppose "global apraxia" characterizes autism: that affected people cannot voluntarily initiate movement. The usage error, again, is that they mean autistic people only experience difficulty. Further, there is no research evidence at all to support the position that people with autism experience such global problems. The usual clinical finding, familiar to any psychologist who routinely works in this area, is that motor impairment and delay is much less prominent than communication disorder and delay (Jacobson and Ackerman 1990). In fact, when playing quietly, or simply walking or using a climbing toy, their relatively smooth and coordinated movement and lack of physical deformity is part of the reason autistic children look so appealingly normal and kindle such general interest and hope.

Surprisingly, although the media and disability advocacy groups have greatly promoted FC, little fundamental opposition by professionals to FC has emerged. Very little criticism has been voiced within developmental disability service agencies or regulatory bodies. Many seem to view FC as the greatest breakthrough of all time, the very breakthrough for which advocates interested in full integration of all handicapped people have been waiting so that no one will, by virtue of inability to express themselves, be deprived of a place in society. Many in disability services, who might be critical, understandably hesitate to be, probably for fear of rejection by those involved with FC every day at their place of work.

Why, at the same time, there has not been an aggressive and visible reaction by professional and scientific societies in medicine, psychology, and neuroscience is more difficult to explain. There are numerous pressing issues facing professionals and scientists in their own fields, many involving funding and competition for resources. It may also be that large portions of these professional and scientific groups have abandoned issues related to people with chronic disabilities, whom they cannot cure, to the human service organizations and government agencies that vocally and confidently present themselves as meeting all disability needs. At the same time, toleration of marginal standards for care and quality by the learned professions serving disabled people does everyone a serious disservice. Ineffective services can be costly, unnecessarily increasing the strain on the nation's health budget. Further, it represents indirect support for the prevalent belief among growing segments of the public that health professionals and scientists are not truly to be trusted.

CONCLUSION

There is good reason to be skeptical of extravagant claims made for FC. Our impression is that, at best, it represents a false ray of hope for many families. Many parents and empathic, concerned paraprofessionals might be especially vulnerable to the appeal of FC because of avoidant coping styles or the action of cognitive denial mechanisms that reduce perception of some features of severe disability in others.

The promotion of FC diverts effort and funding from more plausible longterm strategies that have empirical support. The theoretical confusion gratuitously injected into the research and professional literature by FC proponents is damaging to accumulation of knowledge about handicapping conditions and their causes and detracts from the credibility of sincere efforts to integrate findings about abnormal development. The popular confusion of FC with other nonspeech communication systems that have been used successfully with disabled people will discourage public support for these tried-and-true strategies when the FC bubble bursts. And burst, it will. In the end, regret will provide small solace. The irony of FC is perhaps best revealed by a poem reproduced in the Fall 1992 issue of *The Advocate*, a newsletter of the Autism Society of America (p. 16), a poem that won recognition for its author, identified as a 26-year-old youth with autism. The poem (despite its spooky resemblance to an old Beatles song) was said to have been produced via FC. Two lines, in particular stand out; it begins, "I am you, and you are me...," and goes on, "... We are each other as we are what We can be...."

The professional and scientific communities, as well as government human service and regulatory agencies, should not allow people with handicaps and their families to be used by a few professors and therapists who stoke their hopes with empty promises, regardless of their sincerity, while reaping personal or political rewards and working hard to prevent systematic verification of their claims. Such practices should always be called into question. In our experience, people with handicaps can be valued members of their families and communities without resorting to appeals to miracle cures. There is effective help available, help that makes scientific sense. The genuine efforts of scientifically trained and compassionate professionals surpass all fad treatments, and always will. Advances in treatment and understanding come at the price of rigorous training, dedication to accuracy and scientific standards, and objective verification of all treatment claims.

REFERENCES

Bandura, A. 1973. *Aggression: A Social Learning Analysis.* Englewood Cliffs, N.J.: Prentice-Hall.

Biklen, D. 1992. Typing to talk: Facilitated communication. *American Journal of Speech and Language Pathology*: 15–17, 21–22, January.

———. 1991. Communication unbound: Autism and praxis. *Harvard Educational Review* 60: 291–314.

Biklen, D., M. W. Morton, D. Gold, C. Berrigan, and S. Swaminathan. 1992. Facilitated communication: Implications for individuals with autism. *Topics in Language Disorders* 12: 1–28.

Biklen, D., M. W. Morton, S. N. Saha, J. Duncan, D. Gold, M. Hardardottir, E. Karna, S. O'Connor, and S. Rao. 1991. "I AMN NOT A UTISTIVC OH THJE TYP" (I'm not autistic on the typewriter). *Disability, Handicap & Society* 6: 161–80.

Cory, L. W., P. H. Viall, and R. Walder. 1984. "Computer Assisted Communication Devices." In *Transitions in Mental Retardation*, Vol. 1: *Advocacy, Technology, and Science,* edited by J. A. Mulick and B. L. Mallory, 151–73. Norwood, N.J.: Ablex.

Crossley, R. 1992a. Facilitated communication. Invited address at the Annual Conference of the Northeast Region X American Association on Mental Retardation, Albany, N.Y., October 5.

———. 1992b. Getting the words out: Case studies in facilitated communication training. *Topics in Language Disorders* 12: 46–59.

Crossley, R., and J. Remington-Gurney. 1992. Getting the words out: Facilitated communication training. *Topics in Language Disorders* 12: 29–45.

Cummins, R. A., and M. P. Prior. 1992. Autism and facilitated communication: A reply to Biklen. *Harvard Educational Review* 62: 228–41.

DEAL (Dignity Through Education and Language) Communication Centre. 1992. *Facilitated Communication Training.* Caulfield, Victoria, Australia: Author.

Demasco, P., and R. Foulds. 1982. A new horizon for nonvocal communication devices. *Byte* 7: 166–82, September.

Eberlin, M., G. McConnachie, S. Ibel, and L. Volpe. 1992. A systematic investigation of "facilitated communication": Is there efficacy or utility with children and adolescents with autism? Paper presented at the Annual Conference of Northeast Region X of the American Association on Mental Retardation, Albany, N.Y., October.

Flannelly, K. J., R. J. Blanchard, and D. C. Blanchard, eds. 1984. *Biological Perspectives on Aggression.* New York: Alan R. Liss.

Intellectual Disability Review Panel. 1988. *Report to the Director-General on the Validity and Reliability of Assisted Communication.* Melbourne, Victoria, Australia: Victoria Community Services.

Jacobson, J. W., and L. J. Ackerman. 1990. Differences in adaptive functioning among people with autism or mental retardation. *Journal of Autism and Developmental Disorders* 20: 205–19.

Kiely, M., J. W. Jacobson, N. Schupf, W. B. Zigman, and W. P. Silverman. 1989. *Prevalence of Developmental Disabilities* (Project Report). Staten Island, N.Y.: New York State Institute for Basic Research in Developmental Disabilities.

Lovaas, O. I. 1987. Behavioral treatment and normal education and intellectual functioning in young autistic children. *Journal of Consulting and Clinical Psychology* 55: 3–9.

Margalit, M., A. Raviv, and D. B. Ankonina. 1992. Coping and coherence among parents with disabled children. *Journal of Clinical Child Psychology* 21: 202–209.

Matson, J. L., and J. A. Mulick, eds. 1991. *Handbook of Mental Retardation.* New York: Pergamon.

Meichenbaum, D. 1985. *Stress Inoculation Training.* New York: Pergamon.

Moses, K. 1983. "The Impact of Initial Diagnosis: Mobilizing Family Resources." In *Parent-Professional Partnerships in Developmental Disability Services,* edited by J. A. Mulick and S. M. Pueschel, 11–34. Cambridge, Mass.: Academic Guild Publishers.

Mulick, J. A., and S. M. Pueschel, eds. 1983. *Parent-Professional Partnerships in Developmental Disability Services.* Cambridge, Mass.: Academic Guild Publishers.

Mulick, J. A., F. D. Scott, R. F. Gaines, and B. M. Campbell. 1983. "Devices and Instrumentation in Skill Development and Behavior Change." In *Treatment Issues and Innovations in Mental Retardation,* edited by J. L. Matson and F. Andrasik, 515–80. New York: Plenum.

Pueschel, S. M., J. C. Bernier, and L. E. Weidenman. 1988. *The Special Child: A Source Book for Parents of Children with Developmental Disabilities.* Baltimore, Md.: Paul H. Brookes.

Schreibman, L., L. K. Koegel, and R. L. Koegel. 1989. "Autism." In *Innovations in Child Behavior Therapy,* edited by M. Hersen, 395–428. New York: Springer.

Shapiro, J. P. 1992. See me, hear me, touch me. *U.S. News & World Report,* 63–64, July 27.

Simons, R. 1987. *After the Tears: Parents Talk About Raising a Child with a Disability.* San Diego: Harcourt Brace Jovanovich.

Spake, A. 1992. "In Classroom 210 . . ." *Washington Post Magazine,* 17–22, 28–30, May 31.

Szempruch, J., and J. W. Jacobson. 1992. Evaluating the Facilitated Communications of People with Developmental Disabilities (TR #92-TA2). Rome, N.Y.: Rome DDSO.

Vanderheiden, G. 1982. Computers can play a dual role for disabled individuals. *Byte* 7: 136–62, September.

Veale, T. K.1992. Facilitated communication. Presentation at the 4th Annual State Conference of the Autism Society of Ohio, Toledo, October 23–24.

Wheeler, D. L., J. W. Jacobson, R. A. Paglieri, and A. A. Schwartz. 1993. An experimental assessment of facilitated communication. *Mental Retardation* 31 (1): 49–60.

Whittemore, H. 1992. He broke the silence. *Parade: The Sunday Newspaper Magazine,* pp. 8, 25, September 20.

PART EIGHT

SOCIAL DYNAMICS AND BELIEF

37

WHY CREATIONISTS DON'T GO TO PSYCHIC FAIRS

John H. Taylor, Raymond A. Eve, and Francis B. Harrold

(T) he United States boasts a peerless scientific establishment but is also home to a wide variety of people passionately opposed to many accepted scientific findings or even to the conduct of science itself. This latter category includes many different groups, ranging from certain religious fundamentalists to some New Age adherents. For all their diversity, they share a willingness to dismiss any scientific findings that contradict their beliefs. Their frequent response, for example, to the consistent scientific rejection of their empirical claims is to assert that scientists are guilty of excessive conformity and dogmatism. A truly fascinating element within this essentially antiscience grouping is the subset we will consider here: people who conditionally profess to respect the prestige and authority of science (Cavanaugh 1985).

The beliefs of these antiscience groups, despite their aspiration to scientific status for their claims, are often presented in both the mass media and the academic literature under the rubric of "pseudoscientific."[1] Included in this category are such well-known examples as scientific creationism,[2] certain aspects of UFOlogy and parapsychology, (Immanuel) Velikovskian catastrophism, and diffusionist claims of visits by ancient Celtiberians to Massachusetts or Africans to Mexico. Previous research by two of the authors (Harrold and Eve 1987, 1993; Harrold, Eve, and de Goede 1995; Eve and Harrold 1991) has been largely concerned with describing the patterns and sources of pseudoscientific beliefs about the human past.

Until recently, most of the scientific literature arising in reaction to these claims has been devoted to "debunking" such claims, rather than attempting to understand their origins (Harrold and Eve 1987). Thus when sources are discussed, the advocates of fantastic claims are usually said to be ignorant, stupid, or disordered. These explanations do not go far toward explaining these beliefs—especially among people who are apparently not ignorant, stupid, or disordered.

This article first appeared in *Skeptical Inquirer* 19, no. 6 (November/December 1995): 23–28.

For example, some of the principals of the Institute for Creation Research and its more recent counterpart, the Foundation for Thought and Ethics, hold advanced scientific degrees. Their publications—perhaps most notably the creationist biology textbook *Of Pandas and People* (Davis and Kenyon 1993)—are quite professionally packaged, but pseudoscience nonetheless.

SOURCES OF PSEUDOSCIENTIFIC BELIEFS

One of the grand adventures in studying pseudoscience is the attempt to understand the many social and behavioral scientific perspectives of these belief systems. We have found particularly useful the formulation by social psychologists Singer and Benassi (1981). These authors have suggested four distinct classes of factors that condition for the acceptance of paranormal and pseudoscientific beliefs:

1. *Cognitive Biases:* Natural errors in processes of reasoning exist (Piatelli-Palmerini 1994), such as the tendency to perceive order in random data or to jump to an emotionally attractive conclusion. Unfamiliar lights in the night sky thus may be seen as an alien spaceship.

2. *Uncritical or Erroneous Media Coverage of Science:* All too often, the mass media give sensationalistic coverage to extravagant claims about extraterrestrials or the alleged evidence of Bigfoot. Frequently, however, the media fail to present representative information from mainstream science that rejects these claims.

3. *Inadequate Science Education:* A disheartening series of studies in the 1980s and 1990s (Walters 1995; Miller 1987) indicated that many Americans learn little in school about either the methods or the findings of science. A study by Eve and Dunn (1990), for example, found that more than one-fourth of the nation's biology teachers actually favored the teaching of creationism over the teaching of evolution.

4. *Sociocultural Factors:* People tend to adapt to and maintain the beliefs of the society, social class, ethnic or religious group, and family in which they are socialized. A college student reared in a devout fundamentalist home will be unlikely to find an evolutionary biology course of interest. A black student who finds a chilly reception in a predominantly white school may be drawn to writings and lectures on "Afrocentric" claims that his supposed Egyptian ancestors invented civilization (Ortiz de Montellano 1993).

PURPOSE OF THE STUDY

Readers new to the scholarly study of pseudoscience may well be asking, "Why all the concern about such marginal beliefs?" One answer will become apparent when we discuss survey results, but documenting the alarming numbers of adherents to these fantastic beliefs is only a starting point. Of greater significance is that an emerging body of research has only recently positioned us to frame a number of important questions.

In laying the groundwork for the current study, Harrold and Eve in the mid-

1980s conducted a collaborative project with Kenneth Feder and Luanne Hudson that examined pseudoscientific beliefs among college students (Feder 1987; Harrold and Eve 1987; Hudson 1987). In this work, the researchers identified a minimum of two basic types of such belief—creationism and what was termed *fantastic science*. The latter category consists of a panoply of fanciful claims, including Erich von Däniken's famous "ancient astronauts" hypothesis.[3] Further, the researchers noted that the creationist dimension appeared tied to a conservative Protestant theology and worldview, and was almost entirely unrelated to fantastic science.

Building on these contributions, we empirically test these suggestions in the form of three hypotheses explained below, and present results. We believe that we have begun to be able to show that there exists a certain logic about the "rules" one uses to decide on the "truth" of a thing, and that such rules underlie the adherence to many pseudoscientific beliefs. Finally, it appears likely that such rules provide the mechanism by which one can describe the source of the two main categories of pseudoscientific beliefs, creationism and fantastic science.

CULTURE WARS DRIVEN BY HEURISTIC RULES?

Heuristic methods are the rules of thumb or procedures used to search for solutions or answers. If we believe that these "rules for knowing" allow one to establish truth, how might we conceptualize a measurement of these rules? We propose that the members of one subculture within American society might be termed *cultural traditionalists*. Specifically, we suggest that until recently most U.S. citizens (and probably most other humans) believed a thing to be true because of faith, tradition, revelation, or authority. Such an attitude might be summed up in bumper sticker declarations such as "God said it, I believe it, that settles it."

A second subculture, a thread in Western society since the late 1700s at least, derives from some of the epistemological principles of Enlightenment. Specifically, truth is to be sought by the putting forth of hypotheses that are evaluated and accepted or rejected by empirical testing. Those who use an approach or have a worldview that stresses empiricism and scientific inquiry we term *cultural modernists*.

We suggest that a third, and emerging, subculture (or better, collection of subcultures) within U.S. society opposes a return to traditionalism as defined in the first subculture, and views the modernism in the second subculture as having led to rampant militarism, consumerism, pollution, and global warming. We will term the members of the third subculture *postmodernists*. As yet very loosely organized and multinucleated, this subculture spans orientations as diverse as New Age followers, holistic health practitioners, and even the far left of academe (including certain subsets of rhetoreticians, philosophers, and feminists). These are strange bedfellows, indeed. They can only be lumped together as an entity in the postmodernist subculture because of their common rejection of the validity of cultural traditionalism and cultural modernism.

We expected from our testing that those respondents most inclined toward cultural traditionalism would be the most likely to support religiously based pseudoscientific concepts like creationism, and that such beliefs would be highly correlated with a religiously and politically conservative worldview and social agenda. We also expected that those who appeared to be most inclined toward postmodernism would adhere to fantastic science beliefs.

This leads to the presentation of the following hypotheses to be tested here:

Hypothesis 1: Fantastic science beliefs and religiously based beliefs (such as creationism) can be shown to constitute mutually exclusive empirical domains.

Hypothesis 2: The sources of beliefs concerning matters of fantastic science differ from the sources of beliefs in religiously based pseudoscience.

Hypothesis 3: The differential sources of the two categories of belief center around the rules that believers use to determine what makes a proposition or claim true.

THE SAMPLE AND THE RESEARCH INSTRUMENT

A total of 338 students from the University of Texas at Arlington (enrollment approximately 23,000) completed anonymous questionnaires during the 1993-1994 academic year.[4] The survey instrument consisted of 75 items. These items included a range of statements associated with what may be termed *opinions about facts, opinions about social and moral issues,* and *opinions about reliable rules for determining truth.* Opinions were solicited regarding creationism, paranormal beliefs, and belief in fantastic science.[5] Finally, items were included that measured educational, religious, demographic, and socioeconomic background factors. This sample was not statistically representative of the general population nor of college students in the United States, though the percentages are congruent with many prior student studies. What we are interested in is not the percentages of various beliefs, but their relationships to other variables.

RESULTS OF THE SURVEY

A range of fantastic science beliefs are prominent in our sample. No less than 59 percent agreed with or were unsure whether UFOs are actual spacecraft from other planets. The belief that some people can use psychic power to accurately predict future events was undisputed by 49 percent of respondents. When presented with the statement that time travel into the past is possible, uncertainty crept up somewhat with only 45 percent agreeing or unsure. That séances can communicate with the dead, however, was agreed with by 56 percent of respondents. So much for any optimism on our part that pseudoscientific beliefs are eliminated by exposure to higher education!

When asked about items relating to creationism, 60 percent indicated agreement or uncertainty that "There is a lot of scientific evidence for the Bible's account of mankind's creation." Moreover, 88 percent were unable to dispute the statement that "Noah's Ark has been found on the top of Mount Ararat in Turkey." Finally, 80 percent agreed or were unsure that "Adam and Eve, the first human beings, were created by God."

THE CREATION OF SCALES

Our next step was to ask what influences could be identified in the answers we received. We used the statistical technique of factor analysis and were able to identify two meaningful factors.[6] A "factor" can be interpreted as an underlying force within a data set that causes some items to be highly interrelated with each other. Upon examination, all survey items that were highly correlated with, or "loaded" on the first factor (hereafter, Factor 1) involved such as belief in Bigfoot as a real creature, extraterrestrials visiting ancient humans, psychic powers predicting the future, the Loch Ness Monster as a real creature, and UFOs as spaceships from other planets. We designated Factor 1 the "Fantastic Science" factor.

Similarly, all survey items that loaded on the second factor (hereafter, Factor 2) shared issues such as belief in the existence of physical evidence for Noah's Ark, the existence of scientific proof of creationism, and the actuality of Adam and Eve. Factor 2 was labeled "Creationism." Further statistical analyses established that the variables associated with each factor could be combined to create two separate scales, each of which were quite statistically reliable (Fantastic Science alpha = .72 and Creationism alpha = .71).

THE CORRELATES OF CREATIONISM

Creating the above scales allowed us to ask, "Are there differences in the way various survey items correlate with the scales of Fantastic Science and Creationism?" Four additional scales were created from items in the survey in an attempt to operationalize these concepts. We created a scale named "Biblical Literalism" (alpha = .78), which included statements such as "Everything written in The Bible is literally true, word for word." We then developed a "Vitality" scale (alpha = .76), which measured attitudes regarding the sanctity of life. (For example, the Vitality scale included statements in opposition to genetic engineering, mercy killing, and fetal tissue research.) The third scale, "Abortion," consisted of statements asking for the evaluation of various moral and legal aspects of abortion. Finally, a scale of "Crime" was developed. It included statements attributing an increase in the rate of crime to excessive leniency and advocating more punitive approaches, including the death penalty.

As predicted earlier, the Creationism scale was indeed strongly correlated with scoring high on cultural traditionalism and with holding a conservative social agenda (Figure 1). Creationism was strongly correlated (Pearson's r = .7)[7] with Biblical Literalism and with the conservative views on the preservation of life embodied in the scale of Vitality (r = .6). Those scoring high on Creationism also tended to be strongly opposed to abortion (r = .5). Finally, more traditional beliefs about Crime showed a moderate correlation with Creationism (r = .4).

Figure 1: Correlates of Pseudoscientific Beliefs

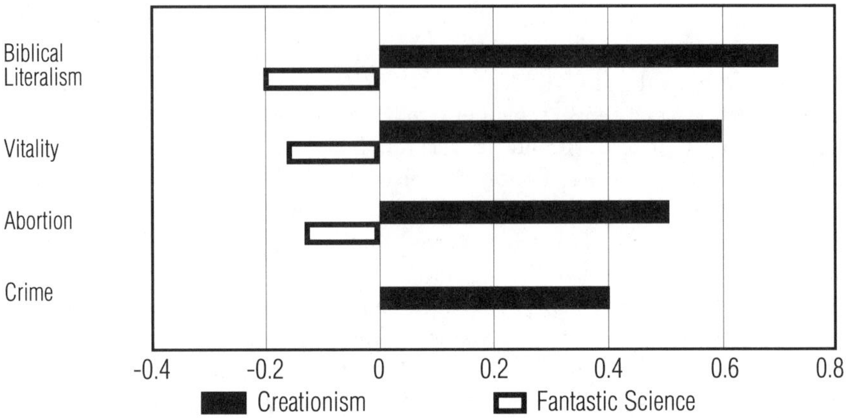

All correlations significant to the .05 level or greater

THE CORRELATES OF FANTASTIC SCIENCE

In significant contrast, the Fantastic Science scale generally produced a weak *inverse* correlation with cultural traditionalism. Biblical Literalism ($r = -.2$), Vitality ($r = -.16$), and Abortion ($r = -.15$) all also showed this reversed, and much weaker, relationship. Crime ($r = -.13$) showed essentially no relationship to belief in fantastic science.

CREATIONISM AS CULTURAL TRADITIONALISM

Interestingly, it can be seen that persons with a deep commitment to creationism and related religiously based pseudoscientific ideas have almost nothing in common with strong believers in fantastic science! If anything, the two thought systems are not just independent of one another, but largely antagonistic. This is just what we would expect on the basis of our initial hypothesis. Remember, we predicted that the rules people use to ascertain the truth would be quite different depending on the subculture one belongs to. It seems reasonable to suggest on the basis of our findings that those believing strongly in creationism are likely to use faith, tradition, revelation, or authority as their major way of establishing the truth.

In order to test this assertion, we performed a statistical analysis using all 26 separate variables related to the Creationism scale. The statistical procedure we used is known as a stepwise regression. It ranks each variable as to its effectiveness in predicting a strong belief in creationism where all other variables in the model are held constant. In an allegorical example, the springtime arrival of storks in Sweden and a correlated increase in birth rate has tempted many to believe that storks bring babies. Of course, a closer analysis reveals a third variable, seasonality, causes the first two items to covary or rise and fall together. The correlation between storks and babies is therefore a noncausal one, often called a spurious correlation. Unfortunately, the general public all too often fails to distinguish that two things that occur together may not involve a causal relationship. So regression techniques provide a better idea of causality among variables. While correlation establishes simultaneity (two or more things occurring at the same time), regression can be used to model causal relations.

We found that among the 26 variables related to creationism, one clearly emerged as having by far the most explanatory power. This variable asked the respondent to assess the statement "It does not matter what scientists say, it is God's word that defines the Truth." There can hardly be any clearer statement that cultural traditionalists don't much care for the rules for truth used by cultural modernists. It also suggests that rules for knowing seem causally prior to all other types of predictor variables.

FANTASTIC SCIENCE AS POSTMODERNISM

Is there, then, any chance that we can support the argument presented in our earlier hypotheses that believers in fantastic science might use postmodernist rules for establishing their own version of the truth?

Postmodernist rules, it seems likely, would eschew both traditional religion and science as ways of knowing. When regression analysis is applied to all 20 items related to fantastic science, an intriguing confirmation of this hypothesis emerged. The variable with the most explanatory power asked whether it was true that "Neither the 'beliefs' of the world's churches today nor scientific 'studies' adequately explain the world around us. An adequate explanation requires other forms of spirituality." The item that ranked second in explanatory power asked for agreement with the statement "Pagan religions from previous times have much to teach us about how to solve today's problems." Again, there could hardly be clearer evidence for support of our hypothesis. Obviously, postmodernists don't care what traditional religion or science determines as the truth.

SUMMARY AND CONCLUSIONS

It can be almost a shocking experience to review survey results and realize that beliefs in pseudoscience are so widespread, even among college students. As with the microscope or telescope, however, statistical analysis is a grand tool of exploration with which to combat excessive reliance on intuition. Within the raw percentages we often find subtle trends that hold promise for a more complete understanding of the sometimes amazing world around us. This study may be a case in point.

We began by noting that, contrary to the intuition of many (including many scientists), pseudoscientific beliefs are not restricted to the domain of the ignorant, stupid, or disordered. Further, we observed that at least two major categories of pseudoscientific belief exist, each being almost causally separate phenomena. The first of these, identified with a religious belief in creationism, seemed strongly influenced by the subculture within which a respondent is located. Such sociocultural factors, however, seem to create their effect largely through influencing the epistemological rules (tradition, authority, etc.) a respondent uses for determining truth.

In contrast, the second major category of pseudoscience appears to be a set of beliefs we have termed fantastic science. These beliefs seem far less influenced by sociocultural factors or subculture than a belief in creationism. Moreover, belief in fantastic science appeared to be highly correlated with the rejection of traditional religion and science as valid methods for establishing the truth. This hints that perceptual errors like those identified at the beginning of this article may play a larger explanatory role in fantastic science beliefs than in the other beliefs examined here. Another fascinating dimension of fantastic science is that the apparently diverse issues of UFOs, monsters, and mysterious mental powers appear highly empirically interrelated.

We hope we have persuaded the reader of the value of exploring the rules used by adherents of pseudoscience to assess the truth. Further, we suggest that the guidelines believers use to establish these rules should prove a productive line of inquiry. Considering the profound role played by sociocultural issues, another intriguing question is whether the choice of rules employed is influenced by variations in social setting. For example, is it possible that one might use cultural traditionalist rules when considering the origins of humanity when one is in a traditional religious community, but unknowingly switch to a different set of rules when one arrives on a college or university campus and is asked to understand molecular biology? Do people use one set of rules for knowing when they are on campus and another set of rules when they return to a traditional community?

This study has looked only at college students. It will not surprise us, though, if these results are supported when surveying a broader population. For example, preliminary analyses of a similar survey collected at a creationist fair earlier this year [1995] appear to confirm relationships discussed in this article. We hope, however, that this study will stimulate further research among other segments of the public (such as UFO-believer groups, pagan religious groups like Wiccans, and so on). In the old days, social scientists who studied deviance adopted a medical perspective. They diagnosed symptoms, declared the subject mentally ill, and prescribed cures. Recently, some have tried a different strategy. They asked the deviants how they perceived the world, and an odd thing happened. Given the assumptions the deviants made about the nature of their own realities, their "sick" behavior appeared quite reasonable, often even logical.

Perhaps for too long we have also used the medical model to process our impression of those who believe in pseudoscience. We have found them stupid, ignorant, or even psychopathic. While this may well be true of some relatively small subsets, we might well be surprised if we ask them to tell us how they make sense of things (including the heuristic assumptions they use). We might well learn again that the temptation to apply pejorative labels to outsiders as an explanation for their behavior and attitude is more successful in satisfying our own emotional needs than in advancing toward greater insight. Without this communi-cation, we will fail to develop effective tactics for responding to the increasing prevalence of pseudoscientific belief in today's world.

POSTSCRIPT

The chapter above was written on the basis of an exploratory sample composed of a student population. Since that time, many of the central ideas presented above have been put to a somewhat more stringent test by asking essentially the same questions of a sample of about 300 individuals attending a "Creationist Fair" in Glen Rose, Texas, and another sample of about 300 neo-Pagans (mostly Wiccans) attending a "Magical Arts" fair and campout near Austin, Texas. It was expected that the creationists would confirm the assertions made above concerning "cul-tural traditionalists" and that the neo-Pagans would give support to the assertions about "postmodernists." It was expected the two groups would differ wildly on a number of major issues as suggested in the article and their respective beliefs would form two generally internally coherent but mutually exclusive domains.

While each sample was small, the results were so dramatically in support of the major assertions made in this chapter that there can be little doubt of the gen-eral validity of most of the predictions extrapolated from the student sample that is the focus of the report in this volume. While these data have not yet been pub-lished, interested parties are invited to contact Ray Eve at eve@uta.edu for a copy of the results of the creationist/neopagan study (which was presented in 1996 at the annual meeting of the American Association for the Advancement of Science meetings as Raymond A. Eve, John Taylor, and Ladorna Goff. 1996. "Postmodern Heuristics and the Rejection of Science").

NOTES

1. Eve and Harrold have discussed elsewhere (1991, 84–86) why we accept the term pseudoscience over alternative terms—often presented as less harshly judgmental—such as *unconventional science* or *parascience*.

2. There are several different types of creationist belief, most of which purport a sci-entific methodology. They are distinguished by tenets such as whether the Earth is a few thousand years old or a few billion years old. Moreover, new variants of creationism are emerging. For example, subsequent to the 1987 United States Supreme Court ruling that defined creation science as religion and banned it from the public classroom, creationist activism in the schools has centered instead on concepts such as "intelligent design" and "abrupt appearance."

3. With apologies to Stephen Williams, whose *Fantastic Archaeology* (1992)—an authori-tative account of North American cult archaeology—provided the inspiration for our term.

4. Classes selected were sections of a variety of upper- and lower-level undergrad-uate sociology and anthropology courses. Survey participation was voluntary, but nearly all students chose to complete the survey. A total of 107 males (32 percent) and 226 females

(67 percent) responded. Black students accounted for 10 percent of the responses, whites 75 percent, and Hispanics 6 percent. The age of respondents included 54 percent who were 22 years old or younger, 28 percent who were from 23 to 29, and 17 percent who were 30 years of age or older. Undergraduate class standing was fairly evenly distributed among seniors (31 percent), juniors (28 percent), sophomores (21 percent), and freshmen (18 percent). Social science majors comprised 44 percent of respondents, humanities majors 16 percent, science majors 16 percent, and anthropology majors 6 percent.

5. Opinion items were designed as Likert-type scales. Many items were selected because they were prominent in previous research efforts. Other items were entirely new. A copy of the questionnaire is available from the authors upon request. Space limitations have prevented reporting in detail the wording of all the items used in the analysis.

6. Details of the statistical analyses are available from the authors upon request.

7. Numbers within the parentheses refer to statistical measures of association. Pearson's r here refers to simple product-moment correlation used when all measures are interval level data. While some relationships were more significant than others, all correlations were significant at the .05 level or better. A .05 level of significance means that among random data such a correlation could not occur more often than one time out of twenty.

8. The authors would like to express their appreciation to Julia Lam for her assistance in formulating the questionnaire and in coding and analyzing the data.

REFERENCES

Cavanaugh, Michael. 1985. Scientific creationism and rationality. *Nature* 315: 185–89.

Davis, William P., and Dean H. Kenyon. 1993. *Of Pandas and People: The Central Questions of Biological Origins,* 2d ed. Dallas: Haughton Publishing.

Eve, Raymond A., and Dana Dunn. 1990. Psychic powers, astrology and creationism in the classroom? Evidence of pseudoscientific beliefs among high school biology and life-science teachers. *American Biology Teacher* 52, January.

Eve, Raymond A., and Francis B. Harrold. 1991. *The Creationist Movement in Modern America.* Boston: Twayne Publishers.

Feder, Kenneth D. 1987. "Cult Archaeology and Creationism: A Coordinated Research Project." In *Cult Archaeology and Creationism: Understanding Pseudoscientific Beliefs About the Past,* edited by Francis B. Harrold and Raymond A. Eve, 34–48. Iowa City: University of Iowa Press.

Harrold, Francis B., and Raymond A. Eve. 1987. "Patterns of Creationist Belief Among College Students." In *Cult Archaeology and Creationism: Understanding Pseudoscientific Beliefs About the Past,* edited by Harrold and Eve, 68–90. Iowa City: University of Iowa Press.

———. 1993. "Scientific Creationism and the Politics of Lifestyle Concern in the United States." In *Religion and Politics in Comparative Perspective,* edited by Bronislaw Misztal and Anson Shupe, 97–109. Westport, Conn.: Braeger.

Harrold, Francis B., Raymond A. Eve, and Geertruida C. de Goede. 1995. "Cult Archaeology and Creationism in the 1990s and Beyond." In *Cult Archaeology and Creationism: Understanding Pseudoscientific Beliefs About the Past,* an expanded edition, edited by Harrold and Eve, 134–75. Iowa City: University of Iowa Press.

Hudson, Luanne. 1987. "East Is East and West Is West? A Regional Comparison of Cult Belief Patterns." In *Cult Archaeology and Creationism: Understanding Pseudoscientific Beliefs About the Past,* edited by Harrold and Eve, 49–67. Iowa City: University of Iowa Press.

Miller, Jon D. 1987. The scientifically illiterate. *American Demographics,* 9(6):26–31.

Ortiz de Montellano, Bernard. 1993. Melanin, Afrocentricity, and Pseudoscience. *Yearbook of Physical Anthropology* 36: 33–58.

Piatelli-Palmerini, Massimo. 1994. *Inevitable Illusions: How Mistakes of Reason Rule Our Minds.* New York: John Wiley and Sons.

Singer, Barry, and Victor A. Benassi. 1981. Occult beliefs. *American Scientist* 69: 49–55.

Walters, Laurel Sharper. 1995. World educators compare notes. *National Times,* December/January, pp. 38–39.

Williams, Stephen. 1992. *Fantastic Archaeology: The Wild Side of North American Prehistory.* Philadelphia: University of Pennsylvania Press.

38

CONSPIRACY THEORIES AND PARANOIA: NOTES FROM A MIND-CONTROL CONFERENCE

Evan Harrington

(T) he debate over "recovered" and "false" memories continues to be one of the most contentious issues in the field of psychology today. The debate is extremely polarized with very little amicable communication among members of the opposing camps. While such a dispute may eventually be beneficial to science, in that both sides are clearly being spurred on to produce original research at a frenetic pace, at the moment the clearest manifestation of this dichotomy is miscommunication and friction between factions. Such miscommunication has been exacerbated by a tendency of some theorists on both sides to make sweeping generalizations and use vague terminology. An example of such miscommunication is the use of the term *recovered memory therapy*, used frequently in books such as *Making Monsters* by Richard Ofshe and Ethan Watters (1994). The term as they used it is not without its critics (e.g., Dalenberg 1995) who complain that the term is overgeneralized. Conversely, in a televised debate, Charles Whitfield, a trauma therapist, stated that there is no such thing as recovered memory therapy. The true state of affairs likely rests somewhere in between.

While some misunderstandings may be rooted in semantics, others are more difficult to trace and harder still to describe adequately. It is very difficult to get quantitative data in the area of the beliefs held by therapists regarding topics that *may* manifest in the form of false memories in their patients. And although some surveys have attempted to obtain quantitative measures of therapists' beliefs, practices, and experiences regarding traumatic memory recovery and therapy (e.g., Poole et al. 1995), such surveys fail to fully inform the reader of the quality of those beliefs. In an attempt to obtain a qualitative analysis of the beliefs of therapists with regard to recovered memories of traumatic events, I have frequently attended sexual- and ritual-abuse conferences. Some of these conferences have afforded me valuable insight into the dynamics of a scientifically informed trauma

This article first appeared in *Skeptical Inquirer* 20, no. 5 (September/October 1996): 35–42.

therapy. At other times I have gained valuable insight into the beliefs of some "fringe" therapists who believe in vast and nefarious conspiracies organized to harm children. My purpose here is not to argue whether such beliefs are accurate or not; rather, I simply wish to outline what some of those beliefs are. The following is not meant to be representative of all therapists in this field. I offer only a description of what *some* therapists believe. The reader will please keep in mind that any qualitative description, such as this one, may not be used to infer anything about the population as a whole, but it may be illuminating in that there is a certain subpopulation that clearly is represented.

This article describes my experiences at a conference held in Dallas, Texas, March 23–26, 1995, by a group calling itself the "Society for the Investigation, Treatment and Prevention of Ritual and Cult Abuse" (SITPRCA). SITPRCA may be reached at P.O. Box 835564, Richardson, TX 75083–5564.

The 1995 SITPRCA conference was titled "Cult and Ritual Abuse, Mind Control, and Dissociation: A Multidisciplinary Dialogue." The word *dialogue* is misleading because there were no skeptics or critics among the speakers and, as will be demonstrated, any dissension from the audience was strongly discouraged—it was essentially a monologue. The 1995 conference offered continuing education credit available through the Texas State Board of Examiners of Licensed Professional Counselors.

The conference was attended by 150 to 200 people. A significant minority of the audience consisted of patients who claimed to have had recovered memories of ritual abuse (several of whom I spoke with) and who were allowed access to even the most advanced professional training sessions, sometimes at the recommendations of their therapists.

The SITPRCA organization was created by Dallas therapist James Randall "Randy" Noblitt, currently the president of the group, and Pamela Perskin, its executive director. Noblitt lectures widely on the existence of ritual cults and mind-control techniques, and has served as an expert witness in a number of child-abuse cases. In the 1992 Austin, Texas, day-care case of Fran and Dan Keller, he helped obtain a conviction by informing the jury that cults across America regularly ritually abuse children through torture and sexual abuse and that the cults make child pornography with these victims. Noblitt stated that these children will often not be able to recall the events because they are so highly traumatized, and that the severity of the abuse causes the amnesia. This testimony, combined with Noblitt's statement that he was "convinced" that the child in this case had experienced extreme trauma, apparently helped convince the jury that the Kellers operated a ritual-abuse cult in their day-care center. At the time of that trial, Noblitt testified that in addition to supervising his own clinical employees he had been sought to consult in 15 similar cases and that he provides supervision for therapists individually and in groups. Noblitt and Perskin (1995) recently released a book outlining their beliefs about ritual abuse. While some mainstream therapists may conclude that those associated with SITPRCA represent a fringe element, I would point out that such organizations are able to have a dramatic influence on society.

OPENING REMARKS

The conference opened with a panel consisting of Walter Bowart (author of *Operation Mind Control*, Dell, 1978), Mark Phillips (who claimed to have inside information on government mind-control techniques), and Alan Scheflin. Scheflin is a lawyer who has for years documented the Central Intelligence Agency experiments with "brainwashing" in the 1950s and 1960s and who spoke on a panel at the 1993 American Psychological Association (APA) meeting with memory researcher Elizabeth Loftus and again at the 1995 annual meeting along with Richard Kluft and several others. Bowart opened the conference with a direct appeal to the therapists. Bowart claimed that "the False Memory *Spindrome* [*sic*] Foundation . . . is a Central Intelligence Agency action. It is an action aimed at the psychological and psychiatric mental health community to discredit you, to keep you in fear and terror." Bowart stated that everyone connected with the False Memory Syndrome Foundation (FMSF) will be shown to be "spooks or dupes." According to Bowart, the CIA is currently conducting a campaign of mind control against the American public and wants to discredit victims of these experiments so that their stories will be seen as false memories. Phillips spoke for a while about how he would reveal the trade secrets of mind control.

Scheflin gave a lengthy talk about how therapists can protect themselves against the lawsuits brought by former patients who retract memories of childhood abuse. These lectures were warmly received, especially Scheflin's. Perhaps because several speakers at the conference had been successfully sued by former clients, the therapists in attendance seemed quite fearful that their clients would retract their memories of abuse and sue them for instilling false memories. I felt that the opening remarks were overtly political for what was purported to be a scientific gathering.

RACIST CONSPIRACY THEORIES AND THE MILITIAS

Doc Marqui, a self-described former "school teacher and witch," lectured about the satanic "Illuminati" conspiracy, which he alleged President Bill Clinton was part of, serving as the "anti-Christ." Marqui assured the audience that this theory is not racist; but the fact is the Illuminati theory is the same one advocated by most members of the American militia movement, and it was utilized by the Nazis in their effort to justify their campaign of genocide against the Jews of Europe (Cohn 1966). *The Protocols of the Elders of Zion* is an anti-Semitic document (based on the Illuminati conspiracy theory) that purports to document plans for Jewish world domination and which first appeared in Russia in 1903 in a newspaper edited by a "noted and militant anti-Semite" (Cohn 1966, 65). The book was instituted as mandatory reading in German schools by the Nazis in 1933 (Cohn 1966). Marqui touted the overall validity of the *Protocols* while replacing the word *Jews* with the word *satanists*. The Illuminati conspiracy holds, in part, that large Jewish banking families have been orchestrating various political revolutions and machinations throughout Europe and America since the late eighteenth century, with the ultimate aim of bringing about a satanic New World Order. Members of the militia movement have said they believe that the United Nations has been infiltrated by these "demonic forces" and is poised for a violent overthrow of the American government, after which American rights to own firearms will be removed and American citizens will be enslaved by the introduction of a cashless society, as foretold in the Bible's book of Revelation (see, e.g., Constantine 1995; Kelly 1995; Springmeier 1995; Stern 1996). Marqui stated that the Illuminati is essentially a shadow government that has controlled the United States since its inception, controls the Masonic order, and commits all manner of occult crime culminating in human sacrifices on eight days of each year. Much of this paranoia was chronicled more than 30 years ago by Richard Hofstadter (1965).

While the Illuminati conspiracy theory is widely endorsed by militia members, it is also embraced by reactionary groups such as: the Lyndon LaRouche organization (political analyst Chip Berlet [1994] stated that in the early 1970s, Lyndon LaRouche "took his followers . . . and guided them into fascist politics"); the John Birch Society (which Berlet [1994] said believes "Insiders" have for years controlled the U.S. and former Soviet Union governments); and the Liberty Lobby. The Liberty Lobby, with its newspaper *Spotlight*, was created by Willis Carto, who also founded the Institute for Historical Review, which asserts that the Holocaust was a hoax (Berlet 1994).

Author Linda Blood, who spoke later in the day, protested that she was "unhappy to be following someone [Marqui] who is pushing the *Protocols of the Elders of Zion*," which she said was anti-Semitic trash. Blood's protest deeply angered some and bewildered others, while about four of Blood's friends clapped in support. Perskin, who moderated the session, announced that although she is Jewish she found nothing offensive in Marqui's lecture. Marqui appeared to me to be connecting existing racist conspiracy theories with the therapists' theories about satanic cults.

Marqui was followed by former Federal Bureau of Investigation agent Ted Gunderson, who highly praised Marqui's lecture. Gunderson is well known for his claims that an archaeological dig under the McMartin preschool showed evidence of tunnels, through which the children were allegedly spirited to other buildings to be prostituted in the community (Summit 1994). The results of this dig have for years gone unpublished while calls for funds to self-publish the results have been issued in newsletters such as the *Survivor Activist* (1994). Meanwhile, the integrity of the dig has been strongly disputed (Earl 1995). Gunderson presented what he called "new evidence" in the 1984 McMartin preschool sex-abuse case in Manhattan Beach, California. He produced a number of photographs of the foundation of a house in the hills above San Bernadino, California, that had burned down, he claimed, the night the charges were filed in the McMartin case. He alleged that the McMartin children were flown to this house and ritually abused, and that the house was torched to destroy evidence. The sum total of the evidence he presented to support this allegation was the existence of spray-painted satanic graffiti on the foundation stones and on boulders on the property. Apparently, several years had gone by between the time of the alleged fire and the time Gunderson snapped the photos. Yet Gunderson was dismissive of the idea that the house foundation on the lot, with its hillside vista of San Bernadino, had been used by teenagers who might have painted the graffiti after the fire. The therapists were enraptured and later asked if Gunderson was planning to publish his photos or if there was any chance of using this evidence in a new trial. The McMartin preschool case resulted in the longest criminal proceeding in American history and failed to produce any convictions (see e.g., Nathan and Snedecker 1995).

Gunderson then described a conversation he had with a witness, Paul Bonacci, from an alleged satanic-ritual abuse case in Nebraska that was detailed by former Nebraska State Sen. John DeCamp (1992), who was also a speaker at this conference. The grand jury of Lincoln described this case as an attack by DeCamp "for personal political gain and possible revenge" (Dorr 1991, 1), a "smear campaign," and a "carefully crafted hoax" (United Press International, September 18, 1990). The grand jury jailed one and indicted two others (including Bonacci) for perjury, and was so critical of DeCamp that he sued the grand jury for ridicule, though he quickly lost (Dorr 1991). A church in the area, the Nebraska Leadership Conference, responded by publishing a tract (no date) named *The Mystery of the Carefully Crafted Hoax*, with a foreword by Gunderson, in which he continued the allegations of satanic-ritual crime. At the conference Gunderson related Bonacci's description of a slave auction in Las Vegas in which 25 to 30 vans pulled up, airplanes landed, and foreign men with turbans bought children and took them away. According to Gunderson: "Nobody knows what happened to those kids. They use them for several things: body parts, they use them for sacrificing, for sex slaves. But this is a big market. Does anybody have any idea what a blue-eyed, blond-haired eleven- or twelve-year-old girl would sell for? Fifty thousand dollars."

Gunderson claimed that there are currently 500 satanic cults in New York City alone, each averaging eight sacrificial murders a year, for a total of 4,000 human sacrifices *every year.* Gunderson did not explain how the cults remove bodies in the asphalt jungle of New York.

Gunderson believes in the threat posed by the New World Order, as do Marqui and militia members. Gunderson has appeared on "Dateline NBC," at militia conferences (Witt 1995), on Michigan Militia member Mark Koernke's shortwave radio program, and on the cover of *Spotlight* (May 13, 1995), stating that the U.S. government intentionally bombed the Oklahoma City federal building in April 1995, in order to remove our rights through anti-terrorism bills. Gunderson informed the audience that *Spotlight* "tells it like it is," and urged audience members to call the subscription number, which he read aloud. On top of this, Gunderson gave an interview to Lyndon LaRouche's *Executive Intelligence Review* (May 25, 1990), in which he described FBI special agent Ken Lanning as "probably the most effective and foremost speaker for the satanic movement in this country, today or at any time in the past." Gunderson and Marqui seem to me to be attempting to introduce therapists to racist conspiracy theories and reactionary propaganda, while at the same time groups such as the LaRouche organization endorse satanic conspiracy theories to draw in new members.

Political analyst Chip Berlet's argument that radical right elements are seducing the left should be taken seriously. In his monograph *Right Woos Left* (Berlet 1994), he describes, among other examples, how the LaRouche organization has persistently destabilized legitimate leftist activist organizations by infiltrating these groups and then claiming that these groups endorse LaRouche. The LaRouchians also gain credibility through their association with legitimate political activists, which enables them to draw new converts. The cult-ritual abuse field is a prime example of such infiltration. Many therapists who specialize in treating ritual or other forms of abuse identify to some degree with feminism and other liberal ideals. When radical right conspiracists get such liberals to believe in the New World Order or "Operation Monarch" (a similar movement, described later) they gain a boost in credibility far beyond what they could expect by printing their stories in *Spotlight* or the *Executive Intelligence Review*.

Former Nebraska State Sen. John DeCamp, mentioned earlier, has been on the ritual-abuse circuit for some time now, talking about his 1992 book *The Franklin Cover-Up*, which purports to document a satanic organization in Nebraska that abused children and prostituted them within the White House. DeCamp gives a favorable mention to a fact-finding mission sponsored by LaRouche (DeCamp 1992, 241). The editors of the *Executive Intelligence Review* repeat DeCamp's claims and praise his book as "important" in their virulently anti-Semitic party tract titled *The Ugly Truth About the ADL* (Anti-Defamation League) (Editors of the *Executive Intelligence Review* 1992). The July 27, 1990, issue of the *Executive Intelligence Review* stated that the FBI in Nebraska covered up child abuse and murder.

On June 15, 1995, DeCamp appeared before a U.S. Senate subcommittee hearing on domestic terrorism chaired by Arlen Spector. DeCamp appeared as a lawyer representing the American militia movement and the four militia leaders testifying that day. At a Washington, D.C., news conference, DeCamp glowingly described the militia movement as "a political movement in the birthing . . . painful, joyous, confusing, and exciting" (Janofsky 1995, 10). DeCamp also has clear ties with the Nebraska Leadership Conference. A call to the church office confirmed that the Nebraska Leadership Conference had "contributed significantly" to DeCamp's book.

DeCamp delighted the therapists at this conference during a luncheon session in which he described the allegations put forth in his book.

CONSPIRACY THEORIES IN ACTION

I struck up a conversation with a woman and her son and learned that the woman claimed to have recovered memories of being abused in a satanic cult. She drove across two states to attend the conference, she said, in the hope that she could learn about Nazi scientists being brought to the United States after World War II. She knew nothing about this topic but seemed to suspect that it had something to do with her. The conversation drifted to the topic of treatment for sex offenders while they are incarcerated. At this point we were joined by a man, whom I'll call Felix, and his companion, who said that treatment for sex offenders is unnecessary because when the New World Order takes control of the country, members are going to shoot all prisoners and also eliminate three-quarters of the world's population. Felix described to us how the New World Order operated, manufacturing multiple personality disorder through torture and creating sex slaves and drug mules under the mind control of the CIA (this is the basis for the alleged "Operation Monarch"). Felix also described how the black helicopters of the New World Order landed in his hometown of Portland, Oregon, and black-suited storm troopers illegally searched all the homes in the neighborhood. There was a total news blackout of this because, Felix said, the media are part of the conspiracy. Later, Felix confided to me that his companion was wrong: the New World Order would not kill all the prisoners, but would use them as slave labor. Felix said he did not like to disagree with her because she was a former "Monarch" mind-control slave.

Felix sold me his newsletter, as big as a book, in which he makes some very strange claims: Charles Manson was programmed by the Illuminati, the Anti-Defamation League is controlled by Jewish satanists, and Marilyn Monroe was a mind-control slave. According to Felix, virtually anyone who disagrees with Felix is a Monarch slave, including prominent militia leader Bo Gritz, who talked Randy Weaver into surrendering at the 1992 incident at Ruby Ridge, Idaho. Most disturbingly, Felix told me that he works as a counselor and has helped "a lot" of people suffering from multiple personality disorder. Felix apparently has no mental-health counseling credentials, and his name badge identified him as "clergy." Nevertheless, he said he counsels dissociative clients and guides them through the intricacies of international cabals.

By this time a crowd had gathered around Felix and me. After Felix's monologue, a social worker from North Carolina informed the group that in the day-care sex-abuse case she was investigating, she *thought* she remembered the kids talking about black helicopters. She said she would look into it.

SECRETS OF MIND CONTROL REVEALED

Felix's claims paled in comparison to what came next. Mark Phillips claimed to be a former government agent involved in mind-control experiments. He was always vague, never giving any information that could be checked. His companion, Cathy O'Brian, claimed to have survived years of torture and abuse at the hands of her CIA handlers in Operation Monarch (these two seem to be the source of most of the Monarch material). O'Brian maintained she had been tortured in unimaginable ways since the time she was a child, and that her cult handlers successfully created dissociative identity disorder in her, which was cured by Phillips, who also managed to hide her from the CIA. She was so savagely tortured, she said, that her back was a complete mass of scar tissue. Phillips added that he had once tried to count the scars but lost count somewhere in the hundreds. We never saw the scars, photos of the scars, or doctors' reports about the scars.

O'Brian stated that she was forced to have sex with a plethora of political figures including George Bush, Ronald Reagan, Jimmy Carter, and Gerald Ford (whom she said she knew as "the neighborhood porn king"). She also said she was abused by Hillary Clinton (but not by Bill). Politicians were not the only ones involved—O'Brian stated that a number of baseball figures were in this satanic/CIA mind-control plot. She told me personally that virtually the entire country music industry is set up by the New World Order to make money. According to O'Brian, most popular country singers are Monarch slaves who had alter-personalities created with good voices for singing. Phillips and O'Brian, along with Bowart and others, claimed that the CIA is currently abusing people through Operation Monarch. Phillips claimed 20 years of experience in genetics and said that the cults would breed slaves selectively to create musical geniuses. To test his vast experience with genetics, I asked him what he thought of the Human Genome Sequencing Project. He had never heard of it. It seems impossible for anyone with even a rudimentary knowledge of genetics to be unaware of the biggest project ever in that field. Nevertheless, one author claims that Phillips is "currently deprogramming at least six Monarch slaves" (Springmeier 1995, 243).

It seems that a number of people in the audience were accepting of Phillips's and O'Brian's claims, although Perskin (of SITPRCA) informed me that this duo will not be asked back in the future because they failed to produce evidence of Operation Monarch. In a personal conversation with me (July 12, 1995), Scheflin stated that he had been able to obtain internal CIA documents corroborating the existence of mind-control experiments in the 1950s and 1960s. (The documents demonstrate that the CIA conducted unethical experiments to try to create multiple personalities in people for the purpose of creating a super spy who could keep vital information submerged in an alter personality [Thomas 1990].) But, he said, the paper trail completely died out by 1976. According to Scheflin, there are no credible reports of mind-control experiments after 1976 and no credible reports of any nature on Operation Monarch.

Catherine Gould gave an advanced workshop in which she described the mechanics of cult mind-control, extensively utilizing the mind-as-computer model. At one point she puzzled over the idea of cult members catching AIDS. She said that no one can figure out why the offenders are not "dropping like flies, because we know they don't practice safe cult sex." With all the blood, cannibalism, and unprotected sex, they ought to be catching a lot of sexually transmitted diseases. Therapist Jerry Mungadze offered a unique explanation. He suggested that mind-control programming boosts the immune system, making the victim resistant to the HIV virus, and that is why children in day-care satanic-ritual abuse cases do not have elevated levels of sexually transmitted diseases.

Well, if they've found a cure for AIDS, why do they bother making money with pornography? Such a cure must be worth several billion dollars! In the grand tradition of conspiracy theories, discrepant information is explained away or, as in this case, incorporated into the scheme. Amazingly, this solution to the AIDS conundrum appeared to be taken seriously by most in the room.

ALTERNATE VIEWS NOT WELCOME

Chrystine Oksana lectured on her experiences of recovering memories of ritual abuse and her subsequent search for corroboration (see Oksana 1994). Oksana stated that she had read some 500 books on the topic of trauma and child abuse. For this reason I asked what she thought of the recent study by Linda Meyer Williams (1994). Oksana said she had not heard of it. The report by Williams is a pivotal study that demonstrated that a substantial minority of adults failed to disclose their documented emergency room visits when they were children, which ostensibly occurred because they had been sexually abused. The study demonstrated that some people may forget such events. There is a mistake in the text of the paper that states the existence of a nonsignificant trend such that, as the amount of force used in the commission of the abuse increases, recall decreases. The trend in the data actually shows that as the amount of force used in the commission of the abuse increases recall *increases,* which is opposite from, and fails to support, the theory of repression of traumatic memory (Harrington 1995). My description of this data set visibly angered several in the audience. One woman voiced disbelief of what I had said (preferring to believe that greater trauma typically was related to nonrecall), while a second woman shouted at me twice to read Lenore Terr's *Unchained Memories.* After a couple more rebuffs, the session ended in a stony silence. Yet another woman approached me and bluntly stated that she did not believe what I had said. I told her that I had a signed letter from Williams affirming my observations. This woman shrugged her shoulders and walked away smiling, as if to say that she still did not believe me. This appears to be an example of the resistant nature of strong beliefs toward discrepant information.

In the final analysis of the Williams data, the nonsignificant trend of force being associated with greater recall is probably a confound wherein both greater force and greater recall are associated with older age at time of abuse. Nevertheless, mine was a legitimate question to raise during a session on traumatic memory where it was stated that events that are more traumatic are more likely to be dissociated from consciousness. The scalding reaction I received from the audience supports the view that group social representations are not amenable to contradiction (Guerin, in press), and indicates that these are not issues open for discussion.

SKEPTICISM AND SATANISM

The next session featured lawyer John Kiker and therapists Noblitt, Michael Moore, and Jan Maclean on the topic of the travails of being sued. Moore described in detail how violated he felt by being sued by former patients. Maclean stated you can *always* believe the stories children tell of being abused—children might make up other things, but they never make up traumatic events. I asked the panel what they thought of Steve Ceci's work. There was a moment of dead silence. None of the four panelists had ever heard of Ceci, who is one of the top developmental psychologists in the country and is well known for his recent experiments demonstrating the suggestibility of children. Ceci's "mousetrap" experiments (Ceci 1993; Ceci and Bruck 1995) demonstrated that repeated interviews regarding a false traumatic event (getting a finger caught in a mousetrap and being taken to the hospital) can result in a portion of children saying (and apparently believing) that the fictional traumatic event occurred. After I described this experiment, the panelists concluded (without reading Ceci's papers) that "these analogue studies" cannot be generalized to the real world.

It seems incredible that a psychological conference could be constructed with a seminar focusing on legal issues and the testimony of children in court, without a single person involved ever having heard of Ceci, who has contributed so much in this area. Indeed, this was the third day of the conference and there had been much talk of children's accusations of abuse, but not one mention of Ceci's research, which was why I felt obliged to pose the question. Often when I attend lectures I ask the speakers what they think of criticisms against them.

Immediately after the session a man connected with the conference demanded to know who I was, where I was from, and why I had asked the question. He was not satisfied with my answers and became visibly agitated when I tried to describe Ceci's experiments in greater detail. He soon gave up and informed me in a brusque tone that "everyone here thinks you are a *plant*." Perturbed, I entered the main hallway where I was confronted by Perskin, who asked if I had set out any literature in the bathroom. Apparently, someone had set out flyers from the Temple of Set, a satanic church, in the men's room!

CONCLUSION

Conspiracy theories have operated in many societies at many times and may be seen from a social-psychological perspective as serving certain functions within society. Conspiracy theories may of course represent real conspiracies, but they may also act in a manner similar to racist stereotyping in which the targeted group is seen as deviant and deeply immoral (Moscovici 1987). Conspiracy scholarship is on the one hand irrational, while on the other "far more coherent than the real world, since it leaves no room for mistakes, failures, or ambiguities. It is, if not wholly rational, at least intensely rationalistic" (Hofstadter 1965, 36). Conspiracy theories offer individuals well-organized enemies against whom the self is defined; this offers them a guiding structure and purpose (Farr 1987).

I frequently observed a categorical rejection of the possibility that there could be "false" memories of traumatic events, and that anyone who made such claims must be "dirty" or a part of the "backlash," and that such claims could be dismissed without serious consideration. There was clearly an assumptive worldview or social representation that unified the audience and speakers, deviation from which would brand one as a spy. Actual debate was an anathema. The assumptions that united the group often veered toward conspiracism, though the particular elements of the conspiratorial plots could change from person to person (satanic cults, New World Order, etc.). Most, though by no means all, of the therapists appeared to be previously unaware of New World Order conspiracy, though some appeared receptive to such ideas. Many seemed to be familiar with and believe in the Operation Monarch conspiracy, despite the lack of credible evidence for this. Of course, belief in conspiracies does not necessarily indicate therapeutic incompetence. However, I would be worried if those therapists interviewing children who are suspected of being victims of sexual abuse believed that the biblical revelation was coming in the form of satanic UN troops sweeping up children in black helicopters.

We cannot know what effect these therapists' conspiratorial beliefs may have on their clients. What we can see from these anecdotes is that strong beliefs are highly resistant to discrepant input and they do have a certain persuasive power. An indication of the influence of this conference can be seen in a quote from Jerry Leonard, a physicist who attended and wrote a review of the conference (Leonard 1995), in which he stated:

> I came away with the opinion that cults are far more prevalent, well connected, sophisticated and dangerous than I had ever dreamed . . . apparently, this type of cult activity is fairly widespread. Police departments have stumbled on well organized nationwide child kidnapping rings. Ted Gunderson . . . described one case in which he personally uncovered an elementary school which had been built on a system of tunnels through which children were taken into neighboring houses . . . to participate in Satanic ritual abuse. . . . It is my personal view that the larger satanic cults are being manipulated by the federal intelligence and law enforcement agencies from behind the scenes.

Leonard informed me that this was his introduction to claims of cult child abuse. This testimonial demonstrates the persuasive power of the rumors that were put forth at this conference, at least to someone who was receptive to hearing them.

We have no way of knowing the percentage of practicing therapists who are represented by this style of thinking. Even if only a very small minority of the therapeutic community is represented, it is troubling to think of the effect these therapists may have on their colleagues, to say nothing of their clients. The theories presented at this conference may at times find wider appeal among more traditional therapists who are searching for evidence of cults, and it appears that such theories have enjoyed fairly wide popular circulation in the recent past (Victor 1993). Sherrill Mulhern (1991, 1994) has outlined the role played by conspiracy theories both historically, and at prestigious gatherings of psychologists. While the majority of psychological trauma specialists are not "conspiracists," they may at times be influenced by conspiracy claims, such as the claim that tunnels existed under the McMartin preschool, because such claims resemble or circumstantially support in some way the memories reported by clients.

The possibility of right-wing racist organizations using the present mental-health dilemma for their political gain is something therapists working in this area should be aware of. Therapists who only seek what is best for their clients may at times be vulnerable to propaganda put out by such groups. In the end it is the client, along with the client's family, who suffers. Whether motivated by such groups, claims that critics are active CIA agents who are engaged in a secret war against the American public, or that they are part of a nationwide backlash against belief in child abuse, only serve to make some therapists antagonistic to all forms of criticism, regardless of the motives of the critic. This is unfortunate because, as trauma therapist and researcher John Briere stated at the 1995 APA meeting, many of the criticisms have merit, and the field will be made better, not worse, because of them.

NOTE

I would like to thank Sherrill Mulhern for comments.

POSTSCRIPT

"Conspiracy Theories and Paranoia: Notes From a Mind-Control Conference" generated considerable reaction, some of it published in the *Skeptical Inquirer* (January/February 1997), some of it sent directly to me. Following was my reply, which was published in the March/April 1997 issue:

It is true that there are few reliable data on what the *typical* member of the militia movement believes. Some have argued that members of the militia movement have been unfairly stereotyped as rabid racists. While it is quite accurate to say that much of the militia movement is rooted in the extremely racist rhetoric of the Christian Identity movement (Aho 1990) or in other racist ideologies (Stern 1996), we still lack the type of portrait that a random survey would give us. I suspect that a significant minority of militia members are simply people fed up with a government they see as hopelessly corrupt and incompetent, are not at all racist, and do not believe in conspiracy theories. However, the premise that the militia movement has a basis in racist ideology and fosters hatred is, I believe, beyond dispute.

The therapists described in my article are in no way representative of the typical therapist who treats recovered memories, and I was somewhat distressed to see that some readers apparently made such an inference. At the NATO Advanced Study Institute held last summer in France, many of the top practitioners from both sides of the memory debate met and seemed to discover that the antagonistic dichotomy that has marked this debate from the beginning seemed to melt away. At least some from each side moved more toward a middle ground, where differing views were seen as reasonable. Some of the top recovered memory researchers stated that they had seen clients with false memories or false *beliefs*, and some of the top false memory researchers stated that forgetting and later recall of traumatic experiences is not only plausible but clearly documented.

The conspiratorial beliefs on display at the SITPRCA conference clearly are not representative of the beliefs of those who attended the NATO conference nor, I think, of the beliefs of recovered memory therapists in general. Furthermore, since recovered memories of satanic ritual abuse (SRA) comprise only a small minority of all recovered memories, failure to substantiate SRA does not logically indicate that all recovered memories are in error. Indeed, the ability to confabulate a memory for an event that never occurred (like getting lost in a shopping mall or seeing grandma boiling babies in a cauldron) and the ability to forget a traumatic event for many years and then remember it are not necessarily mutually exclusive: The existence of one does not logically negate the other.

Here's an example: When I was seventeen years old I was violently mugged in New York City, where I grew up. I was painfully injured and developed a slight phobia for the spot on which the attack had taken place. I thought about the event a great deal for two years and then forgot about it. More than ten years later I bragged to colleagues about how I had grown up in the city without ever being a victim of violent crime. The next day I remembered the event in full. Some would call this a recovered memory. I certainly have no doubt that the event took place, in part because I remember the pain of my injuries, but I never told anyone so I have no corroboration. The possibility that false memories may occur does not constitute evidence that I was not mugged, but it might leave the question open for those who do not have access to my memory.

Finally, note should be made of a recent event illustrating the crossover of militia and SRA conspiracy theories. The *New York Times* on October 2, 1996, reported that the prominent militia leader James "Bo" Gritz was arrested in Connecticut and charged with criminal attempt to commit kidnapping in the first degree. According to the *Times,* Gritz sought to rescue two boys he believed were being abused in a satanic cult. A letter I obtained off the Internet that was distributed by the Militia of Montana contains a statement that the lawyer John DeCamp had been an advisor in the case prior to the arrest. I described DeCamp and his claims of a widespread satanic child-abusing cult in Nebraska at some length in my article. Additionally, at the 1996 SITPRCA conference, the trends exhibited in 1995 continued in that a wealth of militia-oriented conspiracy literature (such as the anti-Semitic *Spotlight* newspaper) was on display, and one book that was being sold by a participant at the conference put forth the theory that all homeless people are being mind-controlled by the CIA. New Orleans therapist Valerie Wolf stated that a pattern exists wherein patients first recover memories of incest, then of SRA, and finally they recover memories of mind-control abuse by the CIA. The idea that the CIA and the satanic cults are in cahoots was repeated frequently. As the cross-fertilization of ideas between the two social movements increases, we might expect more behavior like that exhibited by Gritz and more recovered memories of abuse in government mind-control projects.

REFERENCES

Berlet, C. 1994. *Right Woos Left: Populist Party, LaRouchian, and Other Neo-Fascist Overtures to Progressives, and Why They Must Be Rejected.* Political Research Associates, 678 Massachusetts Ave., Suite 702, Cambridge, MA 02139.

Berlet, C. 1995. *Armed Militias, Right Wing Populism, and Scapegoating.* Cambridge, Mass.: Political Research Associates.

Berlet, C., and J. Bellman. 1989. *Lyndon LaRouche: Fascism Wrapped in an American Flag.* Cambridge, Mass.: Political Research Associates.

Blood, L. 1994. *The New Satanists.* New York: Warner.

Burke, W. K. 1995. The wise use movement: Right-wing anti-environmentalism. In *Eyes Right: Challenging the Right Wing Backlash,* edited by C. Berlet, 135–45. Boston: South End Press.

Ceci, S. J. 1993. "Cognitive and Social Factors in Children's Testimony." Master lecture presented at the 101st annual meeting of the American Psychological Association, August.

Ceci, S. J., and M. Bruck. 1995. *Jeopardy in the Courtroom: A Scientific Analysis of Children's Testimony.* Washington, D.C.: American Psychological Association.

Cohn, N. 1966. *Warrant for Genocide: The Myth of the Jewish World-Conspiracy and the Protocols of the Elders of Zion.* New York: Harper and Row.

Collette, L. 1995. Encountering holocaust denial. In *Eyes Right: Challenging the Right Wing Backlash,* edited by C. Berlet, 246–65. Boston, Mass.: South End Press.

Constantine, A. 1995. *Psychic Dictatorship in the U.S.A.* Portland, Ore.: Feral House.

Dalenberg, C. J. 1995. The war against recovered memories of trauma. *Contemporary Psychology* 40: 1065–67.

DeCamp, J. W. 1992. *The Franklin Cover-Up: Child Abuse, Satanism, and Murder in Nebraska.* Lincoln, Neb.: A.W.T., Inc.

Dorr, R. 1991. DeCamp suit against grand jury dismissed. *Omaha World Herald,* January 5, p. 1.

Earl, J. 1995. The dark truth about the 'dark tunnels of McMartin.' *Issues in Child Abuse Accusations* 7: 76–131.

Editors of the *Executive Intelligence Review.* 1992. *The Ugly Truth About the ADL.* Washington, D.C.: Executive Intelligence Review.

Farr, R. M. 1987. Self/other relations and the social nature of reality. In *Changing Conceptions of Conspiracy,* edited by C. F. Graumann and S. Moscovici. New York: Springer-Verlag.

Guerin, B. In press. Some recent and future developments in the study of social representations. *Japanese Journal of Experimental Social Psychology.*

Hardisty, J. 1995. Constructing homophobia: Colorado's right-wing attack on homosexuals. In *Eyes Right: Challenging the Right Wing Backlash,* edited by C. Berlet, 86–104. Boston: South End Press.

Harrington, E. 1995. Research note. *FMS Foundation Newsletter* 4(2): 9–10.

Hofstadter, R. 1965. *The Paranoid Style in American Politics.* New York: Alfred A. Knopf.

Janofsky, M. 1995. Paramilitary group leaders try to burnish their image. *New York Times,* May 26, section A, p. 10.

Kelly, M. 1995. The road to paranoia. *The New Yorker,* June 19.

Leonard, J. 1995. Mind control conference reviewed on the Internet. *Free Thinking* 1(6). Newsletter of the Freedom of Thought Foundation, P.O. Box 35072, Tucson, AZ 85740.

Moscovici, S. 1987. The conspiracy mentality. In *Changing Conceptions of Conspiracy,* edited by C. F. Graumann and S. Moscovici. New York: Springer-Verlag.

Mulhern, S. 1991. Satanism and psychotherapy: A rumor in search of an inquisition. In *The Satanism Scare,* edited by J. T. Richardson, J. Best, and D. G. Bromley, 145–72. New York: Aldine de Gruyter.

———. 1994. Satanism, ritual abuse, and multiple personality disorder: A socio-historical perspective. *International Journal of Clinical and Experimental Hypnosis* 42: 265–88.

Nathan, D., and M. Snedeker. 1995. *Satan's Silence: Ritual Abuse and the Making of a Modern American Witch Hunt.* New York: Basic Books.

Nebraska Leadership Conference. No date. *The Mystery of the Carefully Crafted Hoax: A Report.* Nebraska Leadership Conference, Box 30165, Lincoln, NE 68503.

Noblitt, J. R., and P. S. Perskin. 1995. *Cult and Ritual Abuse: Its History, Anthropology, and Recent Discovery in Contemporary America.* Westport, Conn.: Praeger.

Ofshe, R., and E. Watters. 1994. *Making Monsters: False Memories, Psychotherapy, and Sexual Hysteria.* New York: Charles Scribner's Sons.

Oksana, C. 1994. *Safe Passage to Healing: A Guide for Survivors of Ritual Abuse.* New York: Harper Perennial.

Poole, D. A., D. S. Lindsay, A. Memon, and R. Bull. 1995. Psychotherapy and the recovery of memories of childhood sexual abuse: U.S. and British practitioners' opinions, practices, and experiences. *Journal of Consulting and Clinical Psychology* 63: 426–37.

Springmeier, F. 1995. Project Monarch: How the U.S. creates slaves of Satan. In *Cult Rapture,* edited by A. Parfrey. Portland, Ore.: Feral House.

Stern, K. S. 1996. *A Force upon the Plain: The American Militia Movement and the Politics of Hate.* New York: Simon and Schuster.

Summit, R. C. 1994. The dark tunnels of McMartin. *Journal of Psychohistory* 21: 397–416.

Survivor Activist. 1994. Announcements. *Survivor Activist* 2(4): 12. Available from Frank Fitzpatrick, 52 Lyndon Road, Cranston, RI 02905-1121.

Thomas, G. 1990. *Journey into Madness: The True Story of Secret CIA Mind Control and Medical Abuse.* New York: Bantam.

Victor, J. S. 1993. *Satanic Panic: The Creation of a Contemporary Legend.* Chicago: Open Court.

Williams, L. M. 1994. Recall of childhood trauma: A prospective study of women's memories of child sexual abuse. *Journal of Consulting and Clinical Psychology* 62: 1167–76.

Witt, H. 1995. Amid Oklahoma mysteries, conspiracy ideas win hearing. *Chicago Tribune,* May 9.

39

THE WEST BANK COLLECTIVE HYSTERIA EPISODE

James R. Stewart

$\left(\mathbf{E}\right)$ pisodes of mass hysteria have captivated researchers for years, but have thus far eluded the efforts of social scientists to provide a comprehensive, systematic explanation of the causes and processes involved. Mass hysteria typically involves the "contagious" spread of physical symptoms (such as fainting, convulsions, nausea, dizziness, and headaches) along with the adoption of a belief system that attributes causation to a toxic agent (such as insects or gases). Although the victims are firmly convinced their illness is "real," extensive medical and environmental studies fail to identify a chemical or biological cause for the symptoms. With no evidence to support the "real illness" explanation, the authorities typically label the episode as an example of mass hysteria and the whole thing is quietly forgotten. The mass-hysteria explanation, however, is usually met with opposition from members of the afflicted group, who remain convinced of the legitimacy of their illness.

Investigations of these outbreaks commonly proceed from an assumption that an accumulation of stress from various sources acts as the underlying cause of the episodes. The stress builds and eventually causes one person to "snap" under the pressure and exhibit an idiosyncratic form of behavior (i.e., symptoms of physical illness). This behavior then spreads throughout the immediately affected group, most commonly occurring in schools or workplaces. The behavior is believed to represent a form of release or escape from unpleasant situations and, as such, most closely resembles an episode of panic in the literature of collective behavior (Smeller 1962; Kerckhoff and Black 1968; Klapp 1972; Rose 1982).

Other social scientists, however, have suggested a competing model of explanation. Instead of focusing upon the fear and anxiety victims are supposedly escaping from in a "panicky" fashion, they look at the possible rewards the participants receive. Rather than viewing participants as involved in an escape/panic,

---

This article first appeared in *Skeptical Inquirer* 15, no. 2 (Winter 1991): 153–61.

they perceive the victims to be participating in an illness craze in an effort to glean rewards (Schuler and Parenton 1943; McGrath 1982; Gehlen 1977). This depiction of mass hysteria as a craze seems to be the more compelling of the two models and was used to analyze the following episode.

On March 21,1983, scores of schoolgirls in the village of Arrabah on the Israeli-occupied West Bank were stricken with a mysterious illness. The symptoms included nausea, headache, dizziness, and fainting spells. The first girl stricken was sent home. Within the next few hours dozens of other girls were similarly afflicted. The local public-health officer was called after the initial cases were reported to the school administrators, but was unable to locate the cause of the illnesses. The health official's investigation seemed to precipitate an alarming increase in the number of cases, and the school was forced to close later that morning (Landrigan and Miller 1983, 17).

Public-health officials returned to the scene in the evening and noted an unusual odor of gas. Although the source of the gas could not be pinpointed, it was generally concluded that it emanated from an open latrine pit located nearby. Only a very few cases were reported during the next 72 hours and the situation calmed down.

During this episode more than 60 girls were hospitalized with severe symptoms. Those with milder cases were treated by local physicians. Medical personnel conducted a barrage of screening tests on the hospitalized victims, but they were unable to offer conclusive results regarding a cause of the complaints. Public-health personnel conducted a comprehensive investigation of the environment in and around the school. These studies also yielded negative results. The incident was reported by the local media and accusations were made against both the occupying Israeli Defense Forces and recent Israeli settlers, who were accused of using poisons to terrorize the local Arab population. The Israelis vehemently denied the charges and, in turn, suggested that local Arabs were attempting to politicize what was actually an episode of mass hysteria. Their conclusions were based upon studies of similar episodes reported and researched in the United Kingdom and the United States. Since neither side was able to offer enough convincing evidence to support their respective cases, an uneasy stalemate developed as the situation quieted down during the next few days.

The prevailing calm was short-lived. A new outbreak flared up very quickly when 300 schoolgirls were stricken in the neighboring community of Jenin on March 26, 1983. The outbreak was similar to the preceding episode in the patterned symptoms of the victims, but it was significantly larger and involved more of a cross-section of the local population. Almost 400 persons were reported to have developed symptoms (Landrigan and Miller 1983, 19–20). The vast majority were again schoolgirls; but a significant number of adult females also developed symptoms, and some of the adult males who helped transport the afflicted children to hospitals as well as four soldiers of the Israeli Defense Forces also came down with symptoms. Local hospitals were inundated. Many patients had to be transferred to hospitals in surrounding communities.

Once again medical personnel from the local community and area public-health officials conducted tests. The results again failed to discover an organic or environmental cause of the illness. By this time the episode was receiving world-wide news coverage, including accusations by the Palestinian Liberation Organization that "schools had been sprayed with a poisonous gas" by the Israelis (*New York Times,* March 29, 1983). The United Nations called for an investigation. The World Health Organization and the International Committee of the Red Cross sent teams of medical personnel to the area to conduct independent investigations. The Israeli Ministry of Health requested that a medical team from the Centers for Disease Control in Atlanta, Georgia, be dispatched to the West Bank to conduct an investigation (Landrigan and Miller 1983, 2).

During this time the episode was marred by violence. Angry mobs of Arab demonstrators protested the "poisonings." During one confrontation, Arab teenagers were injured by gunfire from frightened Israeli settlers (Shipler 1983). There were also isolated incidents of vigilanteism on both sides as tensions escalated. This second outbreak ended abruptly on March 28. No new hospital admissions were reported during the next week.

The third and final wave of illnesses occurred April 3. Two different areas were simultaneously struck. One was a neighboring community to the site of the first two outbreaks; the other was located one hundred miles away in the extreme southern part of the West Bank area. More than 500 persons were involved in this phase of the episode. The pattern was similar to that of the first two waves. In each of these stricken areas the illness started among girls in a school and then spilled over into the surrounding community, and each incident appears to have been triggered by children smelling odors that they described as "sulfurlike" or "gaslike" (Landrigan and Miller 1983, 21). Many of the victims were hospitalized and, like the participants in the preceding episodes, released in a few days. A few isolated reports followed during the next two weeks, but these involved single individuals and did not spread to others.

The final tally indicated that almost a thousand persons had been involved in this series of incidents. Most were hospitalized, although only briefly. There were no fatalities and no reported lingering effects.

Similar occurrences of "mass hysteria" have been frequently reported in medical and social-science journals. These, however, were generally much smaller in size than the West Bank outbreaks. This episode captured the journalistic fancy of the world news media, which featured stories of the illness on an almost daily basis. Never before had an outbreak of mass hysteria been so widely covered or so heavily politicized. Speculation abounded regarding the cause of the mass illness and seemed to reflect more political than medical considerations. Those supporting the "poison gas" explanation thought that either Israeli soldiers or recent Israeli settlers in the West Bank area were terrorizing the local Arab population with poison gas. Schools were considered to be favorite targets because of their high concentration of population and the vulnerability of schoolchildren. On the other side were the advocates of a "mass hysteria" explanation. These persons attributed the illnesses to a pre-existing emotionally charged situation (Israeli-Arab hostilities) in which some gullible schoolgirls had succumbed to an environmental triggering stimulus (sewer gas), and the resulting spread of symptoms was a textbook example of mass psychogenic illness.

Persons who subscribed to the poison-gas interpretation included local Arab public-health officials (who maintained the illness was real), the Palestinian Liberation Organization and their spokesman, Yasser Arafat (who labeled it a planned and systematic crime against Arabs), Radio Moscow (which issued a statement condemning the Israelis for their use of gas), and most third world delegates to the World Health Organization (who voted to condemn Israel for glossing over and rejecting the poison-gas theory in favor of the mass-hysteria explanation). Other fervent supporters included local "radical" PLO supporters who were purported to have encouraged the episode by hanging around the schools and hospitals (*New York Times,* May 21, 1983).

Proponents of the mass-hysteria explanation included Israeli public-health officials and a team of two physicians from the Centers for Disease Control in Atlanta (Landrigan and Miller 1983). Unlike previous documented examples of this phenomenon, the West Bank episode involved relatively large numbers and a geographically dispersed population. Yet these differences could not detract from sound medical and scientific evidence that clearly supported a psychogenic explanation. Landrigan and Miller (1983, 1) concluded:

> ... This epidemic of acute illness was induced by anxiety. It may have been triggered initially either by psychological factors or by subtoxic exposure to hydrogen sulfide. Its subsequent spread was mediated by psychogenic factors. Newspapers and radio reports may have contributed to this spread.

Drawing from its extensive investigation, the CDC team cited a number of findings strongly supporting the mass-hysteria explanation. First and foremost were the findings from comprehensive laboratory studies, which produced no evidence of the existence of any toxic agents in the air, soil, or water of the stricken areas. Tests conducted on blood and urine samples of the hospitalized patients similarly produced no evidence of toxic etiology. Second, the great majority of the victims were adolescent girls, a group previous investigations had demonstrated to be the most vulnerable to these outbreaks. Finally, the form and duration of the clinical symptoms were consistent with previous mass-hysteria findings. The episodes typically began slowly, flared-up quickly, and terminated after a few days. Although the symptoms were suggestive of a severe illness, all of the patients recovered quickly and suffered no physical aftereffects (Landrigan and Miller 1983, 26–28).

In a limited, parallel investigation, a team dispatched from the World Health Organization reached similar although watered-down conclusions (WHO 1983). Its support for a "mass hysteria" explanation was largely implicit and based upon the inability of their findings to indicate any specific toxic cause or set of causes. The WHO report does, however, make mention of the tension and anxiety under which the people of the occupied territories lived and the compoundingly stressful developmental period of adolescence in females. Both of these observations support a mass-hysteria explanation even though the WHO report doesn't come right out and say so.

The International Committee of the Red Cross also sent an investigator to the area and, based upon his findings, called for a full-fledged investigation to be performed by a competent body (ICRC 1983). They reached no conclusions about the cause of the episode and have refused to release the findings of their investigations, citing the need for confidentiality in their inquiries.

Landrigan and Miller (1983, 29–30) offered a concluding explanation:

> 1. The initial outbreak at Arrabah appears to have been triggered either by psychological factors or possibly by the smell of escaping H_2S from an outdoor latrine. Although the concentrations of H_2S at the school in Arrabah are not at all likely to have reached toxic levels, there may have been sufficient concentration of gas following flooding of the latrine during heavy rains in Spring, 1983 to have produced a foul odor and consequent upper respiratory irritation in a few students. Previous studies of outbreaks of psychogenic illness have emphasized that a perception of strange odors or gases by affected individuals have frequently preceded onset of illness (Colligan and Murphy 1979).
>
> 2. The subsequent spread of the outbreak was due to psychogenic factors. That spread may have been facilitated by newspaper and radio reports which described the symptoms in detail and strongly suggested that a toxic gas was the cause of the outbreaks.
>
> 3. The termination of the outbreak was probably related to the closing of West Bank schools. The closing of the schools dispersed the students and helped to minimize the opportunity for spread of symptoms among students clustered together in the classroom environment.

An analysis using the craze model rests upon identifying the valued goals or rewards that may have motivated the participants to adopt the sick role. For some it may have been a release or escape from a recently arisen avoidance dilemma. During the week immediately prior to the first flareup, "radical Palestinian factors" had been active in encouraging schoolchildren to strike and join demonstrations commemorating Land Day (March 30 was the anniversary of the killing of six Arabs in anti-Israeli demonstrations in 1976). Masked men were reported to have visited several schools in the area and called for the students to skip classes and participate in the planned demonstrations. Officials found a pamphlet in one of the Jenin schools threatening the students and strongly urging them to go on strike (Shipler 1983). These forces of provocation may have caused an intensive avoidance situation for students who were caught between the untenable choice of cutting their classes and joining the strike or being labeled as pro-Israel. The sick-role behavior would have provided a welcome and legitimate way out. One could actually participate in a pseudostrike (mass hysteria) without fear of being held accountable by either parents or authorities. Nor would the students have to fear being stigmatized as pro-Israel.

Another reward was notoriety. Most of the stricken persons were taken to hospitals for treatment. While in the hospitals, the patients were given special care.

Some were treated like celebrities. They received the best available care by local medical personnel and were accorded a special status unavailable to patients with mundane conditions. They were viewed as the innocent victims of heinous chemical warfare perpetrated by the Israeli settlers (*Newsweek,* April 18, 1983).

The schoolgirls also found themselves at the center of attention from the representatives of the World Health Organization, the International Red Cross, and the U.S. Centers for Disease Control. These teams questioned large numbers of the afflicted girls in their attempts to establish an etiology of the illness. In addition, the world media showered them with attention. There was even one instance in which a CBS news team was arrested by Israeli authorities for encouraging "hospitalized school girls to act ill for the camera" (*New York Times,* April 5, 1983).

American television networks carried extensive stories on the hospitalized victims, highlighted by examples of schoolgirls acting in a frenzied, convulsive manner, almost on cue, every time the cameras panned in their direction. The hospitalized schoolgirls depicted on television also gave the 'v' for victory sign for the benefit of cameras at what appeared to have all of the earmarks of a fun-filled hospital slumber party. They exhibited a nonchalant concern over their illness. Given the seriousness of the symptoms, the patients' responses seemed totally inappropriate. Their attitude could best be described as *la belle indifférence*—a lack of concern that psychiatrists have noted in individual cases of conversion hysteria. There were also reports from Israeli sources claiming that some of the girls had admitted faking their symptoms (*Time,* April 18, 1983).

I have noted a somewhat similar attitude characterizing "believers" in episodes of cattle mutilations and Bigfoot sightings. While conducting field research and interviews, I was struck by the enthusiasm with which people adopted an explanation fraught with terror and fright. If people truly believed that extraterrestrials or satanic cults were killing and mutilating cattle or that big, hairy monsters were making mysterious footprints in their areas, they should have been exhibiting fear or at least high levels of anxiety. Instead they seemed to be enthusiastically embracing the most fearsome and gory versions of the monster explanations. Rather than being motivated by panic, they seemed to show great eagerness to participate in these collective delusions. In fact, interviewees willingly performed mental gymnastics to believe the bizarre explanation instead of the more empirically supportable, scientific account of the cause of dead cattle or mysterious, enlarged footprints.

Another way the participants can gain from assuming the sick role is that it gives them the power to manipulate persons in positions of authority. Previous studies of mass hysteria have shown that victims have higher rates of absenteeism or poorer relationships with their supervisors than control-group members (Colligan and Murphy 1982, 43–45). Those who fall victim may have found a way to embarrass or retaliate against immediate authorities. Surveys administered by the CDC team showed no significant differences between the affected and nonaffected schoolgirls with respect to grades, previous use of sick days, or relationships with their teachers. However, these were self-administered questionnaires and the research team had no independent measures to check the respondents' veracity. As a result, no evidence can be presented that the victims used the illness as a weapon against school authorities, but participation in the hysteria episode certainly brought embarrassment to the occupying Israeli soldiers and settlers. There is no need to present a detailed history of Israeli-Arab hostilities in the West Bank; it is obviously one of the most politically sensitive areas in the world today. Tension has been extremely high since the Israeli occupation after the Six-Day War in 1967. Establishment of Israeli settlements in the area during the past two decades have greatly exacerbated an emotionally charged situation that was already at the breaking point. Israelis were accused of using both poisonous gas and a mysterious yellow powder (later identified as pollen) to scare the Palestinians from their rightful homeland in the West Bank. Newspaper headlines and lead stories on radio and television readily reported the accusation of mass poisoning. And as is often the case in episodes of hysteria and delusions, the media tended to share the sensational aspects of the story and give less coverage to the medical refutations of the original accounts.

Smelser (1962, 175–88) has identified four positive goals or media of exchange that motivate people to participate in crazes. They are money/property, power, prestige, and psychic gratification. There is no evidence that any of the West Bank victims realized monetary gains from being stricken with the illness, but there is a strong case that they did enjoy gains in the other three dimensions.

The power to manipulate others could have been beneficial to the stricken patients. A two- or three-day medical vacation from school and family responsibilities with no fear of retribution, along with the pampered care received while in the hospital, may have been a powerful inducement for developing symptoms. The acquisition of power was also evidenced by the ability of stricken patients to embarrass, humiliate, and punish the occupying Israelis in front of the whole world. The illness provided vivid "proof" of the heinous nature of the Israelis, who were pictured as having used poisoned gas against helpless schoolchildren. The opportunity of powerless persons suddenly finding themselves in a position with enough political clout to hurt their "enemies" could have been a powerful motivator. The advantageous position that victims find themselves in illustrates why mass psychogenic illness is considered to be a political as well as a behavioral phenomenon (Colligan and Murphy 1982).

The temporary fame enjoyed by the victims also may have served as a method of gaining prestige among peers and the larger community. The illness could have provided instant notoriety and a method of becoming the center of attention. Phoon (1982, 31) has drawn similar conclusions in a study of outbreaks of occupational mass hysteria. Victims, far from being ridiculed or condemned, often receive expressions of concern and sympathy from parents, workmates, and even management. Finally the pleasantness associated with the carnival atmosphere surrounding the episode could have encouraged others to join those already sick. The "fun" associated with spending a few days in a hospital ward with friends would have been a psychologically gratifying experience. It would have been a way of establishing in-group ties among the afflicted and creating a cohesive "we" feeling. It has been suggested that participation in episodes of mass hysteria is a rewarding experience (Stahl 1982). The victims share their experiences by displaying the "stereotyped" symptoms and make the behavior a collective rather than individual experience. The actors mutually reinforce one another.

It's time that social scientists take a hard look at the underlying assumptions upon which previous investigations of mass hysteria were based—that of unwitting victims being involuntarily stricken by psychophysiological symptoms because of stressful environments. Rather than escaping, the victims may be drawn to the possibility of getting a couple of days off from work or school without fear of reprisals. Rather than suffering, victims may actually be enjoying the illness. Rather than being repulsed by some stress stimulus, victims may be attracted to the benefits and rewards of becoming ill. Contracting the disease may represent a form of collective wish-fulfillment and, as such, should be analyzed from the perspective of a craze.

REFERENCES

Colligan, Michael J., and Lawrence Murphy. 1982. A review of mass psychogenic illness in work settings. In *Mass Psychogenic Illness,* edited by Michael J. Colligan, James W. Pennebaker, and Lawrence Murphy. Hillsdale, N.J.: Lawrence Erlbaum.

———. 1979. Mass psychogenic illness in organizations: An overview. *Journal of Occupational Psychology* 52: 77–90.

Colligan, Michael J., James W. Pennebaker, and Lawrence Murphy. 1982. *Mass Psychogenic Illness: A Social Psychological Analysis.* Hillsdale, N.J.: Lawrence Erlbaum.

Gehlen, Frienda L. 1977. Toward a revised theory of hysterical contagion. *Journal of Health and Social Behavior* 18: 27–35.

ICRC (International Committee of the Red Cross). 1983. Israel and occupied West Bank: An ICRC recommendation. ICRC press release, April 7.

Kerckhoff, Alan C., and Kurt Black. 1968. *The June Bug.* New York: AppletonCentury-Crofts.

Klapp, Orrin E. 1972. *Currents of Unrest: An Introduction to Collective Behavior.* New York: Holt, Rinehart and Winston.

Landrigan, Philip J., and Bess Miller. 1983. Epidemic acute illness—West Bank. *Report to the Centers for Disease Control,* April 20.

McGrath, Joseph E. 1982. Complexities, cautions, and concepts in research on mass psychogenic illness. In *Mass Psychogenic Illness,* edited by Michael J. Colligan, James W. Pennebaker, and Lawrence R. Murphy. Hillsdale, N.J.: Lawrence Erlbaum.

Newsweek. 1983. A "poison" scare on the West Bank. April 18: 42.

New York Times. 1983. Israel finds no poison at 6 Arab schools, March 29; U.N. calls for investigation, April 5; World health agency condemns Israel over West Bank epidemic, May 13; Israelis dismiss Arab official in West Bank illness dispute, May 21; Editors' notes, May 21.

Phoon, W. P. 1982. Outbreaks of mass hysteria at workplaces in Singapore: Some patterns and modes of presentation. In *Mass Psychogenic Illness,* edited by Michael J. Colligan, James W. Pennebaker, and Lawrence Murphy. Hillsdale: N.J.: Lawrence Erlbaum.

Rose, Jerry D. 1982. *Outbreaks: The Sociology of Collective Behavior.* New York: Free Press.

Ross, E. A. 1908. *Social Psychology.* New York: Macmillan.

Schuler, Edgar, and Vernon J. Parenton. 1943. A recent epidemic of hysteria in a Louisiana high school. *Journal of Social Psychology* 17: 221–35.

Shipler, David K. 1983. Hundreds fall sick in West Bank; poison suspected. *New York Times,* March 28.

Smelser, Neil J. 1962. *Theory of collective behavior.* New York: Free Press.

Stahl, Sidney M. 1982. Illness as an emergent norm or doing what comes naturally. In *Mass Psychogenic Illness,* edited by Michael J. Colligan, James W. Pennebaker, and Lawrence Murphy. Hillsdale, N.J.: Lawrence Erlbaum.

Time. 1983. Ailing schoolgirls. April 18: 52.

WHO (World Health Organization). 1983. Health conditions of the Arab population in the occupied Arab territories, including Palestine. Report to the Director General, May 11.

40

POLICE PURSUIT OF SATANIC CRIME: THE SATANIC CONSPIRACY AND URBAN LEGENDS

Robert D. Hicks

L aw-enforcement officials throughout the United States flock to training seminars about satanic cults and crime. In Virginia alone, cult-crime officers gave at least 50 seminars in 1988. The seminars, orchestrated by a loose network of investigators, ex-police officers (now cult consultants), therapists, and clergy, offer a world-view that interprets the familiar and explainable—and the unfamiliar and poorly understood—in terms of increasing participation by Americans in satanic worship. The seminars further claim that satanism has spawned gruesome crimes and aberrant behavior that *might* presage violent crime. In particular, law-enforcement officials have developed a model of "the problem," a scheme widely disseminated through police-training seminars as well as through networks of investigators, newsletters, and public presentations. Is this concern justified?

I argue that the current preoccupation of law-enforcers with satanism and cults has not been prompted by anything new: the phenomenon has a firm historical and cultural context. Further, I suggest that the news media are largely responsible for the law-enforcement model of cult activity, since the evidence officers cite for cult mayhem is generally based on nothing more than newspaper stories. Frequently, these news stories do not even attribute nasty incidents to cults, but the police infer causality anyway. I suggest that for police the actual problem with cults, in terms of their threat to public order, is very small, has non-supernatural explanations, and requires no new law-enforcement resources. Newspaper accounts substantiate and fuel police interest in cults. For instance:

• In July 1988, the Myrtle Beach, Florida, *Sun-News* reported that police arrested four teenagers for vandalizing a cemetery, allegedly trying to remove body parts from a grave for satanic rituals (Edge 1988). The judge denied bond because she felt that, once freed, the boys would run amok during upcoming

This article first appeared in *Skeptical Inquirer* 14, no. 3 (Spring 1990): 276–85.

Walpurgisnacht (April 30), a satanic holiday. The boys, the judge felt, couldn't control their own behavior once in thrall to Satan.

• In February 1989, the *Wall Street Journal* reported that police arrested two brothers for trying to kill a judge through a hoodoo spell. (Hoodoo, a variant of voodoo, survives in the American South primarily among impoverished blacks.) The brothers had arranged with a Jamaican "voodoo priest" to cast a death spell using a photograph of the judge and a lock of his hair. Although the brothers got caught by ingenuously asking the judge's wife for the hair and the photograph, the police nevertheless charged them with conspiracy to commit murder based on the hex alone (McCoy 1989).

• In Virginia, in October 1988, a *Style News* article described cult paraphernalia left at a popular riverside park, but a park official wasn't worried: the paraphernalia could not be the work of dangerous satanists because "real satanists don't leave any traces," he observed (Bacon 1988).

• The *Kansas City Times* reported in March 1988 that a Chicago police investigator said of cult crime, "I think it's going to be a growing problem as we go into the nineties." He further noted that, although there are no national statistics on the problem, a network of satanists does exist—people who perform child molestation and murder as a form of worship. A deputy sheriff warned that satanists are responsible for as many as 50,000 human sacrifices a year, "mainly transients, runaways, and babies conceived solely for the purpose of human sacrifice" (Berg 1988).

• The same *Kansas City Times* article, surveying the law-enforcement interest in cults, recounted the first of several preschool or day-care-center cases in which children's uncorroborated testimony caused indictments of many adults for sexual abuse. The children said that adults dressed in robes performed ceremonies involving not only rape but even murder, cannibalism, and mutilation (Berg 1988).

The readers of such accounts, given the impassioned testimony of police officers, therapists,

"Signs of Satan"

Horned hand

Symbols representing the moon goddess Diana and the morning star of Lucifer.

Lightning bolt

Swastika

teachers, and concerned parents, may well presume the existence of a problem of national proportions. The lack of criminal convictions for these crimes has not deterred satanic-conspiracy proponents.

In my role as a law-enforcement specialist with the Virginia Department of Criminal Justice Services, I have a professional interest in what has become a trendy topic on the police seminar circuit: cult crime. At first alarmed by what I learned at the seminars, I became progressively more skeptical, then even more alarmed by the cult experts' anti-intellectual and anti-rationalist stance. The law-enforcement model of cult crime appeared shoddy, ill-considered, and rife with errors of logic (faulty causal relationships, false analogies, lack of documentation, unsupported generalizations) and ignorance of an anthropological, psychological, and historical context.

One cannot easily analyze the law-enforcement concern with cults, for two reasons: first, the sources of information are irregular, sometimes obscure or not verifiable (e.g., no public access to ongoing criminal investigations); second, the eclectic nature of the law-enforcement model of cult crime makes focused criticism difficult. The information law-enforcers use to document cult activities derives largely from newspaper articles. Reporters often cater to the lurid and the macabre, frequently implying cause-effect relationships or hinting at dark deeds.

For example, articles on teen suicides sometimes note that the victim was known to listen to heavy metal rock music or to play "Dungeons and Dragons," a fantasy role-playing game. Some law-enforcers and concerned parents perceive a cause-effect relationship: "Dungeons and Dragons" introduces young lives to the occult and may prompt suicides. Patricia A. Pulling, founder of the Virginia-based group Bothered About Dungeons and Dragons (BADD), implies such a relationship when she claims that many teen suicides are linked to the game, giving only newspaper articles (including even the *Weekly World News*) as her sources.

Law-enforcement literature makes the same kind of mistakes. For example, an article in *Law-Enforcement News,* a publication of the John Jay College of Criminal Justice in New York, began: "A 14-year-old Jefferson Township, N.J., boy kills his mother with a Boy Scout knife, sets the family home on fire, and commits suicide in a neighbor's back yard by slashing his wrists and throat. Investigators find books on the occult and Satan worship in the boy's room" (Clark 1988). But did the boy have a collection of spiders? A stack of pornographic literature under his bed? A girlfriend who just jilted him? Newspaper accounts don't often mention such other possible explanations.

THE CULT–CRIME MODEL

Fundamentalist Christianity drives the occult-crime model. Cult-crime officers invariably communicate fundamentalist Christian concepts at seminars. They employ fundamentalist rhetoric, distribute literature that emanates from fundamentalist authorities and sometimes offer bibliographies giving many fundamentalist publications, and they sometimes team up with clergy to give seminars on satanism. The most notable circular among cult-crime investigators, *File 18 News-letter,* follows a Christian world-view in which police officers who claim to sepa-rate their religious views from their professional duties nevertheless maintain that salvation through Jesus Christ is the only sure antidote to satanic involvement, whether criminal or noncriminal, and point out that no police officer can honor-ably and properly do his or her duty without reference to Christian standards.

At seminars, cult-crime officers distribute handouts showing symbols to iden-tify at crime scenes, accompanied by their meanings. The handouts typically attribute no sources, but many derive from Christian material. For example, the peace symbol of the 1960s is now dubbed the "Cross of Nero." Someone decided that the upside-down broken cross on the symbol somehow mocks Christianity. (In fact, common knowledge has it that the symbol was invented in the 1950s using semaphore representations for the letters "n" and "d" for nuclear disarmament.)

Fundamentalist Christianity motivates the proponents of cult-crime con-spiracy theories in other ways. Apparently, arguing against their theory is, to them, attacking their world-view. To some cult-crime officers, arguing against the model denies the existence of Satan as a lurking, palpable entity who appears to tempt and torture us. Satan becomes the ultimate crime leader: the drug lord, the Mafia don, the gang boss.

Chicago police investigator Jerry Simandl doesn't just investigate crimes, he also interprets cult behavior, particularly that which threatens Christians. He apparently can tell whether an act of church vandalism was committed mindlessly by kids or purposefully by a cult group: "For example, an organ might be vandal-ized by having its keys broken. That means that the vandals were seeking to deny a congregation the ability to 'communicate with God' through music" (Clark 1988). Simandl draws amazing inferences about these crimes, although they have the lowest clearance rate because they frequently leave no suspects and no evi-dence beyond the destroyed property. The church vandalism so shocks religious sensibilities that some cult-crime officer—armed with the world-view that cults cause crime—can only interpret the crime as satanic.

FACING FACTS

Evil is, indeed, the operative word. Law enforcers who meld cult-crime theories with their professional world-views have transformed their legal duties into a moral confrontation between good and evil. Larry Jones, a police officer from Boise, Idaho, edits the *File 18 Newsletter.* Jones believes that a satanic network exists in all strata of society and maintains extreme secrecy to shroud its program of murder.

Defensive about the lack of physical evidence of cult mayhem, Jones writes:

> Those who deny, explain away, or cover up the obvious undeniably growing mountain of evidence often demand statistical evidence or positive linkages between operational suspect groups. At best, this demand for positive proof of a "horizontal conspiracy" is naive.... Consider the possibility that the reason sup-

posedly unrelated groups in different localities over various time periods acting-out in a similar manner, is that consistent directives are received independently from higher levels of authority. Instead of being directly linked to each other, these groups may be linked vertically to a common source of direction and control. Those who accept this theory as a reasonable possibility need to rethink the meaning, scope and effects of the term conspiracy! (*File 18 Newsletter,* 4[1], 1989)

In other words, if the evidence doesn't seem to fit a particular conspiracy theory, just create a bigger conspiracy theory. Jones and other cult-crime officers impose their model on a pastiche of claims, exaggerations, or suppositions. For example, cult investigators would have us believe that cult practitioners develop skills in the vivisection of livestock and household pets. One investigator, retired police captain Dale Griffis from Ohio, says: "Occultists will stun the animal on his back with an electric probe. Then they will spray freon on the animal's throat.... The heart's still pumping, and they will use an embalming tool to get the blood out. It's fast and efficient. Hell, the farmer heard the animal whine, and he was there within five minutes" (Kahaner 1988, 146).

A sheriff's investigator, in a memorandum about cattle mutilations, interviewed a young woman who claimed to be a former satanic-cult member who had mutilated animals. Her cult, which consisted of "doctors, lawyers, veterinarians," were taught by the vets how to perform the requisite fatal surgery. The animal's blood and removed organs, it seems, were used for baptismal rites. She further related:

> When using the helicopter [the cult members] sometimes picked up the cow by using a homemade ... sling ... and they would move it and drop it further down from where the mutilations occurred. This would account for there not being any footprints or tire tracks.... When using the van trucks they would also have a telescoping lift which ... was about 200 feet long mounted outside the truck and would use that to extend a man out to the cow, and he would mutilate it from a board platform on the end of the boom and he would never touch the ground.... They sometimes do three or four cows. (Kahaner 1988, 148)

It seems that the cult members went to such lengths because they delighted in baffling the police.

The sheriff's investigator reported to his supervisor each detail of this story, but obviously he was unacquainted with Occam's Razor. Trucks with 200-foot booms are not plentiful and would appear conspicuous in rural America, particularly when helicopter air support is called in. The investigator never considered the work of a predator, or even the action of a vandal. Of course, news accounts of such livestock deaths, particularly if related by cult-crime officers, most often attribute deaths to cultists and claim the animals were killed and the organs "surgically" removed. Did a surgeon do the work? Can a police officer tell the difference between a hole in a cow's belly put there by a scalpel and one caused by a predator's sharp teeth?

A comprehensive investigation of cattle-mutilation claims, carried out by former FBI agent Kenneth M. Rommel Jr., made exactly these criticisms, and many others. Rommel (1980) concluded that virtually all reported livestock mutilations are due to natural actions of scavengers and predators. He cautioned law-enforcement officers not to use the term "surgical precision" and not to be misled by colorful statements by people interested in spreading rumors, theories, and fears.

The pentagram is another "sign of Satan."

Cult-crime officers may deny facts that contradict their theories. For example, one of the recent murders they dubbed "satanic" was that of Stephen Newberry, a teenager from Springfield, New Jersey, whose friends bashed him to death with a baseball bat. Even though Larry Jones quotes local investigators, a prosecutor, a psychologist, and an academic cult expert who claimed that no satanic sacrifice of Newberry occurred but instead blamed drug abuse, Jones nevertheless offers the opinion that the experts "do not give credit to the strong influence of the tenets of the satanic belief system over its initiates. In some cases the subjects become involved with satanism . . . prior to the onset of family problems. . . . The only true and lasting solution to 'devil worship' or satanic involvement is a personal encounter with true Christianity" (*File 18 Newsletter*, 4[1], 3).

The police have found no evidence to support Jones's earlier suggestion that a "vertical conspiracy" might exist—a higher leader directing groups to do murderous business within an authoritarian cult led by a charismatic leader.

THE CULT-CRIME MODEL: A DESCRIPTION

Characteristically, law-enforcement cult seminars all promote the same model of satanic cults, although largely without any substantiation or documentation. The model persuades because it takes phenomena familiar to the officers and imbues them with new meanings: officers learn a new lexicon to describe old phenomena and therefore see the cult problem as a new threat to public order. So what is the model?

The model, now [1990] almost seven years old, loosely postulates various levels of satanic or cult involvement (see box). Characteristically, the officers— the self-proclaimed experts who teach the seminars—do not define the object of their concern. They use the terms *cult, occult*, and *satanism* interchangeably but with the connotation of disruption, coercion, mind-control by a charismatic leader, plus, of course, criminality. The terms are extended to religious practices dubbed "nontraditional," e.g., voodoo, Santeria, Native American practices. The label masks an

The Cult-Crime Model

This model is promulgated widely at law-enforcement cult seminars, largely without any substantiation or documentation.

Traditional Satanists: The first, and highest, level of satanists includes transgenerational family satanism, the cult survivors' tales, and daycare-center ritual abuse. Such satanists comprise an international underground, tightly organized and covert, responsible for upwards of 50,000 human sacrifices a year (some of which are babies bred for sacrifice).

Organized Satanists: The second level of satanic involvement includes public groups, such as the Church of Satan and the Temple of Set. Cult-crime officers' definition of this level as dangerous is ambiguous since organized groups formally proscribe acts of violence. But cult seminars imply *a fortiori* that such groups promote self-indulgence to the point of attracting psychopaths or criminals. Thus it is the perceived likelihood of public satanic organizations attracting bad people that justifies law-enforcement surveillance.

Self-styled Satanists: The third level of involvement includes self-styled satanists, such as mass murderers like John Wayne Gacey and Henry Lee Lucas. These men, also social isolates, invented ideologies to affirm their behavior. Some cult-crime officers even maintain that these criminals do their evil deeds as a form of satanic sacrifice to give them power, but other law-enforcers, such as Griffis, believe that self-styled satanists borrow from the occult because satanic ideology permits or encourages their crimes. This idea is the most plausible component of the model: sociopaths or psychopaths, already distanced from common standards of behavior, may choose an ideology that helps them reconcile their crimes with their consciences.

Dabblers: Dabblers, those in the outer, or fourth level of cult involvement, are mostly children, teenagers, or very young adults who, in unsophisticated fashion, play with satanic bits and pieces. At this level, "Dungeons and Dragons" (D&D) and like games rope kids into the occult, as does heavy metal rock music with satanic lyrics. Some investigators here introduce the implantation (backmasking) of satanic messages in music. But the real *bête noir* of youth is the fantasy role-playing game, usually D&D. Says one investigator, "Every kid that plays D&D will not get into satanism, but how many kids do we lose before we have a problem?" (Richmond Bureau of Police seminar, Virginia, September 13, 1988). But through playing the game, "some kids cross over an imaginary line and start connecting their D&D world with the real world." In the seminars, cult-crime officers give estimates of 95 to 150 documented deaths of children directly linked to the game (based on news articles).

implicit bias that Christianity is the traditional belief, the norm. The same law-enforcers who employ the term *nontraditional* apparently don't see the irony of introducing their seminars with the caveat that officers must respect First Amendment rights and not interfere with noncriminal religious practices. Larry Jones also advises his readers not to interfere with constitutionally protected civil liberties, yet nonetheless judges nontraditional groups or cults according to his standards.

In a discourse on Wicca (as some witches call themselves), he posits, for example, that any belief system must set absolute standards of conduct. Relative ones won't do because they "open the door to excesses" (*File 18 Newsletter*, 3[3], 7). He can only find fault with Wicca by abstracting this standard that measures the legitimacy of belief systems. While concluding that Wicca is benign and its practitioners claim no connection with satanism, Jones nevertheless describes much of Wicca as derived from "Luciferian" Aleister Crowley, who allegedly had ties to satanist and black-magic organizations.

The cult-crime model begins with a brief, disjointed history of satanic practices. Typically, lecturers usher in pagans and witches as part of the satanic extended family, sometimes accompanied by specific details of how cultists ply their beliefs. One investigator generalizes that witches pray to 300 deities, and not for benign purposes: witches pray for something smacking of self-interest (from a Richmond Bureau of Police seminar, Virginia, September 13,1988). (Cult-crime officers imply that Christians never do this.) When satanists pray, they demand, he says. "Satanism is a self-indulgent religion" based on two themes: "All humans are inherently evil," and "Life is a struggle for the survival of the fittest."

THE SATANISTS: LAVEY AND CROWLEY

The historical discourse continues by pegging two twentieth-century satanists who have molded the contemporary philosophy of their movement: Aleister Crowley and Anton LaVey. Crowley, described in police seminars as an "influential satanist," although indulging in pagan shenanigans during the early twentieth century, became involved with (although cult officers mistakenly say that he founded) the Order of the Golden Dawn and the Ordo Templi Orientis, "the largest practicing Satanic cult operating today." Further, say the police, the main belief fostered by groups deriving from Crowley's legacy involves "sexual perversion."

LaVey, on the other hand, a former police photographer and circus performer, founded the Church of Satan in San Francisco in 1966 at the zenith of Haight-Ashbury hippiedom. Police officers teach that LaVey's two books, *The Satanic Bible* and *The Satanic Rituals Book,* can be dangerous, and they observe incredulously that both can be found in shopping-mall bookstores. In particular, law enforcers cite LaVey's nine dicta of the Church of Satan, which include (LaVey 1969):

Satan represents indulgence, instead of abstinence! . . .

Satan represents vengeance, instead of turning the other cheek! . . .

Satan represents all of the so-called sins, as they lead to physical or mental gratification! . . .

Cult officers maintain that LaVey's dicta foster in his followers the attitude, "If it
feels good, do it," thus justifying criminal acts.

Aleister Crowley is said to have added a more wicked dimension to this philosophy, for in his *Book of the Law* (written before World War I) he stated, "Do what
thou wilt shall be the whole of the law" (Crowley 1976, 9). The statement quoted
by the law-enforcement officers out of context implies to them license for murder.
In context, however, one reads a metaphorical jaunt through the ancient Egyptian
pantheon full of erotic and occasionally Masonic allusions. One might infer from
context that the law officers' quotation, too, is figurative speech.

The Book of the Law, as dictated by a shadowy prophet to Crowley, contains a
damning quotation: "Love is the law, love under will. There is no law beyond Do
what thou wilt." But the text even explains the credo by pointing out that people
move through their lives according to their destinies, that people act according to
experience, impulse, and the "law of growth." In short, people are controlled by
destiny: they cannot act apart from it. "Do what thou wilt" means "Do what
accords with your destiny." Crowley's most recent biography points out that he did
not intend the phrase to mean, "Do what you like," but rather, as Crowley later
wrote, "Find the way of life that is compatible with your innermost desires and
live it to the full" (King 1978, 36). The same biographer adds that an exegesis of
the work may be impossible because Crowley himself claimed that he didn't
understand all of it.

Nevertheless, say law-enforcers, deviant people use Crowley's prescription to
justify sex crimes, child-molesting, and murder. To add to the mystery, one police
investigator held up a copy of Crowley's book at a seminar, stating that one can
obtain it only from a certain Pennsylvania occult bookstore or from the Ordo
Templi Orientis, and that he himself could not reveal how he obtained his copy.[1]

LaVey, on the other hand, operates without a deity. To the Church of Satan,
the Evil One is no deity but rather a symbolic adversary. The Church of Satan,
then, pulls a clever trick:

> "What are the Seven Deadly Sins?" LaVey is fond of asking. "Gluttony, avarice,
> lust, sloth—they are urges every man feels at least once a day. How could you set
> yourself up as the most powerful institution on earth? You first find out what
> every man feels at least once a day, establish that as a sin, and set yourself up as
> the only institution capable of pardoning that sin." (Lyons 1988, 111)

Since people's guilt, apprehension, and anxiety about such urges are worse than
experiencing the urges themselves, the Church of Satan offers people a release:
Indulge yourselves, says the church, as long as you abide by the law and harm no
one. Lyons (1988) reports examples of Church of Satan psychodramas that engineer people's confrontations with their own fears, such as a woman afraid of her
domineering husband who role-plays him to help reduce his menacing effect on

her. Further, the rituals of the Church of Satan frequently invoke fictional deities. "In joining the Church of Satan, these people not only managed to inject a little mystery and exoticism into their otherwise banal lives, they achieved a mastery of their own fates by the practice of ritual magic," Lyons wrote (1988, 116). Lyons's point was confirmed by the participant observation experiment of anthropologist Edward Moody, who found the Church of Satan therapeutic (1974).

If LaVey's ideology is contrived of fiction, symbolism, and a deliberate anti-dote to Establishment Christianity, and Crowley retailed in the metaphysical (what one would now call New Age), why the law-enforcement interest? Cult offi-cers focus on these two because they have been published and their philosophies are within easy reach. No other "satanic" ideologies exist that have so openly and publicly philosophized. They make easy targets. One of the first articles on this subject in a law-enforcement journal even pointed out that LaVey uses a symbolic Satan and noted in context that LaVey's church even condemns sex crimes, including bestiality, but nevertheless stated, "It seems contradictory for a group to encourage all forms of sexual expression, and at the same time place parameters on that activity" (Barry 1987, 39).

By touting certain books as evil and pernicious, cult-crime officers have appointed themselves conservators of our libraries. Cult consultant Dale Griffis has recommended that officers contact public libraries for names of patrons who have borrowed books on the occult (according to an Office of Intellectual Free-dom Memorandum, American Library Association, January/February 1988, p. 7).

The cult-crime officers not only cite numerous books a la LaVey and Crowley as bona fide compendia of occult knowledge rising from the dim horizon of ancient history, but also cite as dangerous the occult symbols on rock music albums, the songs' lyrics, and the fantasy characters that appear in the advanced levels of "Dungeons and Dragons." Yet the game does not invoke any Mephistoph-eles from the arcanum arcanorum of medieval alchemists: the D&D gods, in fact, derive largely from the imaginations of the game designers and the encyclopedia (Michael Stackpole, game designer, personal communication, 1989).

NOTE

1. Taking the officer's statement as a challenge, I examined his copy, the title page complete with an ISBN number and the reprinting publisher's name, Samuel Weiser, Inc. With help from directory assistance, I contacted the publisher's customer-relations repre-sentative. I discovered that the company, which publishes many New Age books, still prints Crowley in paperback, so I placed an order for *The Book of Law*. I alarmed the representa-tive by explaining what the officer had said about the impossibility of obtaining a copy, to which the surprised woman said, "But we'll sell it to anyone who asks!" I received my copy within ten days. (The officer who created the mystery over the book was an investigator for the Richmond, Virginia, Bureau of Police, lecturing on September 13, 1988.)

REFERENCES

American Library Association (Office of Intellectual Freedom). 1988. *Memorandum,* January/February.

Bacon, L. A. 1988. Cult activity in James River Park. *Style Weekly* 6 (42): 6. October 18.

Barry, R. J. 1987. Satanism: The law enforcement response. *National Sheriff* 38 (1): 38–42.

Berg, M. 1988. Satanic crime increasing? Police, therapists alarmed. *Kansas City Times,* March 26.

Clark, J. R. 1988. The macabre faces of occult-related crime. *Law Enforcement News* 14: 279–80. October 31 and November 15.

Crowley, A. 1976. *The Book of Law.* Reprint. York Beach, Me.: Samuel Weiser.

Cult Crime Impact Network. 1988. *File 18 Newsletter* 3(3).

———. 1989. *File 18 Newsletter* 4 (1).

Edge, C. 1988. Four Satan worshippers arrested in attempt to rob Conway grave. *Sun News* (Myrtle Beach, Fla.), July 28.

Kahaner, L. 1988. *Cults That Kill.* New York: Warner Books.

King, Francis. 1978. *The Magical World of Aleister Crowley.* New York: Coward, McCann & Geoghegan.

LaVey, Anton. 1969. *The Satanic Bible.* New York: Avon.

Lyons, A. 1988. *Satan Wants You: The Cult of Devil Worship in America.* New York: Mysterious Press.

McCoy, C. 1989. Mississippi town is all shook up over voodoo plot. *Wall Street Journal,* February 24.

Moody, E. J. 1974. Magic therapy: An anthropological investigation of contemporary Satanism. In *Religious Movements in Contemporary America,* edited by I. I. Zaretsky and M. P. Leone. Princeton, N.J.: Princeton University Press.

Rommel, Kenneth M., Jr. 1980. Operation animal mutilation. Report of the District Attorney, First Judicial District, State of New Mexico, June 1980.

Stackpole, M. 1988–89. Personal communications.

41

THE SPREAD OF
SATANIC–CULT RUMORS

Jeffrey S. Victor

$\left(\mathbf{S}\right)$ cientists sometimes discover something really interesting by accidentally stumbling onto completely unexpected findings. This is what happened to me after I decided to investigate the causes of a rumor-panic about satanic cults that occurred in Jamestown, New York, on Friday, the 13th of May, 1988. It caught my curiosity because I knew a few of the teenagers who were victims of the malicious gossip and vigilante harassment created by the stories.

I have been trained as a sociological researcher, which means that I investigate the causes of group behavior. A rumor-panic is much like the stampede of a herd of buffalo: It is a product of group forces, rather than of the personal motives of individuals.

I used a wide variety of methods to collect information relative to the rumor stories and behavior in response to those stories. My goal was to obtain an "ecological" view of the community of Jamestown as a social system affected by social forces in the larger social system of the national society.

My research methods included interviews I conducted with a wide variety of community authorities, including police, school officials, youth group workers, ministers, psychotherapists, and newspaper reporters. The Jamestown Police Department was exceptionally helpful in providing me with nonconfidential information regarding their own investigations of the various rumor stories. I also interviewed newspaper reporters from other towns in the region who covered the story. Students from one of my classes conducted interviews with 49 local area teenagers, parents, and informal authority figures (such as teachers and ministers) shortly after the rumor-panic occurred. One student, on an independent study project, did a research study of teenage peer group conflict in Jamestown in reaction to the rumors, interviewing 30 teenagers from different youth subcultures. I also had information of my own as a member of the community and as the father of a teenage son in the local

This article first appeared in *Skeptical Inquirer* 14, no. 3 (Spring 1990): 287–301.

high school. As a teacher in a community college, most of my students (youth and adults) are from the local area. Many of them talked to me at length about the rumors. I also obtained useful information from school attendance records, reports from local government agencies, and other documents.

I used a variety of sources to put my local research findings into a national context, including recent books by crime reporters, fundamentalist Christian proselytizers, and scholars studying cult behavior. I found the most useful information about satanic panics and satanist activity around the country in newspaper articles. The most thoroughly researched overview was published in a series of articles in the *Memphis Commercial Appeal* (Charlier and Downing 1988).

In researching the Jamestown rumors, I was surprised to find that very similar rumor-panics in response to fears of dangerous satanic cults have been occurring all around the country since 1984. One of them, for example, occurred on Friday the 13th over most of the state of South Carolina in 1987, a year before it happened in Jamestown.

I have so far located 21 sites of past rumor-panics about satanic cults. These have almost all occurred in economically declining small towns and rural areas of the country. The rumors are also amazingly similar, for example, "A satanic cult is killing animals at secret ritual meetings in the woods" or "A satanic cult is planning to kidnap and sacrifice blond, blue-eyed children." In addition, there are many locally inspired variations of these themes, which may include references to killing black cats, or lists of teenagers to be sacrificed, or allusions to drug-induced sex orgies.

The people of Jamestown were fortunate in having an exceptionally professional police force that did not let itself succumb to the rumor-mongering hysteria. Instead, they reacted with proper concern for verifiable facts and with responsible actions designed to discourage vigilante violence. My research found that this was not the case in many other locations. In some cases, police reacted by passing on rumors as facts, by arresting the innocent victims of the rumors, and by engaging in irresponsible witch-hunts.

In one case in New Hampshire, for example, police publicly claimed to have found evidence of ritually slaughtered animals, which were later determined to be only road kills cleaned up by state road workers and deposited in the woods. In another case, in Cobleskill, New York, in response to reports of a "satanic cult ritual meeting" and impending human sacrifice, police rushed with guns drawn into a wooded site. They found only some college students in hooded garments practicing a medieval play and using wooden daggers and swords as props. Actions like these are not only embarrassing but also create danger where none exists.

My research found that many people in the Jamestown area believed that the police, the editors of the local newspaper, and Jamestown school officials, who dismissed the stories as the empty rumors they were, were all engaged in some kind of grand cover-up of the real facts. This should not be surprising since so many of our political and corporate leaders have so often lied to the public about unpleasant goings on. Other research has shown that the American people's confidence in the ability of their leaders and institutions to solve problems is at an all-

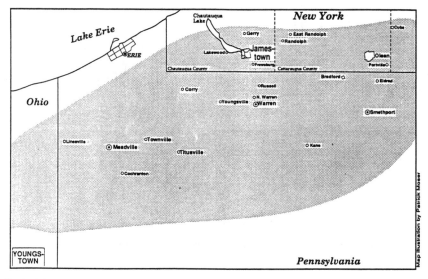

The shaded area is where satanic rumor-panics occurred during May 1988. Locations shown are where rumor activity was most intense according to newspapers reports. (Map by Patrick Moser.)

time low. This may be one reason the satanic-cult rumors are taken seriously by so many people.

When I began my research, I assumed that the rumors were limited to the Jamestown area. However, I was surprised to find that similar stories were circulating at the same time over a 250-mile area of southern New York and northwestern Pennsylvania. Moreover, since similar rumor-panics had occurred across the country before the one in Jamestown, I realized that the underlying causes of the satanic-cult scare must be national in scope.

Nevertheless, my local investigation provided important clues about what is happening nationally. I found, for example, that these rumors were not a sudden outburst. Instead, they were the result of a gradual process that took months to develop. Local gossip "snowballed" into elaborate stories, as layer after layer of fiction were added and the stories were repeated over and over by hundreds of different people. I reasoned that whatever "triggered" the rumor process had to reach a wide audience simultaneously. I found evidence that the satanic-cult rumors first appeared after ideas from television talk-show programs about satanism were distorted to fit into local gossip in different communities in the region.

I also found that various groups of people were affected differently by the rumors. A great many people did not take them seriously and paid little attention. On the other hand, hundreds of parents became alarmed, to the point of keeping their children home from school, out of fear that they might be kidnapped by the "cult." My evidence indicates that less-educated and lower-income parents were more likely to take the stories seriously.

Determining the underlying causes of these satanic-cult rumors is difficult, and it is really a matter of interpretation. Past studies of rumor-mongering have

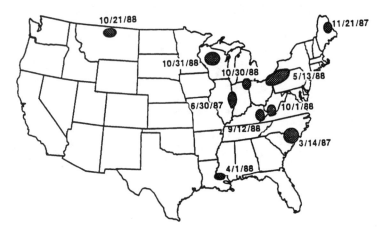

Locations of some recent rumor panics and the dates they were reported in newspapers.

found that such activity increases in times of widespread social stress, when many people are seeking explanations for frustrating life experiences. I believe that economic difficulties and the resultant breakdown in family relations is a source of that frustration. Other research has found that small-town and rural areas all over the country are suffering from a loss of well-paid blue-collar jobs. This stressful situation is most strongly affecting poorly educated young parents in rural areas, about one-third of whom now live in poverty.

Another clue to the causes of these rumors can be found in the stories themselves. Rumors that claim that some kind of alien group is threatening the people in a society are called "subversion myths." Subversion myths about Jews, or communists, or witches are common when people feel that their traditional values are being threatened by social problems. I believe that the satanic-cult scare is an expression of the desire of Americans for clear and consistent values that can give stability to their lives and those of their children.

A great many parents today are seeking scapegoats for their fears and frustrations. Parents in small-town America have to worry about threats to their children that few of their grandparents had to face: the drug plague, violent juvenile crime, disorder in the schools, and even teenage suicide. The increasing breakdown of family structure has served to deepen these fears and intensify the need to blame somebody for the mess.

I found one thing especially disturbing. It appears that some people are going around the country making thousands of dollars from lecture fees by cultivating fear about satanic cults. Many of these same Satan-hunters have broadcast their claims to national audiences on the television talk-shows of Geraldo Rivera, Oprah Winfrey, and Sally Jessy Rafael. Their wild, unsubstantiated claims about satanism have the effect of inflaming passions. Many innocent people can be victimized by their appeals to the scapegoating hysteria.

Kenneth Lanning, the head of the FBI's special unit in charge of investigating claims about satanic-cult crimes, finished a report of his findings in June 1989. In his conclusion, he cautions: "Until hard evidence is obtained and corroborated, the American people should not be frightened into believing that babies are being bred and eaten, that 50,000 missing children are being murdered in human sacrifices, or that satanists are taking over America's day care centers.... An unjustified crusade against those perceived as satanists could result in wasted resources, unwarranted damage to reputations, and disruption of civil liberties." Unfortunately, I found evidence that these harmful effects of rumor-mongering are already being felt around the country.

POSTSCRIPT

The occurrence of satanic cult rumor-panics peaked in the United States during the years of 1988 and 1989, and has diminished since then. By the time that my research was completed and published in my book *Satanic Panic*, I had found 62 areas of the country where these strange phenomena had taken place. These rumor-panics also occurred in other societies after beginning in the United States; especially where English is the primary or a second language. My research found that satanic cult threat stories were given credibility and transmitted, in part, by the popular culture mass media (particularly, TV talk shows), which benefit from the sensational stories by attracting audiences. In addition, other groups gained ideological benefits from these stories, by using them in their propaganda, including some fundamentalist and radical feminist organizations.

Although rumor-panics have diminished, other manifestations of the broader satanic cult scare have persisted. Most important are accounts given by adult psychotherapy patients of childhood ritual sex abuse by their parents, who they claim were participants in secret, criminal satanic cults. Unfortunately, quite a few psychotherapists have attributed credibility to these "recovered memories." Currently, satanic cult survivor accounts are important anomalies in the growing scientific controversies involving the issues of whether or not "repressed memories" actually exist, whether or not recovered memories of childhood sexual abuse are reliable, and whether or not recovered memory psychotherapy techniques are scientifically valid. The interesting research question for sociologists and psychologists is how imaginary stories from popular culture become woven into people's accounts of their memories. Perhaps it is still necessary to report that no convincing scientific or legal evidence has been found to confirm claims about the existence of secret, criminal organizations guided by an ideology of Devil worship.

REFERENCES

Charlier, Tom, and Shirley Downing. 1988. Facts, fantasies caught in tangled web, *Memphis Commercial Appeal,* January 17; Allegations of odd rites compelled closer look, January 17; Links to abuse hard to prove, January 18.

Victor, Jeffrey S. 1989. A rumor-panic about a dangerous satanic cult in Western New York. *New York Folklore* 15 (1–2): 23–49.

42

"LIGHTS OUT": A FAXLORE PHENOMENON

Jan Harold Brunvand

On Saturday August 14, 1993, a small news item—about six column inches —debunking a "faxed warning about gang initiation," appeared in the *Memphis Commercial Appeal.* The headline was "Officials Deny Faxing Gang Warnings," and the subhead read, "Untrue documents promote hysteria." The documents were described as "heavily faxed" and claimed to be official police bulletins. They stated, "This is the time of year for gang initiation," and described the specific threat as follows:

> One of the methods used this year will be for gang members to drive with their lights off at night. When you blink your lights or flash your high beams, they will follow you home and attempt to murder you.

At a press conference, officials of Memphis and of Shelby County, Tennessee, said that these "fax driven rumors criss-crossed the county this week," that police were inundated with calls about them, but that no local law-enforcement office had sent out the faxes, nor was there any evidence for such gang initiation practices.

Although it is clear in this news item that the "Lights Out!" rumor was already in active circulation, this is the earliest example I have located. The next time I encountered the rumor was in an e-mail message from Chicago sent on September 9, and the next *published* report I found was in a suburban Chicago newspaper on September 11, four weeks after the Memphis report. But soon the "Lights Out!" rumor emerged in many cities and became a matter of national concern. The flap continued through autumn 1993 and even into the new year in some places. The rumor was driven nationwide largely by facsimile transmissions and thus provides a unique example of a faxlore phenomenon, likely the trend of the future in the circulation of urban rumors and legends.

This article first appeared in *Skeptical Inquirer* 19, no. 2 (March/April 1995): 32–37.

I collected three kinds of information about "Lights Out!": (1) copies of the warnings themselves, (2) letters from people reporting on the warnings, and (3) further news stories.

My September 9 e-mail report came from someone at the University of Chicago who forwarded a memo credited to the Security Division of the First National Bank of Chicago. The first line, "Beware," was followed by two exclamation points; the capsule description of the initiation ritual, said to be planned for "the Chicagoland area," was followed by the bare elements of an illustrative story: "To date, two families have already fallen victim to this senseless crime." The memo concluded with an appeal for readers to "inform your friends and family not to flash their car lights at anyone!" This early example of the "Lights Out!" warning displays typical features of all subsequent versions: it uses somewhat sensational language, repeats the basic rumor (naming it as "Lights Out!"), and provides a validating reference (i.e., the bank's memo format); it also localizes the supposed criminal plans, alludes in a vague way to specific cases, and recommends a course of action.

Later versions of the warning—usually computer-written and often laser-printed—tend to have more exclamation points (sometimes coming before as well as after sentences), to have more words or whole lines printed in all caps or underlined (sometimes double-underlined, and sometimes both all caps and underlined), and often to have handwritten additions like "*This is not a joke*," or "*Be careful out there*," or "*Urgent!*" A few examples were hand-lettered and photocopied or were retyped on a company's letterhead (including the City of Detroit Water and Sewerage Department). Most examples that I collected had been faxed; others were e-mailed, sent by post-office mail, or posted as hard-copy memos on bulletin boards.

Dates on these warnings range from early September to early December 1993, most coming in mid-September. Other cities mentioned in them are St. Louis, Detroit, Dallas, Atlanta, Norfolk (Virginia), New York, Baltimore, Los Angeles, Sacramento, and Honolulu. (Articles in the press added Memphis, Toledo, Columbus, Pittsburgh, Philadelphia, Washington, D.C., Minneapolis, Little Rock, Tulsa, Houston, San Antonio, Lubbock, Denver, Salt Lake City, San Jose, and San Francisco. Letters from readers added cities in New Jersey, Florida, Missouri, and the Northwest. Clearly, the "Lights Out!" rumor was flying coast to coast.) The institutions circulating the fliers were banks, businesses, law firms, universities, military posts, hospitals, and day-care centers. Several of the warning notices contain routing stamps or slips indicating that they were circulated throughout a whole office or company.

About mid-September the warnings began to name "Grady Harn of the Sacramento Police Department" as the source of the information. Also, the weekend of September 25 and 26 was pinpointed as "Blood [or Bloods'] Initiation Weekend," when, supposedly, gangs, including the notorious Bloods' Gang, would hold their murderous initiation.[1] A few of the fliers, although unsigned, made first-person reference to "my stepfather" or to another relative who had "called me" with this important information. Some fliers admitted that "this information has not been confirmed," but most fliers still advised readers not to flash their lights at anyone, just in case. One person sending me a printed flier added in a note: "One of our stupidvisers [at work] stood up and read this to us." After the supposed "Bloods' Weekend" came and went without incident, the fliers—now minus the dates—continued to appear, mostly in fax machines.

During the 1993 autumn semester at Indiana University at Kokomo a folklore class studied the "Lights Out!" rumor. The instructor, Susanne Ridlen, forwarded a packet of the 20 items resulting from the project, and these display the variations of the story in one community and of the means by which it was transmitted.

The Kokomo students documented transmission of "Lights Out!" via word of mouth, telephone (including long-distance calls), on both commercial and CB radio, via e-mail, fax, printed memos (distributed in schools and workplaces), and in publications. Several Kokomo versions combined "Lights Out!" with other car-related legends (Brunvand 1981, 19–46, 1984, 50–68; 1986, 49–67; 1989, 89–128), and some people claimed that the gang initiations included the cutting off of feet or ears. The specific gangs involved were said to be the Crypts [*sic*] and the Bloods, groups said to be moving to Kokomo either south from Chicago or north from Indianapolis. Other versions claimed an origin of the information from Missis-sippi, where such initiations supposedly had occurred. Motorists would be tar-geted either if they blinked (or "flickered") their lights *or* if they honked at a gang car driving without lights. Police were said to be warning motorists—via gas-sta-tion attendants—not to flash their lights. Despite this, two (or three) people, according to the Kokomo rumors, had already died.

A story in the *Kokomo Tribune* on October 14, ("Police Shoot Hole in Gang Ini-tiation Rumor") described "Lights Out!" as a baseless rumor that was thought to have reached Kokomo from Jackson, Mississippi, on September 19 in a message broadcast on CB radio. Both Indiana and Illinois state police gang-crimes units were quoted in the news story as denying the rumor.

(Similarly, the Salt Lake City Police Department on September 24 issued a news release calling the "Lights Out!" story "unfounded . . . unsubstantiated and without merit." Many police departments across the nation issued similar denials.)

The letters (including e-mail, snailmail, and faxed correspondence) that I received from readers of my books commenting on the "Lights Out!" rumor add to the sketchy story a few more details that must have circulated orally, since they do not appear in the printed fliers. These letters were dated from mid-September 1993 to April 11, 1994, and came from people who knew or suspected that the story was false. Most letters reported a chain of informal communication of the story—i.e., a co-worker whose girlfriend, a nurse, had heard about someone working on the other shift at the hospital or at a different hospital who had treated victims of the ritual. The initiations, according to some writers, "were supposed to test the mental toughness" of gang recruits. Sometimes the gang recruits would trace the offending cars via their license plates. The police were not publicizing the crimes, according to rumors, in order to lull gangs into a false sense of secu-rity in hopes that they could be caught in the act. A letter dated March 29, 1994, from New Cumberland, Pennsylvania, reported a variation in the story: "Cars Full of Satan Worshippers Driving Around Without Their Headlights." The writer had overheard his secretary warning her sister about the threat, and she had heard it from the driver of her van pool, who had heard it from a friend who saw a sign warning people about it in a store window.

A Seattle reader in a letter dated September 24 (the Friday before "Bloods' Weekend") mentioned that on Thursday the 23rd the flier appeared at her workplace and "to my distress, every one of my twelve co-workers fell for it . . . despite my insistence that it couldn't possibly be true." Meanwhile, in Salt Lake City in my own department at the university, here's what happened—or *nearly* happened—on the same days. On the 23rd, a reader from Fort Lauderdale, Florida, faxed me a letter saying he was also faxing a copy of a memo that he suspected was based on a rumor; page 2 of the fax transmission was the "Lights Out!" warning itself, typed on the letterhead of a Florida Jewish community center. The secretary of my department pulled page 1 of the message and put it into my mailbox without reading it beyond the address. Page 2—the warning memo—however, she read with alarm and then took it with her on Friday to a full departmental meeting; or, not *quite* a full meeting, since I myself, as a partial retiree, had decided to skip town that afternoon. The secretary intended to read the notice to the assembled faculty in order to alert them not to flash their lights when driving home, but she was dissuaded from doing so by my colleague, folklorist Margaret Brady. Brady convinced her that the warning was just "one of those legends that people are always sending Jan." Our secretary still felt the story might be true, since her own daughter had also heard about it in Salt Lake City a few days earlier.

The news stories about "Lights Out!" may be taken in chronological order, insofar as dates can be determined from the clippings.

As mentioned earlier, the first clipping I have is from the *Memphis Commercial Appeal* of August 14. The second is from the *Daily Southtown*, a suburban Chicago paper, on September 11 ("Police Say Area Story of Gang Rite Is Hoax"). The lead reads: "It's a great story—just like most so-called 'urban legends.'" Besides describing the same "anonymous handbills," this news story quotes police officials who deny it and adds that it sounds like something that might happen "in the frozen wastelands of Russia." The news story also suggests the possibility of copycat crimes, and it concludes with the threat of prosecution against spreaders of the tale. Every facet of this fairly obscure little news item turns out to be typical of what most papers wrote.

Four days later, on September 15, Mary Schmich, a *Chicago Tribune* columnist, wrote about the rumor, calling it "urban faxlore." Schmich identified the Black Gangster Disciple Nation as the Chicago street gang blamed for the outrage, but she emphasized that the rumor was unsubstantiated and strongly denied by Chicago police. Schmich listed radio stations, colleges, hospitals, churches, and video stores as places where the warning had appeared. Because she had phoned me to discuss the rumor, Schmich was able to cite the earlier Memphis example of the story and also to compare "Lights Out!" to other urban lore about crime and violence. (Mary Schmich said that she was calling me at the suggestion of folklorist Alan Dundes of U.C. Berkeley, whom she had called earlier. He had not yet heard about the warnings but thought that I might have.)

At this point my phone began ringing off the hook, and I was preparing to leave on an extended trip; so I put a message on my answering machine saying, in effect, "No interviews, please." I decided to stay out of the spotlight and see what reporters might make of the rumor on their own, or with the help of other folklorists whom they might contact.

Shortly after taking this vow of silence I started to see results: on September 18 two papers ran stories—the Little Rock *Arkansas Democrat-Gazette* ("Gang Initiation Rumor Baseless, LR Police Say") and the *San Jose Mercury News* ("Cops Try to Arrest Rumor of Gang Rite"). The enterprising Arkansas reporter located three of the Brunvand books and compared "Lights Out!" to other urban lore, including an earlier local outbreak of "The Mutilated Boy" legend (Brunvand 1984, 78–92; 1986, 148–56). The even more enterprising San Jose reporter managed to get a Memphis Police spokesman and Alan Dundes on the phone. From Tennessee she learned that the story was known in Knoxville and Chattanooga as well; from Dundes (who now had heard the story and had a comment ready for it) she picked up the term *faxlore* and also the ideas that a fear of teenagers—especially gang members—was reflected in the story and that it might be inspired by acts of violence against tourists in Florida. In an unusual twist, the *Mercury News* withheld details of the warnings, fearing, they said, to suggest copycat crimes. This article also referred to the speed with which the rumor had blanketed Silicon Valley, where thousands of people are linked to e-mail and to computer bulletin boards.

In Salt Lake City (where I also dodged the press) another gang rumor was flying—that gangs were planning to rape a cheerleader as part of an initiation. Although police and school authorities denied the rumor, a September 18 article in the *Deseret News* reported widespread concern among students and parents, especially on the more affluent east side of the Salt Lake Valley. One television station was criticized for publicizing the story, even though the broadcast had made it clear that the rumors were unconfirmed.

Meanwhile, back in Chicago, on September 21, the student newspaper of the University of Illinois branch there was debunking the story again. One version of the warning photocopied on a state police letterhead was described as having a typeface that did not match the rest of the letterhead, a clue to its unreliability.

On September 22, the *Houston Post*, the *Houston Chronicle*, and the *Toledo Blade* ran articles on the rumor. The *Post* article ("Scary Gang Rumor Has Its Fax All Wrong") was pretty standard, except for calling the warning "an electronic chain letter" and "an urban myth[!]" (Maybe I should have answered the phone after all.) The *Chronicle* article ("HPD Tries to Stop Rumor About Gang Rites") had a novel touch—a color photograph of a grinning police chief holding up one of the fliers; the "Dragnet"-inspired caption read, "Just a phony fax, ma'am." The chief was quoted as saying, "I'd rather deal with the gangs than talk to any more hysterical people on the phone."

A sidelight of the Texas circulation of the rumor was revealed September 23 in a column in the *San Antonio Express-News*. Columnist Roddy Stinson, who a week earlier had dubbed his own city "the wide-eyed booby capital of the nation" for panicking over the "Lights Out!" urban legend (Stinson's term), now gave the booby prize to Houston, where the story had circulated earlier.

During "Bloods' Weekend" itself—September 25 and 26—the newspaper stories increased: I have clippings from papers in central New Jersey, San Francisco, Los Angeles, Salt Lake City, Minneapolis, and Pittsburgh. The New Jersey story mentions dozens of calls to police from fearful people, spurred by warnings traced to the Johnson & Johnson headquarters in New Brunswick and also circulated in area colleges and high schools and in a large insurance company. The San Francisco article ("Rumors Fly on Computer Networks") also names hospitals as hotbeds of the story, but emphasizes computer rather than faxed transmission. Alan Dundes is quoted anew, this time with a different slant on the story. "There is an element of the car, which represents power, mobility, and sex," the Berkeley professor comments, adding that the rumor was probably fueled by recent shootings of foreign tourists in Florida.

The *L.A. Times* article of September 24 is the first publication to mention the mysterious Grady Harn of the Sacramento Police Department, who is named on many fliers; the *Times* checked with Sacramento and found that no one of that name serves on the police force there. Another innovation in the *Times* story is an unattributed claim that "the warnings are the work of a computer hacker who used facsimile telephone numbers to disseminate the messages." The fliers are described as "clumsily worded," which suggests, perhaps, poor social or communicative skills of the assumed hacker. Finally, the *L.A. Times* mentions that these phony warnings are "the latest examples of what sociologists call an urban myth." Although Los Angeles gang investigators are quoted in the story, the reporters who wrote the piece did not attempt to interview any gang members themselves.

Both Salt Lake City newspapers now covered the story, each relating it to the earlier rumors about gang intentions to rape cheerleaders. One county sheriff in Utah said he had not been able to get off the phone for more than 30 seconds at a time while fielding calls about "Lights Out!" from citizens and from the press. In a technically garbled explanation of how the story may have started, a police official suggested that "the hoax note was faxed into a billboard computer system."

Probably "Bloods' Weekend" was when the Associated Press circulated an article debunking "Lights Out!" that was widely reprinted and quoted. (I do not have a dated copy of the AP release.) The AP story quoted Ralph Rosnow and Gary Fine, coauthors of a book on rumor and gossip. Fine voiced the opinion that "*gang* in this particular rumor is a code word for poor, young black men."

The Minneapolis news article of September 25 again described the suspected villain in the case as a "hacker," but the paper associated "hacking" with *faxing* messages rather than with using computerized e-mail or a computer bulletin board. Articles published in Pittsburgh on September 25 and 27 added little to the picture except for speculating that the rumor came from "an unidentified originator in California."

In an interesting juxtaposition, the *Baltimore Sun* on September 27 ran a long story debunking the "Blue Star Acid" rumor (Brunvand 1984, 162–69; 1989, 55–64), which had cropped up there again recently with a sidebar story on "Lights Out!" The Baltimore version was that members of "the Los Angeles-based gang the Bloods were randomly shooting motorists as part of a nationwide initiation campaign."

At last the rumor was mentioned in the national press. In the September 27 issue of *U.S. News & World Report*, a column by John Leo mentioned "Lights Out!" in passing as "a new bit of urban folklore." The column was mostly devoted to comments on the shootings of foreign tourists in Florida. *Newsweek*, in a brief notice headlined "Big Fax Attack," mentioned Houston, Los Angeles, and Atlanta as sites of the rumor. In a syndicated column of September 29, William Raspberry of the *Washington Post* mentioned the rumor, although he indicated that just "people" (gangs were not mentioned) were driving with their lights off, hoping to entice victims to flash their headlights at them and thus become targets for violence.

The last weekend of September the magazine section of the *Dallas Observer* ran a detailed account titled "Anatomy of a Rumor: How the Gang-Initiation Story Terrified Dallas." The author called it "something of a minor urban myth in several cities," and he did a creditable job of tracking local versions back to individuals who had picked it up while traveling, on the phone, via fax, and so on. According to the *Observer*, one Dallas businessman said he was handed the warning "when he boarded an American Airlines return flight from Tampa."

From October to mid-December only a few further news stories appeared; the cities that were late to catch the rumor were Denver; Philadelphia and Allentown, Pennsylvania; Columbus, Ohio; and Lubbock, Texas. With Lubbock, on December 14, my clipping file bottoms out. The Denver story was mainly about rumors of gang rapes at a mall, with "Lights Out!" mentioned only in passing. The *Philadelphia Inquirer* story was unrestrained in calling the rumor "a big, fat, stinking, Pinocchionose kind of lie," but added, "and everybody believes it." Ralph Rosnow, who teaches at Temple University, was again quoted. The *Allentown Morning Call* story mentioned that the rumor had been retyped on Thomas Jefferson University security letterhead; both Patricia Turner, a professor at U.C. Davis, and Alan Dundes were quoted in this story, evidently from original interviews by the Allentown writer. Turner made some interesting observations about the theme of the "demonized outsider" threatening a Good Samaritan on the road. Dundes repeated some of his earlier remarks, and he too touched on the "Good Samaritan" theme.

In Columbus, "Lights Out!" got only a short paragraph in a story that was primarily about another rumor—that groups are trying to raise funds to buy firearms for the homeless. Similarly, in Lubbock, Texas, "Lights Out!" rated only a short mention in a full-page feature story about "Those pesky rumors." The only unusual touch here was the remark of a local police sergeant about how he deals with the panicky public in such cases; he said, "I just try to sound authoritative and reassuring." By mid-December, then, the story was effectively dead in the nation's media, although a few individuals continued to write me letters about it.

Unexpectedly, "Lights Out!" emerged one more time in the national media in "Traps," a new CBS-TV police drama, in which George C. Scott plays Joe Trapcheck, a former homicide investigator who comes out of retirement to serve as a consultant in his old department. In the premiere episode, aired on Thursday, March 31, 1994, he has joined the "Headlight Killer Taskforce," which is working to solve a series of shootings that occurred after people gave a "courtesy blink" to a car driving with its headlights out. Curiously, the episode also included another piece of urban folklore, where a photocopy machine is used to fool a suspect into revealing needed information; he is told that he's actually hooked up to a new type of polygraph (Brunvand 1993, 139–45).

Reviewing the "Lights Out!" phenomenon, briefly, as *folklore*, it seems evident that although modern technology and communications facilitated the rapid dissemination of the rumor over an extremely broad area, the variations that developed were similar to those typically caused by oral transmission alone. Details were added and altered; the story was localized; authorities were invoked to support the story; supposed actual cases were cited, and so on. "Lights Out!" was similar to other recent rumors and legends, including "Blue Star Acid" (also fax- and photocopy-delivered), "The Assailant in the Backseat" and other car horror stories, "The Attempted Abduction" and other mall-crime stories, and the like. *Faxlore* is a catchy term, but it is not clear that it should imply any essentially different kind of modern tradition.

Despite the sensational tone of the warnings, their informal channels of distribution, their poor English usage, and their lack of any specific verification, their message was taken to heart by many Americans. Thousands of people must have duplicated and distributed the fliers for them to have penetrated so widely. I believe the folklorists and sociologists who commented on "Lights Out!" were right in identifying themes like teenage and gang crimes, racism, urban problems, and attacks on foreign motorists as topics underlying the "Lights Out!" hysteria. (I'm not sure, however, where the sex, mentioned by Alan Dundes, comes in.) A well-known gang name, the Bloods, lent itself perfectly to intensifying the hysteria, as did another name, the Cryps (when misunderstood as "Crypts"). In Chicago, a gang with the very word "Black" in its name—the Black Gangster Disciple Nation—fed into this same racial fear, and in Kokomo the idea that the warning had come from Mississippi implies the same stereotype.

Press reports reveal other views of the rumor flap. Journalists tended to confuse terms like *myth, rumor, legend, chain letter,* and *hoax,* as well as to ignore the technical differences between fax, e-mail, and computer bulletin boards or computer newsgroups. Several newspapers mentioned copycat crimes, although none of the papers ever reported any actual "Lights Out!" crimes *or* copycats. In hope of determining the precise origin of the rumor—something folklorists seldom seek or find—journalists focused upon (indeed, may have *invented*) the *hoaxing hacker* character, presumably someone from California. No evidence for the existence of such a hoaxer was ever presented. An idea proposed by both law-enforcement and newspaper sources is that the originator of the story should be found and prosecuted. Despite having this serious plan in mind, newspapers felt free to pun shamelessly in the wording of their stories and headlines: "just the fax," "a fax attack," "has its fax all wrong," "guns down the rumor," and others.

One persistent fear that both the general public and newspapers responded to was the idea of dangerous gang activity entering one's own community from some outside source, usually a nearby larger city that is thought to be awash with gang crimes. It still amazes me (as mentioned above) that no investigative reporter seems to have bothered to have contacted any local gang members to ask about initiation rituals in general and about "Lights Out!" or other car-related customs in particular.

Finally, I reemphasize one theme of the warnings pointed out in only a single news article among the reports I collected. This is the "Good Samaritan" theme mentioned by Turner and Dundes. Flashing one's headlights at a car driving toward you at night without its lights on is a simple, common act of courtesy and safety. It's something most drivers have been taught to do, either by a driver's education teacher or simply by customary example from other drivers. (I'll be so bold as to call this a folk custom.) Some of the warning fliers allude to this custom by referring to it as a "courtesy flash." (I'll concede, however, that light flashing may also imply negative criticism in the minds of some drivers who flash or are flashed at.) Another strong message of "Lights Out!" then, is that something you, as a driver, have learned to do as a good and socially useful thing may, in this crime-ridden modern world, have become an act of aggression toward another driver who represents an outsider to your standards of behavior—perhaps a younger person of a different race who belongs to a street gang—and this outsider will go so far as to murder you for daring to be courteous. In other words, the "Lights Out!" rumor says, "Forget about being a courteous driver, because *it might kill you!*"

NOTE

1. Folklorist Ed Kahn of Berkeley, California, pointed out to me the similarity to the much-feared "Michelangelo" computer virus, supposedly set to strike DOS-based machines worldwide on March 6 (the Renaissance master's birthdate), 1992. Like the "Data-Crime" (or "Friday the 13th") virus of 1989, this threat proved to be much less destructive than predicted.

POSTSCRIPT

My article when originally published prompted two letters, published in the July/August 1995 *Skeptical Inquirer.*

Keith Trexler of Milford, New Hampshire, wrote: "The tone of Jan Harold Brunvand's 'Lights Out!' was a tad condescending toward the general public, which by and large fell for the fable. If we were all authors of books on urban legends (or even readers of such books), perhaps we might have been more suspicious at the outset. However, most of us *are* aware of certain documented aspects of gang behavior: some of them *do* have initiation requirements for membership, many are prone to committing extremely violent acts with little provocation, and more than a few have little or no regard for the sanctity of life, human or otherwise. Thus to hear that inner-city gangs from a distant state were using the 'Lights Out' method of initiation would not have the same effect on us as an alert about alien abductions could. Nor would it be unreasonable to assume that a local gang from a nearby metropolis might adopt the idea and start cruising our quiet suburban streets looking for a Good Samaritan sap to waste. Because of the reports from south Florida, if someone rams my rental car while I'm on my way to Disney

World, I'm not going to stop to exchange insurance information, and this policy could save my life. Until I had received more information on the 'Lights Out' phenomenon, I intended to err on the side of caution with an extra dose of credulity."

Al Christians of Lake Oswego, Oregon, wrote: "A true incident may have inspired the rumors and faxlore described in your 'Lights Out!' story. Around the beginning of 1993, a driver in Stockton, California, signaled to the driver of a lightless vehicle in the customary fashion. An occupant of the unlit car responded with gunfire and killed a schoolteacher in the back seat of the other vehicle. The story received prominent coverage in the region's media."

This was my reply:

"I admit to being just a tad condescending toward the general public, which was perhaps understandably gullible about 'Lights Out!' But many journalists and law-enforcement personnel — quickly recognizing the warnings as bogus—were much more outspoken in their criticisms, as my article points out.

"I learned about the Stockton, California, incident when my article was in press: after examining news stories, I concluded that it had no direct connection to the 'Lights Out!' warning. On September 18, 1992, a driver there used a hand signal to alert a driver behind him at a stoplight that his headlights were out. The driver of this second car, taking this as a sign of disrespect, shot at the first car, killing Kelly Freed, 29, a passenger, who was a longtime popular employee of a local school system, although not a teacher. In August 1993, at the sentencing, the prosecutor suggested that the convicted killer had acted 'to save face with his fellow gang members.' However, the defense attorney denied that his client belonged to a gang. Never was a gang initiation mentioned in connection with this tragic case, and in February 1995, in articles about efforts to establish a Kelly Freed Teen center in Stockton, no reference was made to gangs.

"The differences in details between this actual crime and the supposed planned attacks described in the warnings, plus the fact that never in any news stories about the warnings—nor in any of the faxed warnings themselves—was the Stockton case mentioned, all suggest that the 'Lights Out!' story developed independently."

REFERENCES

Brunvand, Jan Harold. 1981. *The Vanishing Hitchhiker: American Urban Legends and Their Meanings.* New York: W. W. Norton.

———. 1984. *The Choking Doberman and Other "New" Urban Legends.* New York: W. W. Norton.

———. 1986. *The Mexican Pet: More "New" Urban Legends and Some Old Favorites.* New York: W. W. Norton.

———. 1989. *Curses! Broiled Again! The Hottest Urban Legends Going.* New York: W. W. Norton.

———. 1993. *The Baby Train and Other Lusty Urban Legends.* New York: W. W. Norton.

43

STAKING CLAIMS: VAMPIRES OF FOLKLORE AND LEGEND

Paul Barber

(P) eople who learn that I wrote a book on vampire lore often say, "Oh, you mean like Vlad Drakul?"

"Not actually," I tell them. "Vlad Drakul was a figure in Romanian history whose only association with the vampire lore is that Bram Stoker named the character Dracula after him. Until *Dracula* came out, no one ever associated the historical figure with the vampire lore." This has been pointed out many times, and the Romanians have often expressed their dismay over the way we have expropriated their national hero and made him into a vampire. But in the media the sensational always has an edge on the prosaic, and by being associated with vampires—even if only via fiction—Vlad Drakul has become the only figure in Romanian history that Americans have ever heard about. If the Romanians began to make movies portraying George Washington as a ghoul, we would know what they feel like.

Here we see fiction becoming "historical fact," while the scholars who try to correct the "facts" find that they have no hope of getting equal time with the people who purvey mythologies. One of these is Stephan Kaplan, who I think—but I'm never sure—is a notoriety freak who is putting us on and having a wonderful time doing it. For example, he was quoted recently as saying that vampires can come out in the daytime, they just need to wear a sunblock of 15 or higher. As wit, this ranks among the best things I've heard recently, right up there with the story that the Florida citrus industry is trying to get O. J. Simpson to change his first name to Snapple. I suspect that Kaplan will one day call a press conference, wearing a silly hat, and say, "I was just fooling, and you fell for it!" I got a call from the BBC a while back asking me for my reaction to Kaplan's announcement that Los Angeles is awash in vampires. To me this is like an adult asking me what Santa Claus brought me this year: The question had better be ironic, and the answer may

This article first appeared in *Skeptical Inquirer* 20, no. 2 (March/April 1996): 41–44.

as well be. So I told the interviewer that it was true that vampires are everywhere in Los Angeles, but because of the muggers they're afraid to go out at night.

The folklore of the vampire has only a slight connection with the fiction, much the way the folklore of ghosts has little to do with the movie *Ghostbusters*. Most people aren't aware that, throughout European history, there have been extensive and detailed accounts of bodies in graveyards being dug up, declared to be vampires, and killed. I took some years out of my life to study these accounts and find out what in the world could have caused people to set out to kill dead bodies. And here we encounter our first real/non-real boundary: the digging up of the bodies was unquestionably real—indeed, beyond any doubt. We know this because we have a vast array of evidence to that effect, both archaeological and documentary, including highly detailed accounts written by literate outsiders, who gave information that they could not possibly have made up. For example, unless you are a forensic pathologist, you probably don't know that decomposing bodies may undergo a process called "skin slippage," in which the epidermis flakes away from the dermis. The following account, from the eighteenth century, tells of the exhumation of a man named Peter Plogojowitz and remarks on this phenomenon: "The hair and beard—even the nails, of which the old ones had fallen away—had grown on [the corpse]; the old skin, which was somewhat whitish, had peeled away, and a new fresh one had emerged under it. . . . Not without astonishment, I saw some fresh blood in his mouth, which, according to the common observation, he had sucked from the people killed by him." When we see remarks about skin slippage, we know that the author has either (a) read a text on forensic pathology or (b) looked at, or heard about, a decomposing corpse.

Yet here we are confronted with a predicament: If our source is right about skin slippage, what are we to make of his evidence that the dead body had been drinking blood from the living? The answer, of course, is that we are not obliged to believe our informant's interpretations, let alone those of *his* informants, just because he is giving us an accurate description of a corpse. Scholars have always thrown out the observations because they didn't believe the interpretations. This is not as odd as it might seem, for often description and interpretation are run together, as in such a statement as "the body came to life and cried out when it was staked." But we'll get to that in a moment.

For now, let's slow down and look carefully at the observations in the account we have quoted:

1. "The hair and beard have grown on the corpse." Sorry, this just doesn't happen, even though many people believe it even today. It can *appear* to happen, however, because the skin may shrink back after death and make hair and beard more visible.

2. "The nails have fallen off and new ones have grown." The nails do in fact fall off as a body decomposes. The Egyptians were aware of this and dealt with it either by tying the nails to the fingers and toes or by putting metal thimbles over the tip of each finger or toe. The "new nails," according to Thomas Noguchi, former medical examiner for Los Angeles, were probably an interpretation of the nail bed.

3. "The old skin has peeled away and new skin has emerged under it." This is skin slippage: epidermis and dermis. Many accounts remark also on the "ruddy" or "dark" color of the corpse, a phenomenon that may be caused by decomposition and a variety of other things as well. Contrary to popular belief, the face of a corpse is not necessarily pale at all, since pallor results from the blood draining from the tissues. If the person was supine when he or she died, the face of the corpse may be pale; if prone, the face may be dark. Those parts of the corpse that are lower than the rest may be gorged with blood that, having lost its oxygen, is dark and causes the skin to appear dark as well. And the parts that are under pressure—where the weight of the body is distributed—may be light in color because the (now dark) blood has been forced away from the tissues. The dark coloration resulting from the saturation of the tissues with blood is called "livor mortis" or "lividity." It is this phenomenon that allows medical examiners to determine whether a body has been moved after death: If lividity is present where it shouldn't be, or not present where it should, then the body has been moved.

4. "There is fresh blood at the mouth." The adjective "fresh" is less puzzling if we suppose that the author hasn't actually tested the blood for freshness. What he was surely observing, and confused by, was the fact that the blood was *liquid*. This was remarked on many times by people who observed such exhumations. It is simply not unusual. In fact, blood normally coagulates at death, then either remains coagulated or becomes liquid again.[1] The reason the blood migrates to the mouth is that the body, as it decomposes, bloats from the gases produced by decomposition, and this bloating puts pressure on the lungs, which are rich in blood and deteriorate early on, so that blood is forced to the mouth and nose.

And did you notice that we were just told why people believed that the dead sucked blood from the living? The standard theory about death was that it came from the dead, and when people dug up the first victim of an epidemic and found that he had blood at his mouth, they concluded that he had sucked the blood from the other people who had died. "Not without astonishment," says our author, "I saw some fresh blood in his mouth, which, according to the common observation, he had sucked from the people killed by him." Moreover, the bloating of the body was taken for evidence that it was full to bursting with the blood of its victims.

So we have cleared up an old mystery merely by paying attention to the people who, centuries ago, tried to tell us about it. From here on things will be easier: If our informants tell us that the vampire "came to life and cried out" when they drove a stake through him, we shall accept the observation and reject the conclusion: Yes, a body would "cry out" if you drove a stake into it, because doing so forces air past the glottis—but this is not because the body is still alive. Among modern medical examiners, there is remarkable agreement on both points.

The vampire lore did not die when people worked out forensic pathology: by that time it had become part of literature. The folkloric vampires had been peasants, but in the eighteenth century, authors were still reluctant to make peasants into major characters in stories, so the fictional vampire was moved into the upper classes. By the time of Bram Stoker's *Dracula* (1897), he had became a pallid count, rather than the ruddy peasant of the folklore. Along the way, Linnaeus named a

Central American bat after the European vampire, since the bat lived on blood, and the fiction writers, noting this, added the bat to the store of their motifs. This is why, in modern movies, vampires are apt to turn into bats in the night, when they need to go somewhere quickly.

Oddly, when this material became fiction, it once again became "fact," for nowadays the media keep digging up not just scholars and pseudoscholars who talk about the folklore but also people who actually claim to *be* vampires. The scholars and the vampires are brought together by their common fate: The media trot them out every year around Halloween. The modern "vampires" derive their inspiration not from the perfectly good material from folklore, which in fact has been sadly neglected, but from the fiction, perhaps because it is more dramatic and coherent. The folklore is about cantankerous peasants who come back as spirits to torment their nearest and dearest, and this simply doesn't translate into a glamorous lifestyle. So our modern "vampires" drive hearses, cap their canine teeth, and wear cloaks when they go out at night. None of these things has anything whatever to do with the folklore of the vampire—even the canines are an artifact of the fictional tradition. Some modern "vampires" claim a taste for blood and tell stories of raids on bloodbanks and of obliging friends who let them open a vein.

The baffling part of this is that the modern "vampires" are claiming kinship not with the vampire that our ancestors actually believed in but with the *fictional* vampire derived from that one. This is like somebody claiming to be related to Rhett Butler in the movie *Gone with the Wind*. "You mean Clark Gable," you say. "No, no: Rhett Butler. You know, the character in the movie. He's my cousin." And, lacking anything further to say, you ask, "Do you and Rhett talk a lot?" But in its way, theirs is a successful lifestyle, for those of us who study the folklore have long since become accustomed to getting two minutes on television programs that then give ten minutes to a ditsy lady who sleeps in a coffin. And anyone can get media attention who will bring up Vlad Drakul or even the moribund porphyria theory, which supposes that people really *were* drinking blood to cure their rare disease, even though we have no evidence either that drinking blood would alleviate the symptoms of porphyria or that any live people were accused of drinking blood—it was always corpses. This theory never got beyond the wild hypothesis stage but has historical interest for following the trend that confuses folklore with fiction. I describe it as "moribund," but such theories seemingly never die in the media, no matter how often they are demolished by evidence and argument. By now you couldn't kill the porphyria theory with a stake.

The peculiarities of this subject have a way of compounding themselves with time. We have seen how confusing it is to have data in which accurate observation and inaccurate interpretation are all balled up together. As the discipline of anthropology formed and took shape, it looked back on its earlier indiscretions and made a firm resolution not to view other cultures as inferior to that of the anthropologist. Indeed, it took us many decades to figure out that "primitive" cultures aren't any younger than "advanced" ones. But their attempt at dispassion discouraged anthropologists from making distinctions: Now you're not supposed to notice when someone from another culture is simply wrong about something.

Indeed, it's no longer politically correct to make distinctions at all between right and wrong ideas, unless of course they are the ideas of our own culture. So it doesn't bother us to say that Copernicus corrected Ptolemy, but it does bother us if I point out that nonliterate cultures typically misunderstand the events of decomposition. What is odd about our modern view is that it appears to be the very kind of patronizing that we are trying to get rid of.

One review of my book complained about my applying scientific discourse to my subject. The reviewer did not suggest an alternative mode of interpretation—intuition, perhaps? But the reason I studied this particular aspect of the folklore is that it is replete with evidence, and evidence lends itself to analysis better than hunches or intuition. One objective of the serious scholar, it seems to me, is to find likely subjects, ones where there is enough evidence to base an argument on. I have had several fruitless discussions with television directors who wanted me to tell them not just more about the vampire lore than I know, but more than can even *be* known. "What about the really early stuff?" one woman kept asking. "What about the Paleolithic?"

But we simply don't have any clear evidence from the Paleolithic. The literary evidence, going from present to past, continues to change subtly until finally you would be hard put to identify the "vampire" phenomenon at all. Early Greek views of the dead have much in common with the later vampire lore, but no one would identify Patroclus as a "vampire" simply because he appears to Achilles after his own death. And the early archaeological evidence is often ambiguous: People may put slabs of stone over graves either to keep the dead from returning or to keep animals from digging into a grave.

The fact is, no one leaves documents around explaining the things that everyone knows. It is only much later that it occurs to anyone to wonder about those things—when it is too late, and they are no longer known. So we will almost surely never know anything about the origins of the vampire lore. The most we can know is that by the eighteenth century the vampire was a certifiably dead body that was believed to retain a kind of life and had to be "killed" in order to prevent it from killing other people. And, of course, we now know that the misconceptions about the folklore have proved to be more viable than the folklore itself.

NOTE

1. There are other correlations here that I've dealt with in detail in a book: *Vampires, Burial, and Death: Folklore and Reality.* New Haven: Yale University Press, 1988.

PART NINE

THE MALLEABILITY OF MEMORY

44

REMEMBERING DANGEROUSLY
Elizabeth F. Loftus

We live in a strange and precarious time that resembles at its heart the hysteria and superstitious fervor of the witch trials of the sixteenth and seventeenth centuries. Men and women are being accused, tried, and convicted with no proof or evidence of guilt other than the word of the accuser. Even when the accusations involve numerous perpetrators, inflicting grievous wounds over many years, even decades, the accuser's pointing finger of blame is enough to make believers of judges and juries. Individuals are being imprisoned on the "evidence" provided by memories that come back in dreams and flashbacks—memories that did not exist until a person wandered into therapy and was asked point-blank, "Were you ever sexually abused as a child?" And then begins the process of excavating the "repressed" memories through invasive therapeutic techniques, such as age regression, guided visualization, trance writing, dream work, body work, and hypnosis.

One case that seems to fit the mold led to highly bizarre satanic-abuse memories. An account of the case is described in detail by one of the expert witnesses (Rogers 1992) and is briefly reviewed by Loftus and Ketcham (1994).

A woman in her mid-seventies and her recently deceased husband were accused by their two adult daughters of rape, sodomy, forced oral sex, torture by electric shock, and the ritualistic murder of babies. The older daughter, 48 years old at the time of the lawsuit, testified that she was abused from infancy until age 25. The younger daughter alleged abuse from infancy to age 15. A granddaughter also claimed that she was abused by her grandmother from infancy to age 8.

The memories were recovered when the adult daughters went into therapy in 1987 and 1988. After the breakup of her third marriage, the older daughter started psychotherapy, eventually diagnosing herself as a victim of multiple-personality disorder and satanic ritual abuse. She convinced her sister and her niece to begin therapy and joined in their therapy sessions for the first year. The two sisters also

This article first appeared in *Skeptical Inquirer* 19, no. 2 (March/April 1995): 20–29.

attended group therapy with other multiple-personality-disorder patients who claimed to be victims of satanic ritual abuse.

In therapy the older sister recalled a horrifying incident that occurred when she was four or five years old. Her mother caught a rabbit, chopped off one of its ears, smeared the blood over her body, and then handed the knife to her, expecting her to kill the animal. When she refused, her mother poured scalding water over her arms. When she was 13 and her sister was still in diapers, a group of Satanists demanded that the sisters disembowel a dog with a knife. She remembered being forced to watch as a man who threatened to divulge the secrets of the cult was burned with a torch. Other members of the cult were subjected to electric shocks in rituals that took place in a cave. The cult even made her murder her own newborn baby. When asked for more details about these horrific events, she testified in court that her memory was impaired because she was frequently drugged by the cult members.

The younger sister remembered being molested on a piano bench by her father while his friends watched. She recalled being impregnated by members of the cult at ages 14 and 16, and both pregnancies were ritually aborted. She remembered one incident in the library where she had to eat a jar of pus and another jar of scabs. Her daughter remembered seeing her grandmother in a black robe carrying a candle and being drugged on two occasions and forced to ride in a limousine with several prostitutes.

The jury found the accused woman guilty of neglect. It did not find any intent to harm and thus refused to award monetary damages. Attempts to appeal the decision have failed.

Are the women's memories authentic? The "infancy" memories are almost certainly false memories given the scientific literature on childhood amnesia. Moreover, no evidence in the form of bones or dead bodies was ever produced that might have corroborated the human-sacrifice memories. If these memories are indeed false, as they appear to be, where would they come from? George Ganaway, a clinical assistant professor of psychiatry at the Emory University School of Medicine, has proposed that unwitting suggestions from therapy play an important role in the development of false satanic memories.

WHAT GOES ON IN THERAPY?

Since therapy is done in private, it is not particularly easy to find out what really goes on behind that closed door. But there are clues that can be derived from various sources. Therapists' accounts, patients' accounts, and sworn statements from litigation have revealed that highly suggestive techniques go on in some therapists' offices (Lindsay and Read 1994; Loftus 1993; Yapko 1994).

Other evidence of misguided if not reckless beliefs and practices comes from several cases in which private investigators, posing as patients, have gone undercover into therapists' offices. In one case, the pseudopatient visited the therapist complaining about nightmares and trouble sleeping. On the third visit to the therapist, the investigator was told that she was an incest survivor (Loftus 1993). In another case, Cable News Network (CNN 1993) sent an employee undercover to the offices of an Ohio psychotherapist (who was supervised by a psychologist) wired with a hidden video camera. The pseudopatient complained of feeling depressed and having recent relationship problems with her husband. In the first session, the therapist diagnosed "incest survivor," telling the pseudopatient she was a "classic case." When the pseudopatient returned for her second session, puzzled about her lack of memory, the therapist told her that her reaction was typical and that she had repressed the memory because the trauma was so awful. A third case, based on surreptitious recordings of a therapist from the Southwestern region of the United States, was inspired by the previous efforts.

INSIDE A SOUTHWESTERN THERAPIST'S OFFICE

In the summer of 1993, a woman (call her "Willa") had a serious problem. Her older sister, a struggling artist, had a dream that she reported to her therapist. The dream got interpreted as evidence of a history of sexual abuse. Ultimately the sister confronted the parents in a videotaped session at the therapist's office. The parents were mortified; the family was wrenched irreparably apart.

Willa tried desperately to find out more about the sister's therapy. On her own initiative, Willa hired a private investigator to pose as a patient and seek therapy from the sister's therapist. The private investigator called herself Ruth. She twice visited the therapist, an M.A. in counseling and guidance who was supervised by a Ph.D., and secretly tape-recorded both of the sessions.

In the first session, Ruth told the therapist that she had been rear-ended in an auto accident a few months earlier and was having trouble getting over it. Ruth said that she would just sit for weeks and cry for no apparent reason. The therapist seemed totally disinterested in getting any history regarding the accident, but instead wanted to talk about Ruth's childhood. While discussing her early life, Ruth volunteered a recurring dream that she had had in childhood and said the dream had now returned. In the dream she is 4 or 5 years old and there is a massive white bull after her that catches her and gores her somewhere in the upper thigh area, leaving her covered with blood.

The therapist decided that the stress and sadness that Ruth was currently experiencing was tied to her childhood, since she'd had the same dream as a child. She decided the "night terrors" (as she called them) were evidence that Ruth was suffering from post-traumatic-stress disorder (PTSD). They would use guided imagery to find the source of the childhood trauma. Before actually launching this approach, the therapist informed her patient that she, the therapist, was an incest survivor: "I was incested by my grandfather."

During the guided imagery, Ruth was asked to imagine herself as a little child. She then talked about the trauma of her parents' divorce and of her father's remarriage to a younger woman who resembled Ruth herself. The therapist wanted to know if Ruth's father had had affairs, and she told Ruth that hers had, and that this was a "generational" thing that came from the grandfathers. The therapist led Ruth through confusing/suggestive/manipulative imagery involving a man holding down a little girl somewhere in a bedroom. The therapist decided that Ruth was suffering from a "major grief issue" and told her it was sexual: "I don't think, with the imagery and his marrying someone who looks like you, that it could be anything else."

The second session, two days later, began:

> Pseudopatient: You think I am quite possibly a victim of sexual abuse?
> Therapist: Um-huh. Quite possibly. It's how I would put it. You know, you don't have the real definitive data that says that, but, um, the first thing that made me think about that was the blood on your thighs. You know, I just wonder, like where would that come from in a child's reality. And, um, the fact that in the imagery the child took you or the child showed you the bedroom and your father holding you down in the bedroom . . . it would be really hard for me to think otherwise. . . . Something would have to come up in your work to really prove that it really wasn't about sexual abuse."

Ruth said she had no memory of such abuse but that didn't dissuade the therapist for a minute.

> Pseudopatient: . . . I can remember a lot of anger and fear associated with him, but I can't recall physical sexual abuse. Do people always remember?
> Therapist: No. . . . Hardly ever. . . .
> It happened to you a long time ago and your body holds on to the memory and that's why being in something like a car accident might trigger memories. . . .

The therapist shared her own experiences of abuse, now by her father, which supposedly led to anorexia, bulimia, overspending, excessive drinking, and other destructive behaviors from which the therapist had presumably now recovered. For long sections of the tape it was hard to tell who was the patient and who was the therapist.

Later the therapist offered these bits of wisdom:

> I don't know how many people I think are really in psychiatric hospitals who are really just incest survivors or, um, have repressed memories. It will be a grief issue that your father was—sexualized you—and was not an appropriate father.
> You need to take that image of yourself as an infant, with the hand over, somebody's trying to stifle your crying, and feeling pain somewhere as a memory.

The therapist encouraged Ruth to read two books: *The Courage To Heal*, which she called the "bible of healing from childhood sexual abuse," and the workbook that goes with it. She made a special point of talking about the section on confrontation with the perpetrator. Confrontation, she said, wasn't necessarily for everyone. Some don't want to do it if it will jeopardize their inheritance, in which case, the therapist said, you can do it after the person is dead—you can do eulogies. But confrontation is empowering, she told Ruth.

Then to Ruth's surprise, the therapist described the recent confrontation she had done with Willa's sister (providing sufficient detail about the unnamed patient that there could be little doubt about who it was).

> Therapist: I just worked with someone who did do it with her parents. Called both of her parents in and we did it in here. . . . Its empowering because you're stepping out on your own. She said she felt like she was 21, and going out on her own for the first time, you know, that's what she felt like. . . .
>
> Pseudopatient: And, did her parents deny or—
>
> Therapist: Oh, they certainly did—
>
> Pseudopatient: Did she remember, that she—she wasn't groping like me?
>
> Therapist: She groped a lot in the beginning. But it sort of, you know, just like pieces of a puzzle, you know, you start to get them and then eventually you can make a picture with it. And she was able to do that. And memory is a funny thing. It's not always really accurate in terms of ages, and times and places and that kind of thing. Like you may have any variable superimposed on another. Like I have a friend who had an ongoing sexual abuse and she would have a memory of, say, being on this couch when she was seven and being abused there, but they didn't have that couch when she was seven, they had it when she was five. . . . It doesn't discount the memory, it just means that it probably went on more than once and so those memories overlap. . . .
>
> Pseudopatient: This woman who did the confrontation, is she free now? Does she feel freed over it?
>
> Therapist: Well, she doesn't feel free from her history . . . but she does feel like she owns it now and it doesn't own her . . . and she has gotten another memory since the confrontation. . . .

The therapist told Ruth all about the "new memory" of her other patient, Willa's sister:

> Therapist: [It was in] the early-morning hours and she was just lying awake, and she started just having this feeling of, it was like her hands became uncontrollable and it was like she was masturbating someone. She was like going faster than she could have, even in real life, so that she knew, it was familiar enough to her as it will be to you, that she knew what it was, and it really did not freak her out at all. . . . She knew there was a memory there she was sitting on.

Before Ruth's second therapy session had ended, Ruth's mother was brought into the picture—guilty, at least, of betrayal by neglect:

> Therapist: Well, you don't have to have rational reasons, either, to feel betrayed. The only thing that a child needs to feel is that there was probably a part of you that was just yearning for your mother and that she wasn't there. And whether she wasn't there because she didn't know and was off doing something else, or whether she was there and she knew and she didn't do anything about it. It doesn't matter. All the child knew was that Mom wasn't there. And, in that way she was betrayed, you know, whether it was through imperfection on your mother's part or not, and you have to give yourself permission to feel that way without justification, or without rationalization because you were.

Ruth tried again to broach the subject of imagination versus memory:

Pseudopatient: How do we know, when the memories come, what are symbols, that it's not our imagination or something?

Therapist: Why would you image this, of all things. If it were your imagination, you'd be imaging how warm and loving he was.

. . . I have a therapist friend who says that the only proof she needs to know that something happened is if you think it might have.

At the doorway as Ruth was leaving, her therapist asked if she could hug her, then did so while telling Ruth how brave she was. A few weeks later, Ruth got a bill. She was charged $65 for each session.

Rabinowitz (1993) put it well: "The beauty of the repressed incest explanation is that, to enjoy its victim benefits, and the distinction of being associated with a survivor group, it isn't even necessary to have any recollection that such abuse took place." Actually, being a victim of abuse without any memories does not sit well, especially when group therapy comes into play and women without memories interact with those who do have memories. The pressure to find memories can be very great.

Chu (1992, 7) pointed out one of the dangers of pursuing a fruitless search (for memories): it masks the real issues from therapeutic exploration. Sometimes patients produce "ever more grotesque and increasingly unbelievable stories in an effort to discredit the material and break the cycle. Unfortunately, some therapists can't take the hint!"

The Southwestern therapist who treated Ruth diagnosed sexual trauma in the first session. She pursued her sex-abuse agenda in the questions she asked, in the answers she interpreted, in the way she discussed dreams, in the books she recommended. An important question remains as to how common these activities might be. Some clinicians would like to believe that the problem of overzealous psychotherapists is on a "very small" scale (Cronin 1994, 31). A recent survey of doctoral-level psychologists indicates that as many as a quarter may harbor beliefs and engage in practices that are questionable (Poole and Lindsay 1994). That these kinds of activities can and do sometimes lead to false memories seems now to be beyond dispute (Goldstein and Farmer 1993). That these kinds of activities can create false victims, as well as hurt true ones, also seems now to be beyond dispute.

THE PLACE OF REPRESSED MEMORIES IN MODERN SOCIETY

Why at this time in our society is there such an interest in "repression" and the uncovering of repressed memories? Why is it that almost everyone you talk to either knows someone with a "repressed memory" or knows someone who's being accused, or is just plain interested in the issue? Why do so many individuals believe these stories, even the more bizarre, outlandish, and outrageous ones? Why is the cry of "witch hunt" now so loud (Baker 1992, 48; Gardner 1991)? *Witch hunt* is, of course, a term that gets used by lots of people who have been faced by a pack of accusers (Watson 1992).

"Witch hunt" stems from an analogy between the current allegations and the witch-craze of the sixteenth and seventeenth centuries, an analogy that several analysts have drawn (McHugh 1992; Trott 1991; Victor 1991). As the preeminent British historian Hugh Trevor-Roper (1967) has noted, the European witch-craze was a perplexing phenomenon. By some estimates, a half-million people were convicted of witchcraft and burned to death in Europe alone between the fifteenth and seventeenth centuries (Harris 1974, 207–58). How did this happen?

It is a dazzling experience to step back in time, as Trevor-Roper guides his readers, first to the eighth century, when the belief in witches was thought to be "unchristian" and in some places the death penalty was decreed for anyone who burnt supposed witches. In the ninth century, almost no one believed that witches could make bad weather, and almost everyone believed that night-flying was a hallucination. But by the beginning of the sixteenth century, there was a complete reversal of these views. "The monks of the late Middle Ages sowed: the lawyers of the sixteenth century reaped; and what a harvest of witches they gathered in!" (Trevor-Roper 1967, 93). Countries that had never known witches were now found to be swarming with them. Thousands of old women (and some young ones) began confessing to being witches who had made secret pacts with the Devil. At night, they said, they anointed themselves with "devil's grease" (made from the fat of murdered infants), and thus lubricated they slipped up chimneys, mounted broomsticks, and flew off on long journeys to a rendezvous called the witches' sabbat. Once they reached the sabbat, they saw their friends and neighbors all worshipping the Devil himself. The Devil sometimes appeared as a big, black, bearded man, sometimes as a stinking goat, and sometimes as a great toad. However he looked, the witches threw themselves into promiscuous sexual orgies with him. While the story might vary from witch to witch, at the core was the Devil, and the witches were thought to be his earth agents in the struggle for control of the spiritual world.

Throughout the sixteenth century, people believed in the general theory, even if they did not accept all of the esoteric details. For two centuries, the clergy preached against the witches. Lawyers sentenced them. Books and sermons warned of their danger. Torture was used to extract confessions. The agents of Satan were soon found to be everywhere. Skeptics, whether in universities, in judges' seats, or on the royal throne, were denounced as witches themselves, and joined the old women at the burning stake. In the absence of physical evidence (such as a pot full of human limbs, or a written pact with the Devil), circumstantial evidence was sufficient. Such evidence did not need to be very cogent (a wart, an insensitive spot that did not bleed when pricked, a capacity to float when thrown in water, an incapacity to shed tears, a tendency to look down when accused). Any of these "indicia" might justify the use of torture to produce a confession (which was proof) or the refusal to confess (which was also proof) and justified even more ferocious tortures and ultimately death.

When did it end? In the middle of the seventeenth century the basis of the craze began to dissolve. As Trevor-Roper (1967, 97) put it, "The rubbish of the human mind, which for two centuries, by some process of intellectual alchemy and social pressure, had become fused together in a coherent, explosive system, has disintegrated. It is rubbish again."

Various interpretations of this period in social history can be found. Trevor-Roper argued that during periods of intolerance any society looks for scapegoats. For the Catholic church of that period, and in particular their most active members, the Dominicans, the witches were perfect as scapegoats; and so, with relentless propaganda, they created a hatred of witches. The first individuals to be so labeled were the innocently nonconforming social groups. Sometimes they were induced to confess by torture too terrible to bear (e.g., the "leg screw" squeezed the calf and broke the shinbone in pieces; the "lift" hoisted the arms fiercely behind and back; the "ram" or "witch-chair" provided a heated seat of spikes for the witch to sit on). But sometimes confessions came about spontaneously, making their truth even more convincing to others. Gradually laws changed to meet the growth of witches—including laws permitting judicial torture.

There were skeptics, but many of them did not survive. Generally they tried to question the plausibility of the confessions, or the efficacy of torture, or the identification of particular witches. They had little impact, Trevor-Roper claims, because they danced around the edges rather than tackling the core: the concept of Satan. With the mythology intact, it creates its own evidence that is very difficult to disprove. So how did the mythology that had lasted for two centuries lose its force? Finally, challenges against the whole idea of Satan's kingdom were launched. The stereotype of the witch would soon be gone, but not before tens of thousands of witches had been burned or hanged, or both (Watson 1992).

Trevor-Roper saw the witch-craze as a social movement, but with individual extensions. Witch accusations could be used to destroy powerful enemies or dangerous persons. When a "great fear" grips a society, that society looks to the stereotype of the enemy in its midst and points the finger of accusation. In times of panic, he argued, the persecution extends from the weak (the old women who were ordinarily the victims of village hatred) to the strong (the educated judges and clergy who resisted the craze). One indicia of "great fear" is when the elite of society are accused of being in league with the enemies.

Is it fair to compare the modern cases of "de-repressed memory" of child sexual trauma to the witch-crazes of several centuries ago? There are some parallels, but the differences are just as striking. In terms of similarities, some of the modern stories actually resemble the stories of earlier times (e.g., witches flying into bedrooms). Sometimes the stories encompass past-life memories (Stevenson 1994) or take on an even more bizarre, alien twist (Mack 1994).[1] In terms of differences, take a look at the accused and the accusers. In the most infamous witch hunt in North America, 300 years ago in Salem, Massachusetts, three-fourths of the accused were women (Watson 1992). Today, they are predominantly (but not all) men. Witches in New England were mostly poor women over 40 who were misfits, although later the set of witches included men (often the witches' husbands or sons), and still later the set expanded to include clergy, prominent merchants, or anyone who had dared to make an enemy. Today, the accused are often men of power and success. The witch accusations of past times were more often leveled by men, but today the accusations are predominantly leveled by women. Today's phenomenon is more than anything a movement of the weak against the strong. There is today a "great fear" that grips our society, and that is fear of child abuse. Rightfully we wish to ferret out these genuine "enemies" and point every finger of accusation at them. But this does not mean, of course, that every perceived enemy, every person with whom we may have feuded, should be labeled in this same way.

Trevor-Roper persuasively argued that the skeptics during the witch-craze did not make much of a dent in the frequency of bonfires and burnings until they challenged the core belief in Satan. What is the analogy to that core today? It may be some of the widely cherished beliefs of psychotherapists, such as the belief in the repressed-memory folklore. The repression theory is well articulated by Steele (1994, 41). It is the theory "that we forget events because they are too horrible to contemplate; that we cannot remember these forgotten events by any normal process of casting our minds back but can reliably retrieve them by special techniques; that these forgotten events, banished from consciousness, strive to enter it in disguised forms; that forgotten events have the power to cause apparently unrelated problems in our lives, which can be cured by excavating and reliving the forgotten event.

Is it time to admit that the repression folklore is simply a fairy tale? The tale may be appealing, but what of its relationship to science? Unfortunately, it is partly refuted, partly untested, and partly untestable. This is not to say that all recovered memories are thus false. Responsible skepticism is skepticism about some claims of recovered memory. It is not blanket rejection of all claims. People sometimes remember what was once forgotten; such forgetting and remembering does not mean repression and de-repression, but it does mean that some recently remembered events might reflect authentic memories. Each case must be examined on its merits to explore the credibility, the timing, the motives, the potential for suggestion, the corroboration, and other features to make an intelligent assessment of what any mental product means.

THE CASE OF JENNIFER H.

Some writers have offered individual cases as proof that a stream of traumas can be massively repressed. Readers must beware that these case "proofs" may leave out critical information. Consider the supposedly ironclad case of Jennifer H. offered by Kandel and Kandel (1994) to readers of *Discover* magazine as an example of a corroborated de-repressed memory. According to the *Discover* account, Jennifer was a 23-year-old musician who recovered memories in therapy of her father raping her from the time she was 4 until she was 17. As her memories resurfaced, her panic attacks and other symptoms receded. Her father, a mechanical-engineering professor, denied any abuse. According to the *Discover* account, Jennifer sued her father, and at trial "corroboration" was produced: Jennifer's mother testified that she had seen the father lying on top of Jennifer's 14-year-old sister and that he had once fondled a baby-sitter in her early teens. The defendant's sister recalled his making passes at young girls. Before this case becomes urban legend and is used as proof of something that it might not be proof of, readers are entitled to know more.

Jennifer's case against her father went to trial in June 1993 in the U.S. District Court for the District of Massachusetts (*Hoult* v. *Hoult,* 1993). The case received considerable media attention (e.g., Kessler 1993). From the trial transcript, we learn that Jennifer, the oldest of four children, began therapy in the fall of 1984 with an unlicensed New York psychotherapist for problems with her boyfriend and divided loyalties surrounding her parents' divorce. Over the next year or so she experienced recurring nightmares with violent themes, and waking terrors. Her therapist practiced a "Gestalt" method of therapy; Jennifer describes one session: "I started the same thing of shutting my eyes and just trying to feel the feelings and not let them go away really fast. And [my therapist] just said 'Can you see anything?' . . . I couldn't see anything . . . and then all of a sudden I saw this carved bedpost from my room when I was a child. . . . And then I saw my father, and I could feel him sitting on the bed next to me, and he was pushing me down, and I was saying, 'No.' And he started pushing up my nightgown and . . . was touching me with his hands on my breast, and then between my legs, and then he was touching me with his mouth . . . and then it just all like went away. It was like . . . on TV if there is all static. . . . It was, all of a sudden it was plusssssh, all stopped. And then I slowly opened my eyes in the session and I said, 'I never knew that happened to me'" (pp. 58–59).

Later Jennifer would have flashbacks that were so vivid that she could feel the lumpy blankets in her childhood bed. She remembered her father choking her and raping her in her parents' bedroom when she was about 12 or 13 (p. 91). She remembered her father threatening to rape her with a fishing pole in the den when she was about 6 or 7. She remembers her father raping her in the basement when she was in high school. The rape stopped just as her mother called down for them to come to dinner. She remembered her father raping her at her grandparents' home when she was in high school, while the large family were cooking and kids were playing. She remembered her father threatening to cut her with a letter opener, holding a kitchen knife to her throat (p. 113). She remembered him chasing her through the house with knives, trying to kill her, when she was about 13 years old (p. 283).

Jennifer also remembered a couple of incidents involving her mother. She remembered one time when she was raped in the bathroom and went to her mother wrapped in a towel with blood dripping. She remembered another incident, in which her father was raping her in her parents' bedroom and her mother came to the door and said, "David." The father then stopped raping her and went out to talk to the mother. Jennifer's mother said she had no recollection of these events, or of any sexual abuse. An expert witness testifying for Jennifer said it is common in cases of incest that mothers ignore the signs of abuse.

During the course of her memory development, Jennifer joined numerous sexual-abuse survivor groups. She read books about sexual abuse. She wrote columns.

She contacted legislators. Jennifer was involved in years of therapy. She wrote letters about her abuse. In one letter, written to the President of Barnard College on February 7, 1987, she said "I am a victim of incestuous abuse by my father and physical abuse by my mother" (p. 175). In another letter to her friend Jane, written in January 1988, she talked about her therapy: "Well, my memories came out . . . when I would sit and focus on my feelings which I believe I call visualization exercises because I would try to visualize what I was feeling or be able to bring into my eyes what I could see" (pp. 247–48). She told Jane about her Gestalt therapy: "In Gestalt therapy, the sub-personalities are allowed to take over and converse with one another and hopefully resolve their conflicts. Each personality gets a different chair, and when one new one starts to speak, the individual changes into that personality's seat. It sounds weird, and it is. But is is also an amazing journey into one's self. I've come to recognize untold universes within myself. It feels often very much like a cosmic battle when they are all warring with one another" (pp. 287–88; see also page 249).

In one letter, written on January 1 1, 1989, to another rape survivor, she said that her father had raped her approximately 3,000 times. In another letter, dated January 30, 1989, she wrote: "Underneath all the tinsel and glitter was my father raping me every two days. My mother smiling and pretending not to know what the hell was going on, and probably Dad abusing my siblings as well" (pp. 244–45). In a letter written on April 24, 1989, to *Mother Jones* magazine she said that she had survived hundreds of rapes by her father (p. 231).

Before October 1985, Jennifer testified, she didn't "know" that her father had ever put his penis in her vagina, or that he had put his penis in her mouth, or that he put his mouth on her vagina (p. 290). She paid her therapist $19,329.59 (p. 155) to acquire that knowledge.

In sum, Jennifer reported that she had been molested by her father from the ages of 4 to 17 (p. 239); that she was molested hundreds if not thousands of times, even if she could not remember all of the incidents; that this sometimes happened with many family members nearby, and with her mother's "involvement" in some instances; and that she buried these memories until she was 24, at which time they purportedly began to return to her. No one saw.

These are a few of the facts that the Kandels left out of their article. Jennifer was on the stand for nearly three days. She had "experts" to say they believed her memories were real. These experts were apparently unaware of, or unwilling to heed, Yapko's (1994) warnings about the impossibility, without independent corroboration, of distinguishing reality from invention and his urgings that symptoms by themselves cannot establish the existence of past abuse. At trial, Jennifer's father testified for about a half-hour (Kessler 1993b). How long does it take to say, "I didn't do it"? Oddly, his attorneys put on no character witnesses or expert testimony of their own, apparently believing—wrongly—that the implausibility of the "memories" would be enough. A Massachusetts jury awarded Jennifer $500,000.

GOOD AND BAD ADVICE

Many of us would have serious reservations about the kinds of therapy activities engaged in by Jennifer H. and the kind of therapy practiced by the Southwestern therapist who treated pseudopatient Ruth. Even recovered-memory supporters like Briere (1992) might agree. He did, after all, say quite clearly: "Unfortunately, a number of clients and therapists appear driven to expose and confront every possible traumatic memory" (p. 136). Briere notes that extended and intense effort to make a client uncover all traumatic material is not a good idea since this is often to the detriment of other therapeutic tasks, such as support, consolidation, desensitization, and emotional insight.

Some will argue that the vigorous exploration of buried sex-abuse memories is acceptable because it has been going on for a long time. In fact, to think it is fine to do things the way they've always been done is to have a mind that is as closed and dangerous as a malfunctioning parachute. It is time to recognize that the dangers of false-memory creation are endemic to psychotherapy (Lynn and Nash 1994). Campbell (1994) makes reference to Thomas Kuhn as he argues that the existing paradigm (the theories, methods, procedures) of psychotherapy may no longer be viable. When this happens in other professions, a crisis prevails and the profession must undertake a paradigm shift.

It may be time for that paradigm shift and for an exploration of new techniques. At the very least, therapists should not let sexual trauma overshadow all other important events in a patient's life (Campbell 1994). Perhaps there are other explanations for the patient's current symptoms and problems. Good therapists remain open to alternative hypotheses. Andreasen (1988), for example, urges practitioners to be open to the hypothesis of metabolic or neurochemical abnormalities as cause of a wide range of mental disorders. Even pharmacologically sophisticated psychiatrists sometimes refer their patients to neurologists, endocrinologists, and urologists. For less serious mental problems we may find, as physicians did before the advent of powerful antibiotics, that they are like many infections—self-limiting, running their course and then ending on their own (Adler 1994).

When it comes to serious diseases, a question that many people ask of their physicians is "How long have I got?" As Buckman and Sabbagh (1993) have aptly pointed out, this is a difficult question to answer. Patients who get a "statistical" answer often feel angry and frustrated. Yet an uncertain answer is often the truthful answer. When a psychotherapy patient asks, "Why am I depressed?" the therapist who refrains from giving an erroneous answer, however frustrating silence might be, is probably operating closer to the patient's best interests. Likewise, nonconventional "healers" who, relative to conventional physicians, give their patients unwarranted certainty and excess attention, may make the patients temporarily feel better, but in the end may not be helping them at all.

Bad therapy based on bad theory is like a too-heavy oil that, instead of lubricating, can gum up the works—slowing everything down and heating everything up. When the mental works are slowed down and heated up, stray particles of false memory can, unfortunately, get stuck in it.

To avoid mucking up the works, constructive advice has been offered by Byrd (1994) and by Gold, Hughes, and Hohnecker (1994): Focus on enhancement of functioning rather than uncovering buried memories. If it is necessary to recover memories, do not contaminate the process with suggestions. Guard against personal biases. Be cautious about the use of hypnosis in the recovery of memories. Bibliotherapeutic and group therapy should not be encouraged until the patient has reasonable certainty that the sex abuse really happened. Development and evaluation of other behavioral and pharmacological therapies that minimize the possibility of false memories and false diagnoses should be encouraged.

Instead of dwelling on the misery of childhood and digging for childhood sexual trauma as its cause, why not spend some time doing something completely different. Borrowing from John Gottman's (1994) excellent advice on how to make your marriage succeed, patients might be reminded that negative events in their lives do not completely cancel out all the positives (p. 182). Encourage the patient to think about the positive aspects of life—even to look through picture albums from vacations and birthdays. Think of patients as the architects of their thoughts, and guide them to build a few happy rooms. The glass that's half empty is also half full. Gottman recognized the need for some real basis for positive thoughts, but in many families, as in many marriages, the basis does exist. Campbell (1994) offers similar advice. Therapists, he believes, should encourage their clients to recall some positive things about their families. A competent therapist will help others support and assist the client, and help the client direct feelings of gratitude toward those significant others.

FINAL REMARKS

We live in a culture of accusation. When it comes to molestation, the accused is almost always considered guilty as charged. Some claims of sexual abuse are as believable as any other reports based on memory, but others may not be. However, not all claims are true. As Reich (1994) has argued: "When we uncritically embrace reports of recovered memories of sexual abuse, and when we nonchalantly assume that they must be as good as our ordinary memories, we debase the coinage of memory altogether" (p. 38). Uncritical acceptance of every single claim of a recovered memory of sexual abuse, no matter how bizarre, is not good for anyone—not the client, not the family, not the mental-health profession, not the precious human faculty of memory. And let us not forget one final tragic consequence of overenthusiastic embracing of every supposedly de-repressed memory; these activities are sure to trivialize the genuine memories of abuse and increase the suffering of real victims who wish and deserve, more than anything else, just to be believed.

We need to find ways of educating people who presume to know the truth. We particularly need to reach those individuals who, for some reason, feel better after they have led their clients—probably unwittingly—to falsely believe that family members have committed some terrible evil. If "truth" is our goal, then the search for evil must go beyond "feeling good" to include standards of fairness, burdens of proof, and presumptions of innocence. When we loosen our hold on these ideals, we risk a return to those times when good and moral human beings convinced themselves that a belief in the Devil meant proof of his existence. Instead, we should be marshaling all the science we can find to stop the modern-day Reverend Hale (from *The Crucible*), who, if he lived today would still be telling anyone who would listen that he had seen "frightful proofs" that the Devil was alive. He would still be urging that we follow wherever "the accusing finger points"!

NOTE

1. John Mack details the kidnappings of 13 individuals by aliens, some of whom were experimented upon sexually. Mack believes their stories, and has impressed some journalists with his sincerity and depth of concern for the abductors (Neimark 1994). Carl Sagan's (1993, 7) comment on UFO memories: "There is genuine scientific paydirt in UFO's and alien abductions—but it is, I think, of distinctly terrestrial origin."

REFERENCES

Adler, J. 1994. The age before miracles. *Newsweek,* March 28, p. 44.

Andreasen, N. C. 1988. Brain imaging: Applications in psychiatry. *Science* 239: 1381–88.

Baker, R. A. 1992. *Hidden Memories.* Amherst, N.Y.: Prometheus Books.

Briere, John N. 1992. *Child Abuse Trauma.* Newbury Park, Calif.: Sage Publications.

Buckman, R., and K. Sabbagh, 1993. *Magic or Medicine? An Investigation into Healing.* London: Macmillan.

Byrd, K. R. 1994. The narrative reconstructions of incest survivors. *American Psychologist* 49: 439–40.

Campbell, T. W. 1994. *Beware the Talking Cure.* Boca Raton, Fla.: Social Issues Resources Service (SirS).

Chu, J. A. 1992. The critical issues task force report: The role of hypnosis and amytal interviews in the recovery of traumatic memories. *International Society for the Study of Multiple Personality and Dissociation News,* June, pp. 6–9.

CNN. 1993. "Guilt by Memory." Broadcast on May 3.

Cronin, J. 1994. False memory. *Z Magazine,* April, pp. 31–37.

Gardner, R. A. 1991. *Sex Abuse Hysteria.* Creskill, N.J.: Creative Therapeutics.

Gold, Steven N., Dawn Hughes, and Laura Hohnecker. 1994. Degrees of repression of sexual-abuse memories. *American Psychologist* 49: 441–42.

Goldstein, E., and K. Farmer, eds. 1994. *True Stories of False Memories.* Boca Raton, Fla.: Social Issues Resources Service (SirS).

Gottman, J. 1994. *Why Marriages Succeed or Fail.* New York: Simon & Schuster.

Harris, M. 1974. *Cows, Pigs, Wars, and Witches: The Riddles of Culture.* New York: Vintage Books.

Hoult v. Hoult. 1993. Trial testimony. U.S. District Court for District of Massachusetts. Civil Action No 88–1738.

Kandel, M., and E. Kandel. 1994. Flights of Memory. *Discover* 15 (May): 32–37.

Kessler, G. 1993a. Memories of abuse. *Newsday,* November 28, pp. 1, 5, 54–55.

———. 1993b. Personal communication, *Newsday,* letter to EL dated December 13, 1993.

Lindsay, D. S., and J. D. Read. 1994. Psychotherapy and memories of childhood sexual abuse: A cognitive perspective. *Applied Cognitive Psychology* 8: 281–338.

Loftus, E. F. 1993. The reality of repressed memories. *American Psychologist* 48: 518–37.

Loftus, E. F., and K. Ketcham. 1994. *The Myth of Repressed Memory.* New York: St. Martin's Press.

Lynn, S. J., and M. R. Nash. 1994. Truth in memory. *American Journal of Clinical Hypnosis* 36: 194–208.

Mack, J. 1994. *Abduction.* New York: Scribner's.

McHugh, P. R. 1992. Psychiatric misadventures. *American Scholar* 61: 497–510.

Neimark, J. 1994. The Harvard professor and the UFO's. *Psychology Today,* March/April, pp. 44–48, 74–90.

Poole, D., and D. S. Lindsay. 1994. "Psychotherapy and the Recovery of Memories of Childhood Sexual Abuse." Unpublished manuscript, Central Michigan University.

Rabinowitz, Dorothy. 1993. Deception: In the movies, on the news. *Wall Street Journal,* February 22. Review of television show "Not in My Family."

Reich, W. 1994. The monster in the mists. *New York Times Book Review,* May 15, pp. 1, 33–38.

Rogers, M. L. 1992. "A Case of Alleged Satanic Ritualistic Abuse." Paper presented at the American Psychology-Law Society meeting, San Diego, March.

Sagan, C. 1993. What's really going on? *Parade Magazine,* March 7, pp. 4–7.

Stevenson, I. 1994. A case of the psychotherapist's fallacy: Hypnotic regression to "previous lives." *American Journal of Clinical Hypnosis* 36: 188–93.

Steele, D. R. 1994. Partial recall. *Liberty,* March, pp. 37–47.

Trevor-Roper, H. R. 1967. *Religion, the Reformation, and Social Change.* London: Macmillan.

Trott, J. 1991. Satanic panic. *Cornerstone* 20: 9–12.

Victor, J. S. 1991. Satanic cult "survivor" stories. *Skeptical Inquirer* 15: 274–80.

Watson, B. 1992. Salem's dark hour: Did the devil make them do it? *Smithsonian* 23: 117–31.

Yapko, M. 1994. *Suggestions of Abuse.* New York: Simon & Schuster.

45

RMT:
REPRESSED MEMORY THERAPY

Martin Gardner

(I)n March 1992, a group of distinguished psychologists and psychiatrists banded together to form the False Memory Syndrome (FMS) Foundation. The organization is headquartered in Philadelphia under the direction of educator Pamela Freyd. Its purpose: to combat a fast-growing epidemic of dubious therapy that is ripping thousands of families apart, scarring patients for life, and breaking the hearts of innocent parents and other relatives. It is, in fact, the mental-health crisis of the 1990s.

The tragic story begins with Freud. Early in his career, when he made extensive use of hypnotism, Freud was amazed by the number of mesmerized women who dredged up childhood memories of being raped by their fathers. It was years before he became convinced that most of these women were fantasizing. Other analysts and psychiatrists agreed. For more than half a century the extent of incestuous child abuse was minimized. Not until about 1980 did the pendulum start to swing the other way as more solid evidence of child sexual abuse began to surface. There is now no longer any doubt that such incest is much more prevalent than the older Freud or the general public realized.

Then in the late 1980s a bizarre therapeutic fad began to emerge in the United States. Hundreds of poorly trained therapists, calling themselves "traumatists," began to practice the very techniques Freud had discarded. All over the land they are putting patients under hypnosis, or using other techniques, to subtly prod them into recalling childhood sexual traumas, memories of which presumably have been totally obliterated for decades. Decades Delayed Disclosure, or DDD, it has been called. Eighty percent of the patients who are claimed to experience DDD are women from twenty-five to forty-five years old. Sixty percent of their

This article is reprinted from Martin Gardner, *Weird Water and Fuzzy Logic: More Notes of a Fringe Watcher* (Amherst, N.Y.: Prometheus Books, 1996). Copyright © 1996 by Martin Gardner. Reprinted by permission of the author and publisher. It first appeared under the title "The False Memory Syndrome," in *Skeptical Inquirer* 17, no. 4 (Summer 1993): 370–75.

parents are college graduates, 25 percent with advanced degrees. More than 80 percent of their parents are married to their first spouse.

Here is a typical scenario. A woman in her thirties seeks therapy for symptoms ranging from mild depression, anxiety, headaches, or the inability to lose weight, to more severe symptoms like anorexia. Her therapist, having succumbed to the latest mental-health fad, decides almost at once that the symptoms are caused by repressed memories of childhood abuse. Profoundly shocked by this suggestion, the woman vigorously denies that such a thing could be possible. The stronger her denial, the more the therapist believes she is repressing painful memories.

The patient may be hypnotized, or given sodium amytal, or placed into a relaxed, trancelike state. Convinced that a childhood trauma is at the root of the patient's ills, the therapist repeatedly urges the woman to try to remember the trauma. If she is highly suggestible and eager to please the therapist, she begins to respond to leading questions and to less obvious signs of the therapist's expectations.

After months, or even years, images begin to form in the patient's mind. Shadowy figures threaten her sexually. Under continual urging, these memories grow more vivid. She begins to recognize the molester as her father, or grandfather, or uncle. The more detailed the visions, the more convinced both she and the therapist become that the terrible truth is finally being brought to consciousness. To better-trained psychiatrists, these details indicate just the opposite. Childhood memories are notoriously vague. Recalling minute details is a strong sign of fantasizing.

As the false memories become more convincing, the patient's anger toward a once-loved relative grows. The therapist urges her to vent this rage, to confront the perpetrator, even to sue for psychic damage. Stunned by their daughter's accusations, the parents vigorously deny everything. Of course they will deny it, says the therapist, perhaps even suppress their own memories of what happened. The family is devastated. A loving daughter has inexplicably been transformed into a bitter enemy. She may join an "incest survivor" group, where her beliefs are reinforced by hearing similar tales. She may wear a sweatshirt saying, "I survived."

No one doubts that childhood sexual assaults occur, but in almost every case the event is never forgotten. Indeed, it festers as a lifelong source of shame and anger. Studies show that among children who witnessed the murder of a parent, not a single one repressed the terrible memory. Not only do victims of child incest not repress such painful memories (to repress a memory means to completely forget the experience without any conscious effort to do so); they try unsuccessfully to forget them. That traumas experienced as a child can be totally forgotten for decades is the great mental-health myth of our time—a myth that is not only devastating innocent families but doing enormous damage to psychiatry.

In the past, when juries found a parent guilty of child incest, there has been corroborating evidence: photos, diaries, letters, testimony by others, a history of sexual misconduct, or even open admission of guilt. Juries today are increasingly more often judging a parent guilty without any confirming evidence other than the therapy-induced memories of the "victim."

Patients as well as their families can be scarred for life. They are led to believe that bringing suppressed memories to light will banish their symptoms. On the contrary, the symptoms usually get worse because of traumatic breaks with loved ones. Moreover, this treatment can also cause a patient to refuse needed therapy from psychiatrists who have not fallen prey to the FMS epidemic. Pamela Freyd has likened the traumatists to surgeons doing brain surgery with a knife and fork. Others see the epidemic as similar in many ways to the great witch-hunts of the past, when disturbed women were made to believe they were in Satan's grip. The Devil has been replaced by the evil parent.

FMS takes many forms other than parental sexual abuse. Thousands of victims are being induced by traumatists to recall childhood participation in satanic cults that murder babies, eat their flesh, and practice even more revolting rituals. Although there is widespread fascination with the occult, and an amusing upsurge in the number of persons who fancy themselves benevolent witches or warlocks, police have yet to uncover any compelling evidence that satanic cults exist. Yet under hypnosis and soporific drugs, memories of witnessing such rituals can become as vivid as memories of sexual abuse.

Thousands of other patients, highly suggestible while half asleep, are now "remembering" how they were abducted, and sometimes sexually abused, by aliens in spaceships from faraway planets. Every year or so victims of this form of FMS (assuming they are not charlatans) will write persuasive books about their adventures with extraterrestrials. The books will be heavily advertised and promoted on talk shows, and millions of dollars will flow into the pockets of the authors and the books' uncaring publishers. Still another popular form of FMS, sparked by the New Age obsession with reincarnation, is the recovering of memories of past lives.

Pop-psychology books touting the myth that memories of childhood molestations can be suppressed for decades are becoming as plentiful as books about reincarnation, satanic cults, and flying saucers. Far and away the worst offender is a best-seller titled *Courage to Heal* (Harper & Row, 1988), by Ellen Bass and Laura Davis. Although neither author has had any training in psychiatry, the book has become a bible for women convinced they are incest survivors. Davis thinks she herself is a survivor, having recalled under therapy being attacked by her grandfather. A survey of several hundred accused parents revealed that in almost every case their daughters had been strongly influenced by *Courage to Heal.* (See box for some of the book's more outrageous passages.)

From the growing literature of FMS cases I cite a few typical horrors. A twenty-eight-year-old woman accuses her father of molesting her when she was six months old. "I recall my father put his penis near my face and rubbed it on my face and mouth." There is not the slightest evidence that a child of six months can acquire lasting memories of *any* event.

Betsy Petersen, in *Dancing with Daddy* (Bantam, 1991), tells of being convinced by her therapist that she had been raped by her father when she was three. "I don't know if I made it up or not," she told the therapist. "It feels like a story," he replied, "because when something like that happens, everyone acts like it didn't."

In 1986 Patti Barton sued her father for sexually abusing her when she was seven to fifteen months old. She did not remember this until her thirty-second

Passages from *Courage to Heal*, the "Bible" of Incest Survivors

"You may think you don't have memories, but often as you begin to talk about what you do remember, there emerges a constellation of feelings, reactions, and recollections that add up to substantial information. To say, 'I was abused,' you don't need the kind of recall that would stand up in a court of law. Often the knowledge that you were abused starts with a tiny feeling, an intuition.... Assume your feelings are valid.... If you think you were abused and your life shows the symptoms, then you were."

"If you don't remember your abuse you are not alone. Many women don't have memories, and some never get memories. This doesn't mean they weren't abused."

"If you maintained the fantasy that your childhood was 'happy,' then you have to grieve for the childhood you thought you had.... You must give up the idea that your parents had your best interest at heart.... If you have any loving feelings toward your abuser, you must reconcile that love with the fact that he abused you.... You may have to grieve over the fact that you don't have an extended family for your children, that you'll never receive an inheritance, that you don't have family roots."

"If your memories of the abuse are still fuzzy, it is important to realize that you may be grilled for details.... Of course such demands for proof are unreasonable. You are not responsible for proving that you were abused."

"If you're willing to get angry and the anger just doesn't seem to come, there are many ways to get in touch with it. A little like priming the pump, you can do things that will get your anger started. Then once you get the hang of it, it'll begin to flow on its own."

"You may dream of murder or castration. It can be pleasurable to fantasize such scenes in vivid detail.... Let yourself imagine it to your heart's content."

therapy session. She recalls trying to tell her mother what happened by saying, "Ma, ma, ma, ma!" and "Da, da, da, da!"

Geraldo Rivera, in 1991, had three trauma survivors on his television show. One woman insisted she had murdered forty children while she was in a satanic cult but had totally forgotten about it until her memories were aroused in therapy. Well-known entertainers have boosted the FMS epidemic by openly discussing their traumas on similar sensational talk shows. Comedienne Roseanne recently learned for the first time, while in therapy, that she had been repeatedly molested by her parents, starting when she was three months old! Her story made the cover of *People* magazine. Her parents and sisters deny it and have threatened legal action. A former Miss America, Marilyn Van Derbur, has been in the news proclaiming her decades-delayed recollection of abuse by her father, now deceased.

It is an alarming trend that a dozen states have revised their statute-of-limitations laws and now permit legal action against parents within three years of the time the abuse was *remembered*! In 1990 the first conviction based on "repressed memory" occurred. George Franklin was given a life sentence for murdering an eight-year-old in 1969 almost entirely on the basis of his daughter's memory, allegedly repressed for twenty years, of having witnessed his murdering her friend.* A year later a Pennsylvania man was convicted of murder on the basis of a man's detailed account of what he had seen when he was five, but had totally forgotten for sixteen years.

Although therapists usually deny asking leading questions, tapes of their sessions often prove otherwise. If no memories surface they will prod a patient to make up a story. After many repetitions and elaborations of the invented scenario, the patient starts to believe the story is true. One therapist, who claims to have treated fifteen hundred incest victims, explained her approach. She would say to a patient: "You know, in my experience, a lot of people who are struggling with many of the same problems you are have often had some kind of really painful things happen to them as kids—maybe they were beaten or molested. And I wonder if anything like that ever happened to you?" Another traumatist says: "You sound to me like the sort of person who must have been sexually abused. Tell me what that bastard did to you."

The FMS epidemic would not be so bad if such therapists were frauds interested only in money, but the sad truth is that they are sincere. So were the doctors who once tried to cure patients by bleeding, and the churchmen who "cured" witches by torture, hanging, and burning.

Better-trained, older psychiatrists do not believe that childhood memories of traumas can be repressed for any length of time, except in rare cases of actual brain damage. Nor is there any evidence that hypnosis improves memory. It may increase certitude, but not accuracy. And there is abundant evidence that totally false memories are easily aroused in the mind of a suggestible patient.

A two-part article by Lawrence Wright, "Remembering Satan" (*New Yorker*, May 17 and 24, 1993), tells the tragic story of Paul Ingram, a respected police officer in Olympia, Washington, who was accused by his two adult daughters of sexually abusing them as children. Ingram's family are devout Pentecostals who believe that Satan can wipe out all memories of such crimes. Ingram remembered nothing, but after five months of intensive questioning, he came to believe himself guilty. Psychologist Richard Ofshe, writing on "Inadvertent Hypnosis During Interrogation" (*International Journal of Clinical and Experimental Hypnosis* 11 [1992]: 125–55), tells how he fabricated an imaginary incident of Ingram's sexual abuse of a son and daughter. After repeated suggestions that he try to "see" this happening, Ingram produced a written confession!

Jean Piaget, the Swiss psychologist, tells of his vivid memory of an attempted kidnapping when he was two. The thief had been foiled by Piaget's nurse, who bravely fought off the man. When Piaget was in his teens the nurse confessed that she made up the story to win admiration, even scratching herself to prove there had been a struggle. Piaget had heard the story so often that it seeped into his consciousness as a detailed memory.

---

*Franklin's conviction was recently reversed, his daughter's testimony being called unreliable.

Paul McHugh, a psychiatrist at Johns Hopkins University, in "Psychiatric Mis-adventures" (*American Scholar*, Fall 1992), writes about a woman who under therapy came to believe she had been sexually assaulted by an uncle. She recalled the exact date. Her disbelieving mother discovered that at that time her brother was in military service in Korea. Did this alter the woman's belief? Not much, "I see, Mother," she said, "Yes. Well let me think, if your dates are right, I suppose it must have been Dad."

Although the incest-recall industry is likely to grow in coming years, as it spreads around the world, there are some hopeful signs. Here and there women are beginning to discover how cruelly they have been deceived and are suing ther-apists for inducing false memories that caused them and their parents great suf-fering. They are known as "recanters" or "retractors."

Another welcome trend is that distinguished psychologists and psychiatrists are now writing papers about the FMS epidemic. I particularly recommend the book *Confabulations* (Boca Raton, Fla.: Social Issues Research Series, 1992) by Eleanor Goldstein, and the following three articles: "Beware the Incest-Survivor Machine," by psychologist Carol Tavris, in the *New York Times Book Review* (January 3, 1993); "The Reality of Repressed Memories," by Elizabeth Loftus (psychologist, University of Washington), *American Psychologist* 48 (May 1993, pp. 518–37); and "Making Monsters," by Richard Ofshe and Ethan Watters, in *Society* (March 1993). Most of this column is based on material in those articles. Copies can be obtained, along with other literature, from the FMS Foundation, 3401 Market Street, Philadelphia, PA 19104. The phone number is 215-387-1865, Fax: 215-387-1917.

The FMS Foundation is a nonprofit organization whose purpose is to seek rea-sons for the FMS epidemic, to work for the prevention of new cases, and to aid vic-tims. By the end of 1992, only ten months after its founding, more than two thou-sand distressed parents had contacted the Foundation for advice on how to cope with sudden attacks by angry daughters who had accused them of horrible crimes.

I trust that no one reading this column will get the impression that either I or members of the FMS Foundation are not fully aware that many women are indeed sexually abused as children and that their abusers should be punished. In its newsletter of January 8, 1993, the Foundation responded to criticism that some-how its efforts are a backlash against feminism. Their reply: Is it not "harmful to feminism to portray women as having minds closed to scientific information and as being satisfied with sloppy, inaccurate statistics? Could it be viewed as a pro-found insult to women to give them slogans rather than accurate information about how memory works?"

The point is not to deny that hideous sexual abuse of children occurs, but that when it does, it is not forgotten and only "remembered" decades later under hyp-nosis. Something is radically amiss when therapist E. Sue Blume, in her book *Secret Survivors*, can maintain: "Incest is easily the greatest underlying reason why women seek therapy.... It is my experience that fewer than half of the women who experi-ence this trauma later remember or identify it as abuse. Therefore it is not unrea-sonable that *more than half of all women* are survivors of childhood sexual trauma."

As Carol Tavris, author of *Mismeasure of Women*, comments in her article cited above: "Not one of these assertions is supported by empirical evidence."

ADDENDUM

It was no surprise that many defenders of RMT (Repressed Memory Therapy) would send letters blasting my column. Two such letters, and one favorable letter, appeared in the Winter 1994 issue of the *Skeptical Inquirer.*

> Re Martin Gardner's "The False Memory Syndrome" (*Skeptical Inquirer*, Summer 1993): Some traumas for some people clearly are repressed; other traumas for other people are clearly not repressed. Repressions may last for years.
>
> In World War II, there were hundreds of documented cases of battlefield trauma-induced neuroses, characterized by symptoms appearing after the trauma and an inability to remember the trauma. Among helpful treatments used by the military and VA was a brief psychotherapy utilizing hypnosis or sodium pentothal interviews with the aim of quickly restoring a pre-traumatic level of functioning. Heightened suggestibility of the patients was not a problem, since other people knew what had happened.
>
> Patients usually do try to get external confirmation of what they uncover insofar as it deals with externally observable events. With incest, however, the people best able to confirm or deny have a strong motive to lie if it is true.
>
> Patients who accept false events as "real" memories (during badly conducted psychotherapy) do badly in their subsequent life. In psychoanalytic therapy this is not so much a problem; eventually the truth will be discovered as patient and therapist investigate and consider all possibilities. An honest psychotherapist may not know the truth for a long time; it is therapeutic for patients to learn to tolerate uncertainty and to consider that the reconstructed memory might or might not be true. But courts want the "truth" immediately.
>
> Because of the social importance of memories the American Psychological Association has formed a task force to survey all we know about memory from both experimentation and clinical observations.
>
> Bertram P. Karon, Professor, Department of Psychology,
> Michigan State University, East Lansing, Mich.

> For a journal purportedly devoted to "scientific investigation of the paranormal," the inclusion of "The False Memory Syndrome" by Martin Gardner in your summer issue is disturbing, to say the least. This is an area in which I can speak from some knowledge: I am one of those "better trained" (will Harvard do?), "older" therapists, a writer of one book (*Second Childhood*, Norton, 1989) and many journal articles, and a teacher (of memory, among other subjects) on the postgraduate level. More important, however, I keep up with the literature: I would say, offhand, that I have read carefully at least ten full-length professional books and two hundred or more professional articles, as well as lay articles on the subject of delayed memory, including all of those cited in Gardner's contribution. In addition, just this year I have, in the interests of skepticism, curiosity, and open-mindedness, attended a lecture by John Mack, M.D. (surely an "older" Harvard psychiatrist, he of the UFOs), and a conference in which Pamela Freyd (she of the False Memory Syndrome Foundation) played a major part. I was greatly unimpressed—in fact, horrified—by both of them, each with an axe to grind that blinded them to consideration of real data. Unfortunately, I feel the same way about Gardner.

In the interests of brevity, I will focus on a fundamental point made by Gardner: "No one doubts that childhood sexual assaults occur, *but in almost every case the event is never forgotten*" (italics mine). This statement is in direct contradiction to evidence established in myriad serious studies. Let me refer him to just one small journal article that summarizes a retrospective study of two hundred admissions to an emergency room in which the admission records recorded sexual assault of female children under the age of twelve (Linda Meyer Williams, "Adult Memories of Childhood Abuse: Preliminary Findings from a Longitudinal Study," *The Advisor*, Summer 1992): 38 percent of a hundred of those interviewed seventeen years later did not remember the abuse or the hospitalization, which had been objectively documented. I have, myself, had clients who remembered childhood sexual abuse for the first time as adults, in or out of therapy, and, upon talking to the accused, were actually supported in their allegations by confessions and, sometimes, apologies. Still others located witnesses, or other unimpeachable evidence, such as records of a forgotten childbirth (due to forgotten incest) at a home for unwed mothers. Similar findings of corroborated accusations are documented in Herman's recent scholarly book, *Trauma and Recovery*. So, surely Gardner's assertion here is blatantly incorrect. In addition, his spurious comparison of delayed memory of sexual abuse to claims of "past lives" and UFO abductions puts a poor light on his intentions to be objective and fair-minded.

Marian Kaplun Shapiro, Ed.D., Licensed Psychologist, Lexington, Mass.

Martin Gardner's column about the false memory syndrome (*Skeptical Inquirer*, Summer 1993) is an outstanding expose of a pseudoscientific practice with devastating consequences for patients and their families. I commend Gardner for writing it and *Skeptical Inquirer* for publishing it.

Forrest M. Mims III, Seguin, Texas

MARTIN GARDNER REPLIES

I have no objection to anything said by Bertram P. Karon. No one denies that traumatic memories can be repressed. The question is whether the thousands of such memories now being elicited by zealous, untrained, self-deceived therapists are genuine or are the product of dubious techniques abetted by sensational books read by the patients.

There are many ways memories of child abuse can be substantiated. A father capable of raping his four-year-old daughter betrays such psychotic behavior that it is almost impossible for there to be no other records of his mental illness. Wives, siblings, and other relatives seldom have reason to maintain a state of "denial." There can be confirming diaries and letters, or at times a full confession. When there is no evidence at all, coupled with vigorous denials by relatives and a history of the alleged victim's love and admiration for the alleged perpetrators, incalculable harm can be done both to parents and to their innocent families by therapy-induced false memories.

Marian Kaplun Shapiro's doctorate is in education. This does not qualify her to practice psychiatry. She is a self-proclaimed hypnotherapist.

Her praise of Judith Herman's book is misplaced. Herman did indeed find that in fifty-three cases of incest survivors, thirty-nine found corroborative evidence. This seems impressive until you learn that thirty-nine survivors had no loss of memory to begin with. "It wouldn't be surprising," said University of Arizona psychologist John Kihlstrom in a recent lecture, "if these individuals were able to validate their memories. Confirmation of abuse is not the same as confirmation of repressed memories."

In Herman's study, patients with nonrepressed memories reported sex abuse when they were eight or older. Those who recalled abuse only after prolonged therapy reported an average abuse age of four to five years. Why were these memories pushed so far back? Because, Kihlstrom suggests, they had to push them back to a time when they couldn't remember anything else.

Shapiro objects to my comparing repressed memories of child sex abuse with similar memories of Satanic rituals, UFO abductions, and past incarnations. But techniques for evoking such memories are exactly the same in all four areas. In the case of Satanic cult memories, the therapy is just as widespread.

I urge readers to check the sensational two-part article "Remembering Satan," by Lawrence Wright, in the *New Yorker* (May 17 and 24, 1993), Leslie Bennetts' article in *Vanity Fair* (June 1993), Gayle Hanson's article in *Insight* (May 24, 1993), and Claire Safran's piece in *McCall's* (June 1993). Better still, write for literature to the False Memory Syndrome Foundation, 3401 Market Street, Philadelphia, PA 19104. This rapidly growing organization, supported by eminent psychiatrists, was formed in 1993 to combat the greatest witch hunt and mental health scandal of this half-century.

Ms. Shapiro fired back in the Summer 1994 issue:

> I could respond at great length about the content of Martin Gardner's "reply" to my letter regarding False Memory Syndrome. However, in this letter I will, with difficulty, restrict myself to a response to his inaccurate references to me, as these are potentially injurious to my professional reputation.
>
> It is telling that Gardner needs to bolster his arguments with an artful but deceptive end run on my credentials. First, he says that my doctorate does not qualify me "to practice psychiatry"; however, it is only he who looks foolish here, since I clearly identified myself not as a psychiatrist, but as a licensed psychologist. Next, and far worse, he employs character assassination by labeling me a "self-proclaimed hypnotherapist," which is tantamount to being called an unethical clinician, practicing a specialty without appropriate training. I cannot allow that statement to go uncorrected: After about two hundred hours of postdoctoral training in hypnotherapy, I have been certified as a hypnotherapist by the American Society of Clinical Hypnosis (ASCH), a well-established training organization recognized by the American Psychological Association and the American Psychiatric Association. In addition, I have been certified at the ASCH's highest level, that of consultant qualified to train and supervise other licensed clinicians in hypnotherapy. Had Gardner taken the time as a "scientific inquirer" to ask me, before publishing his derogatory and incorrect characterization, I would have been glad to furnish any relevant documents in relation to my qualifications.

Surely a basic of responsible journalism is to check the accuracy of one's assertions—especially potentially defamatory ones—before publishing them.

Marian Kaplun Shapiro, Licensed Psychologist, Lexington, Mass.

MARTIN GARDNER RESPONDS

Freud began his career as a hypnotherapist, but soon abandoned hypnotism as a worthless technique, an opinion shared by most psychiatrists today. Nevertheless, I apologize to Marian Kaplun Shapiro, who has a doctorate in education, for using the adjective "self-proclaimed." She has indeed been thoroughly trained in the use of hypnotism for treating the mentally ill.

46

MEASURING THE PREVALENCE OF FALSE MEMORIES

Ted Goertzel

Budd Hopkins, David Jacobs, and Ron Westrum's (1992) assertion, in *Unusual Personal Experiences*, that 3.7 million Americans are suffering from "UFO abduction syndrome" has been criticized for methodological and logical deficiencies (Stires 1993; Klass 1993; Hall, Rodeghier, and Johnson 1993; Dawes and Mulford 1993). In an attempt to remedy these deficiencies, my research-methods class at Rutgers University undertook to replicate Hopkins, Jacobs, and Westrum's survey. Our questionnaire included the items that the authors believed were indicators of UFO abduction, but added others that were designed to place their items in a broader psychological context.

In March and April 1993, we conducted interviews with 697 residents of southwestern New Jersey. Our sample was partly random and partly accidental. We began with telephone interviews of randomly selected telephone subscribers, but it proved too difficult for the students to obtain an adequate sample in that way, so we relaxed the sampling plan to allow the students to include their friends and family members. This kind of accidental sampling is not appropriate for generalizing to a population at large, but it is generally viewed as adequate for testing the validity and reliability of a measure. There is no reason to believe that systematic sampling biases were introduced that would have distorted the correlations between questionnaire items.

Despite the use of accidental sampling, the frequency of "unusual personal experiences" in our sample was close to that obtained from Hopkins, Jacobs, and Westrum's representative national sample. Table 1 gives the percentages from both samples for the five experiences Hopkins, Jacobs, and Westrum used as their measure of UFO-abduction status. By their criteria—of having suffered four or five of these experiences—3.4 percent of our respondents qualified as "abductees" This number is quite high, especially considering the fact that the interviews were

This article first appeared in *Skeptical Inquirer* 18, no. 3 (Spring 1994): 266–72.

Table 1

Experience	National Sample	South Jersey Sample
1. "Waking up paralyzed with a sense of a strange person or presence or something else in the room."	18%	25%
2. "Experiencing a period of time of an hour or more, in which you were apparently lost, but you could not remember why or where you had been."	13%	17%
3. "Feeling that you were actually flying through the air although you didn't know why or how."	10%	12%
4. "Seeing unusual lights or balls of light in a room without knowing what was causing them."	8%	13%
5. "Finding puzzling scars on your body and neither you nor anyone else remembering how you received them or where you got them."	8%	18%
"Abductee": Having four or five of the above experiences	2%	3.4%

done in the suburbs of Philadelphia, an urban region not known for a high frequency of UFO sightings or abduction reports.

Hall, Rodeghier, and Johnson (1993, 132) note that Hopkins, Jacobs, and Westrum "apparently did not assess the reliability of this measure by using the customary formulas based on internal consistency, though it would be quite simple to do so." We applied these formulas to our data and achieved a reliability coefficient of .68 (coefficient alpha) for the abduction scale, which is quite acceptable for a five-item scale. This and other standard psychometric analyses we conducted suggest that the scale is measuring a consistent phenomenon of some kind, but it tells us nothing about *what* it is that the scale is measuring.

Assessing the *validity* of a measure is much more difficult and is really the crux of the issue. By validity, we simply mean whether the measure measures what it is supposed to measure, in this case UFO abductions. The best method (known as "criterion validity") is to correlate the measure with a known outcome measure. Thus, a measure of "scholastic ability" can be validated if it correlates with school performance (other things being equal). This method cannot be used in this case, however, since we have no generally accepted measure of abduction status. The only viable alternative is to examine the measure's "construct validity." This means examining whether the measure performs as the theory predicts.

In this case, there are at least two alternative theories that can explain why the measure is internally consistent. One is that the respondents are consistently reporting on experiences similar to those of UFO abductees. The other is that the individuals who score high on the scale share a psychological tendency to have

Table 2
Survey Items Correlated with "UFO-Indicator" Experiences

Items	Correlation
6. "Feeling as if you left your body."	.30
7. "Having seen, either as a child or as an adult, a terrifying figure —which might have been a monster, a witch, a devil, or some other evil figure in your bedroom, closet, or somewhere else."	.38
8. "Seeing a ghost."	.26
9. "Seeing a UFO."	.21
10. "Having vivid dreams about UFOs."	.16
11. "Having a sudden feeling that something bad was happening to someone you know. and later finding out that you were correct."	.37
12. "Seeing a halo or ring of light around another person which other people couldn't see."	.23
13. "Being visited by a deceased relative or friend who spoke to you by name."	.31

false memories. Theodore Flournoy (1911) referred to this phenomenon as *cryptomnesia*. Psychologist Robert Baker (1992: 78) states that this phenomenon of "seeing complex visual images in one's head that you cannot remember ever having seen before or . . . suddenly hearing voices from unknown and unrecollected sources is not only a much more common occurrence than is generally known but is also one of the more interesting and intriguing anomalies in the field of 'normal' human behavior."

To distinguish between these two theories, we included three items in our questionnaire about "unusual personal experiences" that are *not* part of the "UFO abduction" syndrome as posited by Hopkins, Jacobs, and Westrum. These three items are numbers 11, 12, and 13 in Table 2. As Table 2 shows, these items correlated (Pearson's *r*) with the "UFO abduction" measure as well as with the other items from Hopkins, Jacobs, and Westrum's study that are also included in Table 2. Three of Hopkins, Jacobs, and Westrum's items (numbers 6, 7, and 8) actually say nothing about UFOs, but can better be thought of as tapping a cryptomnesiac tendency to have "psychic" or "New Age" experiences (Ring 1992). If all these 13 items are combined in a single scale, we obtain a reliability coefficient of .75.

We used a statistical technique known as "factor analysis" to analyze how the responses to the 13 items grouped together empirically. The factor analysis found that a single factor accounts for the consistency of the measure. There is no empirical basis for separating the "UFO" items from the items that do not relate to the abduction phenomenon. Table 3 shows the loadings on this principal factor.

Table 3	
Item Number	Factor Loadings
1	.58
2	.50
3	.47
4	.57
5	.42
6	.49
7	.59
8	.58
9	.33
10	.32
11	.60
12	.43
13	.55

The loadings measure how well each item correlated with the general factor the group of items have in common. As the table shows, the three non-UFO-related items (11, 12, and 13) fitted right in with the other items, which were intended to measure UFO-abduction experiences. An attempt to divide these items into groupings through a technique known as "rotating the factors" was not successful. The only items that clustered together were "seeing a UFO" and "having vivid dreams about UFOs," which are obviously similar. If we treat these two items as belonging to a separate factor, the item most closely related to them is "seeing a ghost." The factor analysis strongly supports the view that these 13 items are a measure of cryptomnesia, not of UFO abduction. We will henceforth refer to them as the *cryptomnesia scale*.

This conclusion is also strongly supported by Dawes and Mulford's (1993) innovative study at the University of Oregon. This study demonstrated that the dual nature of Hopkins, Jacobs, and Westrum's first item, which asked about "waking up paralyzed" *and* about sensing a strange person in the room in the *same* item, actually led to an *increased* recollection of unusual phenomena as compared to a properly constructed single-issue survey item. Textbooks on questionnaire writing universally warn against "double-barreled" questions of this sort because they are known to give bad results. Dawes and Mulford confirm this and further offer the explanation that the combination of the two issues in one item causes a conjunction effect in memory that increases the likelihood of false recollection.

While the Hopkins, Jacobs, and Westrum scale is not a valid measure of UFO abduction, its developers have inadvertently constructed a useful measure of another phenomenon: the tendency to have false memories. Therefore, we will henceforth refer to this measure as the "cryptomnesia scale." Further support for this interpretation can be obtained by correlating the cryptomnesia scale with other items in the questionnaire. To the extent that the scale behaves as Baker's theory would predict, we can be more confident of its validity as a measure of cryptomnesia. Table 4 shows correlations between cryptomnesia and a number of other questionnaire items. All the correlations reported in this study are statistically significant ($p < .05$).

The first three items show that high scores on the cryptomnesia scale are related to belief in conspiracies, including the JFK assassination and AIDS conspiracies, as well as the UFO cover-up conspiracy. Cryptomnesiacs are also more likely to believe in astrology and less likely to trust their neighbors, relatives, and parents. They are more likely to have difficulty dealing with the stress in their

Table 4	
Items Correlated with a High Frequency of	
"Unusual Personal Experiences" (Cryptomnesia)	

Items	Correlation
14. "High government officials were involved in the Kennedy assassination."	.21
15. "The AIDS virus was created deliberately as part of a conspiracy against certain groups in our society."	.15
16. "The Air Force is hiding evidence that the United States has been visited by flying saucers."	.20
17. "I believe in astrology."	.16
18. "I can usually trust my neighbors."	−.16
19. "I can usually trust my relatives."	−.15
20. "When I was growing up, I could usually trust my parents."	−.12
21. "I often wake up in the middle of the night worrying about things."	.16
22. "I sometimes feel that people are conspiring against me."	.22
23. "I sometimes find it difficult to deal with all the stress in my life."	.16
24. "I enjoy reading books about UFOs and other strange phenomena."	.22

lives, to wake up with worries at night, and to feel that people are conspiring against them. They are more likely to enjoy reading about UFOs and other strange phenomena. We did not, however, obtain significant correlations with items measuring anomie (a personal lack of purpose or ideals, or a feeling of not really belonging to society).

Many of these findings are not surprising. David Jacobs (1992) has noted high stress levels and frequency of psychosomatic symptoms among "UFO abductees," which he attributes to the trauma of the abduction experience. The question here is one of distinguishing cause from effect—do the abductions cause stress, or does the stress cause the false memories, which are interpreted as abductions under hypnosis? It is difficult to resolve this kind of cause-and-effect issue with data gathered at one point in time. Since our analysis suggests that we are dealing with cryptomnesia, not UFO abduction, however, we undertook a causal path analysis based on the assumption that this is a psychological phenomenon rooted in early life experiences. The results of this analysis are summarized in the path diagram (Figure 1).

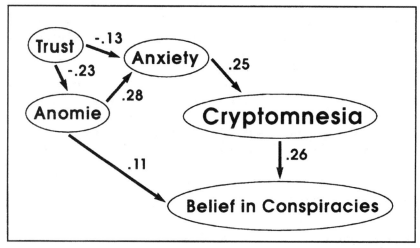

Figure 1

The path analysis suggests that cryptomnesia is rooted in a lack of trusting relationships. This problem may have its origins in early childhood, but it continues into adult life with a lessened feeling of trust of friends and neighbors. This lack of trust leads to feelings of anomie and anxiety that make the individual more likely to construct false memories out of information stored in the unconscious mind. People who think in this way are susceptible to belief in conspiracy theories, since these theories help them to make sense of an otherwise incoherent world.

This causal analysis is consistent with our data, but certainly not proved by it. Path analysis is not a substitute for in-depth psychological case studies. The path coefficients are modest in size, probably because most of the variables were measured with brief three-item scales because of the need to keep the questionnaire brief. This analysis does, however, provide a more plausible interpretation of the data than does the UFO-abduction hypothesis. It is also consistent with the results of detailed psychological testing on "abductees" (Slater 1985; Klass 1989).

David Jacobs was kind enough to speak to our class to familiarize the students with the issue, and I had the opportunity to speak with him informally after the lecture. At that time, I mentioned the "UFO abduction" case discussed in Siegel's (1992) book *Fire in the Brain*. Jacobs was not interested in learning of Siegel's findings and expressed the view that no one was qualified to speak on this issue unless they had done dozens of interviews with abductees under hypnosis, as he had. He clearly fit the profile of the true believer as described in my book *Turncoats and True Believers* (Goertzel 1992). He used numerous ideological defense mechanisms to avoid confronting unwelcome evidence.

The UFO-abduction survey data promise some relief from this subjectivity, thanks to the objectivity of random sampling and statistical analysis. However, even when the data are collected and reported accurately, as they apparently were in this case, the biases of the investigator can enter into the interpretation. In this case, only a strong belief in the prevalence of the abduction phenomenon led the researchers to interpret questions about seeing ghosts or having out-of-body experiences as indicators of abduction. Our data show that there is no objective basis for that assumption. The observed correlations are part of a broader psychological syndrome that includes non-UFO-related phenomena.

Of course none of this proves that UFO abductions have not occurred in a small number of cases. In fact, two of our respondents volunteered that they had experienced such events. (We made no attempt to verify their accounts.) However, the public can rest assured that there is no evidence that millions of Americans are being abducted. Further research on the psychological phenomenon of cryptomnesia, however, is warranted. Both our survey and Hopkins, Jacobs, and Westrum's representative national sample suggest that it is an identifiable syndrome affecting several million people in American society.

REFERENCES

Baker, Robert. 1992. *Hidden Memories.* Amherst, N.Y.: Prometheus Books.

Dawes, Robyn, and Matthew Mulford. 1993. Diagnoses of alien kidnappings that result from conjunction effects in memory. *Skeptical Inquirer* 18: 50–51, Fall.

Flournoy, Theodore. 1911. *Spiritism and Psychology.* New York: Harper.

Goertzel, Ted. 1992. *Turncoats and True Believers.* Amherst, N.Y.: Prometheus Books.

Hall, Robert, Mark Rodeghier, and Donald Johnson. 1992. The prevalence of abductions: A critical look. *Journal of UFO Studies,* n.s. 4: 131–35.

Hopkins, Budd, David Jacobs, and Ron Westrum. 1992. *Unusual Personal Experiences: An Analysis of the Data from Three Major Surveys.* Las Vegas: Bigelow Holding Corporation.

Jacobs, David. 1992. *Secret Life.* New York: Simon & Schuster.

Klass, Philip J. 1989. *UFO-Abductions: A Dangerous Game.* Amherst, N.Y.: Prometheus Books.

———. 1993. Additional comments about the 'Unusual Personal Experiences' survey. *Skeptical Inquirer* 17:145–46.

Ring, Kenneth. 1992. The cosmic connection. Interview by Jonathan Adolph and Peggy Taylor. *New Age Journal,* July/August, pp. 66–69, 122.

Siegel, Ronald. 1992. *Fire in the Brain.* New York: Dutton.

Slater, Elizabeth. 1985. "Conclusions on Nine Psychologicals." In Final Report on the Psychological Testing of UFO "Abductees." Mount Rainer, Md.: Fund for UFO Research.

Stires, Lloyd. 1993. Critiquing the 'Unusual Personal Experiences' survey. *Skeptical Inquirer* 17: 142–44.

RECOMMENDED READING

The past few years have seen an extraordinary number of worthy books published on the themes considered in this volume. This selective listing gives priority emphasis to books published in the 1990s, but I have included several important works of earlier vintage. For more comprehensive lists of those, see the predecessor volumes to this anthology.—The Editor

Alcock, James E. 1990. *Science and Supernature: A Critical Appraisal of Parapsychology.* Amherst, N.Y.: Prometheus Books.

Allen, Steve. 1991. *'Dumbth' and 81 Ways to Make Americans Smarter.* Amherst, N.Y.: Prometheus Books.

Aveni, Anthony. 1996. *Behind the Crystal Ball: Magic, Science, and the Occult from Antiquity Through the New Age.* Times Books, New York.

Baker, Robert A. 1991. *Hidden Memories.* Amherst, N.Y.: Prometheus Books.

Baker, Robert A., and Joe Nickell. 1992. *Missing Pieces: How to Investigate Ghosts, UFOs, Psychics, & Other Mysteries.* Amherst, N.Y.: Prometheus Books.

Blackmore, Susan. 1996. *In Search of the Light.* Amherst, N.Y.: Prometheus Books.

Bulgatz, Joseph. 1992. *Ponzi Schemes, Invaders from Mars, & More Extraordinary Popular Delusions.* Harmony Books.

Carey, Stephen S. 1994. *A Beginner's Guide to Scientific Method.* Belmont, Calif.: Wadsworth Publishing Co.

Cromer, Alan. 1993. *Uncommon Sense: The Heretical Nature of Science.* New York, Oxford: Oxford University Press.

Culver, Roger B., and Philip A. Ianna. 1988. *Astrology: True or False?* Amherst, N.Y.: Prometheus Books.

Dawkins, Richard. 1995. *River Out of Eden.* New York: Basic Books.

———. 1996. *Climbing Mount Improbable.* New York: W. W. Norton.

Dudley, Underwood.1992. *Mathematical Cranks.* Washington, D.C.: Mathematical Association of America.

Edwards, Harry. 1996. *A Skeptic's Guide to the New Age.* Sydney: Australian Skeptics.

Feder, Kenneth L. 1996. *Frauds, Myths, and Mysteries: Science and Pseudoscience in Archaeology.* 2d ed. Mountain View, Calif.: Mayfield Publishing.

Frazier, Kendrick, ed. 1986. *Science Confronts the Paranormal.* Amherst, N.Y.: Prometheus Books.

———. 1991. *The Hundredth Monkey and Other Paradigms of the Paranormal.* Amherst, N.Y.: Prometheus Books.

Friedlander, Michael W. 1995. *At the Fringes of Science.* Boulder, Colo.: Westview Press.

Gray, William D. 1991. *Thinking Critically About New Age Ideas.* Belmont, Calif.: Wadsworth Publishing Co.

Gardner, Martin. 1952, 1957. *Fads & Fallacies in the Name of Science.* New York: Dover Books.

———. 1981, 1989. *Science: Good, Bad, and Bogus.* Amherst, N.Y.: Prometheus Books.

———. 1991. *The New Age: Notes of a Fringe-Watcher.* Amherst, N.Y.: Prometheus Books.

———. 1992. *On the Wild Side.* Amherst, N.Y.: Prometheus Books.

———. 1996. *Weird Water & Fuzzy Logic.* Amherst, N.Y.: Prometheus Books.

George, Leonard. 1995. *Alternative Realities.* New York: Facts on File.

Goldberg, Steven. 1991. *When Wish Replaces Thought: Why So Much of What You Believe Is False.* Amherst, N.Y.: Prometheus Books.

Gross, Paul R., and Norman Levitt. 1994. *Higher Superstition.* Baltimore: Johns Hopkins University Press.

Harrold, Francis B., and Raymond A. Eve, eds. 1987. *Cult Archaeology & Creationism: Understanding Pseudoscientific Beliefs About the Past.* Iowa City: University of Iowa Press.

Hicks, Robert D. 1991. *In Pursuit of Satan.* Amherst, N.Y.: Prometheus Books.

Hines, Terence. 1988 *Pseudoscience and the Paranormal: A Critical Examination of the Evidence.* Amherst, N.Y.: Prometheus Books.

Holton, Gerald. 1993. *Science and Anti-Science.* Cambridge, Mass.: Harvard University Press.

———. 1996. *Einstein, History, and Other Passions: The Rebellion Against Science at the End of the Twentieth Century.* New York: Addison-Wesley.

Hoggart, Simon, and Mike Hutchinson. 1995. *Bizarre Beliefs.* London: Richard Cohen Books.

Huber, Peter W. 1991.*Galileo's Revenge: Junk Science in the Courtroom.* New York: Basic Books.

Humphrey, Nicholas. 1996. *Leaps of Faith: Science, Miracles, and the Search for Supernatural Consolation.* New York: Basic Books.

Hyman, Ray. 1989. *The Elusive Quarry: A Scientific Appraisal of Psychical Research.* Amherst, N.Y.: Prometheus Books.

Kurtz, Paul. 1986, 1991. *The Transcendental Temptation.* Amherst, N.Y.: Prometheus Books.

———. 1992.*The New Skepticism.* Amherst, N.Y.: Prometheus Books.

———, ed. 1985. *A Skeptic's Handbook of Parapsychology.* Amherst, N.Y.: Prometheus Books.

Levy, David A. 1997.*Tools of Critical Thinking.* Boston: Allyn and Bacon.

Loftus, Elizabeth F., and K. Ketcham. 1994. *The Myth of Repressed Memory.* New York: St. Martin's Press.

Matsumura, Molleen. 1995. *Voices for Evolution.* 2d ed. Berkeley, Calif.: National Center for Science Education.

Nickell, Joe. 1995. *Entities: Angels, Spirits, Demons, and Other Alien Beings.* Amherst, N.Y.: Prometheus Books.

Nickell, Joe, ed. 1994. *Psychic Sleuths: ESP and Sensational Cases.* Amherst, N.Y.: Prometheus Books.

Nickell, Joe, and John F. Fischer. 1992. *Mysterious Realms: Probing Paranormal, Historical, and Forensic Enigmas.* Amherst, N.Y.: Prometheus Books.

Ofshe, Richard, and Ethan Watters. 1994. *Making Monsters: False Memories, Psychotherapy, and Sexual Hysteria.* New York: Scribner's.

Randi, James. 1982. *Flim-Flam!* Amherst, N.Y.: Prometheus Books.

———. 1987, 1989. *The Faith-Healers.* Amherst, N.Y.: Prometheus Books.

———. 1995. *An Encyclopedia of Claims, Frauds, and Hoaxes of the Occult and Supernatural.* New York: St. Martin's Press.

Rothman, Milton. 1988. *A Physicist's Guide to Skepticism.* Amherst, N.Y.: Prometheus Books.

Ruchlis, Hy. 1990. *Clear Thinking.* Amherst, N.Y.: Prometheus Books.

Sagan, Carl. 1979. *Broca's Brain.* New York: Random House.

———. 1996. *The Demon-Haunted World: Science as a Candle in the Dark.* New York: Random House.

Schick, Jr., Theodore, and Lewis Vaughn. 1995. *How to Think About Weird Things.* Mountain View, Calif.: Mayfield Publishing Co.

Schnabel, Jim. 1997. *Remote Viewers: The Secret History of America's Psychic Spies.* New York: Dell.

Shermer, Michael. 1997. *Why People Believe in Weird Things.* New York: Freeman.

Spanos, Nicholas. 1996. *Multiple Identities & False Memories.* Washington, D.C.: American Psychological Association.

Stein, Gordon, ed. 1993. *Encyclopedia of Hoaxes.* Detroit: Gale Research.

———. 1996. *Encyclopedia of the Paranormal.* Amherst, N.Y.: Prometheus Books.

Stenger, Victor J. 1995. *The Unconscious Quantum: Metaphysics in Modern Physics and Cosmology.* Amherst, N.Y.: Prometheus Books.

Strahler, Arthur. 1992. *Understanding Science.* Amherst, N.Y.: Prometheus Books.

Victor, Jeffrey S. 1993. *Satanic Panic.* Chicago: Open Court.

Vos Savant, Marilyn. 1996. *The Power of Logical Thinking.* New York: St. Martin's Press.

Williams, Stephen. 1991. *Fantastic Archaeology.* Philadelphia: University of Pennsylvania Press.

Wiseman, Richard. 1997. *Deception & Self-Deception: Investigating Psychics.* Amherst, N.Y.: Prometheus Books.

Wolpert, Lewis. 1992. *The Unnatural Nature of Science.* Cambridge, Mass.: Harvard University Press.

Zimmerman, Michael. 1995. *Science, Nonscience, and Nonsense.* Baltimore: Johns Hopkins University Press.

CONTRIBUTORS

JAMES E. ALCOCK is professor of psychology at Glendon College, York University, Toronto, and author of *Parapsychology: Science or Magic?* (1981) and *Science and Supernature* (1990).

PAUL BARBER is a research associate at the Fowler Museum of Cultural History at UCLA. His speciality is folklore associated with burial and death. He received his Ph.D. from Yale University.

SUSAN BLACKMORE is senior lecturer in psychology at the University of the West of England, Bristol, and author of *Dying to Live, Test Your Psychic Powers* (with Adam Hart-Davis), and *In Search of the Light.*

JAN HAROLD BRUNVAND retired in June 1996 after 30 years as professor of English and folklore at the University of Utah. He is author of five books and numerous articles on urban legends.

ALAN CROMER is a physics professor at Northeastern University in Boston. He is the author of *Uncommon Sense: The Heretical Nature of Science* (1993) and *Connected Knowledge: Science, Philosophy, and Education* (1997).

RAYMOND A. EVE is professor of sociology and anthropology, University of Texas at Arlington.

MANDY FOWLER, KATY McCARTHY, and **DEBBIE PEERS** were at sixth-form college in the UK at the time their study on the Barnum effect was carried out. It won first prize in the British Psychological Society's School Project competition.

KENDRICK FRAZIER, editor of this volume, is a science writer and the editor of the *Skeptical Inquirer.* He is author of four books, including *People of Chaco: A Canyon and Its Culture* and *Solar System,* and editor of three previous general anthologies of *Skeptical Inquirer* articles: *The Hundredth Monkey, Science Confronts the Paranormal,* and *Paranormal Borderlands of Science.* He is also coeditor of a single-subject anthology, *The UFO Invasion* (1997).

CHRISTOPHER C. FRENCH is currently head of the Department of Psychology at Goldsmiths College, University of London, where he teaches a course entitled "Psychology, Parapsychology, and Pseudoscience."

MARTIN GARDNER is author of some sixty books about science, mathematics, philosophy, and literature, including *The Night Is Large, Weird Water & Fuzzy Logic, On the Wild Side, The New Age, Order and Surprise, Science: Good, Bad, and Bogus, The Ambidextrous Universe, The Relativity Explosion,* and *Fads and Fallacies in the Name of Science.* He is also a member of CSICOP's Executive Council.

THOMAS GILOVICH is professor of psychology at Cornell University and the author of *How We Know What Isn't So: The Fallibility of Human Reason in Everyday Life.*

TED GOERTZEL is professor of sociology and director, Forum for Policy Research, Rutgers, State University of New Jersey at Camden.

PAUL R. GROSS is professor of life sciences, emeritus, University of Virginia, and Visiting Scholar, Harvard University.

FRANCIS B. HARROLD is professor and chair of sociology and anthropology at the University of Texas at Arlington.

EVAN HARRINGTON is currently working on his dissertation in social psychology at Temple University. The subject of his research is the nocebo effect (negative placebo) and mass psychogenic illness.

ROBERT D. HICKS is a law-enforcement specialist with the Virginia Department of Criminal Justice Services and author of *In Pursuit of Satan: The Police and the Occult* (1991). His chapter expresses his opinions, not those of his agency or the Virginia state government.

PETER HUSTON is the author of *Scams From the Great Beyond: How to Make Easy Money Off of ESP, Astrology, UFOs, Crop Circles, Cattle Mutilations, Alien Abductions, Atlantis, Channeling, and Other New Age Nonsense* (1997) and *Tongs, Gangs, and Triads: Chinese Crime Groups in North America* (1995). He lives in Schenectady, N.Y.

RAY HYMAN is professor of psychology at the University of Oregon and author of *The Elusive Quarry: A Scientific Appraisal of Psychical Research* (1989). He is also a member of CSICOP's Executive Council.

JOHN W. JACOBSON is an associate planner with the New York Office of Mental Retardation and Developmental Disabilities.

NORETTA KOERTGE is professor of history and philosophy of science at Indiana University, Bloomington, Indiana. She is coauthor of *Professing Feminism: Cautionary Tales from the Strange World of Women's Studies* (1994) and editor of *A House Built on Sand: Exposing Postmodernist Myths about Science* (1998).

FRANK H. KOBE practices clinical psychology in Columbus, Ohio.

PAUL KURTZ is chairman of the Committee for the Scientific Investigation of Claims of the Paranormal and professor emeritus of philosophy at the State University of New York at Buffalo. He is author of more than 30 books, including *The Transcendental Temptation, The New Skepticism, Toward a New Enlightenment,* and *The Courage to Become.*

Contributors

DAN LARHAMMAR is professor of molecular cell biology in the Department of Medical Genetics at Uppsala University, Sweden. He does research on evolution and brain function and is a member of the Swedish branch of CSICOP.

LEON M. LEDERMAN is Pritzker professor of science at the Illinois Institute of Technology (where he teaches freshman physics), director emeritus of the Fermi National Accelerator Laboratory, and founding board member of the Illinois Math Science Academy in Aurora and the Teachers Academy for Math and Science in Chicago. He received the Nobel Prize in physics in 1988.

JAMES LETT is professor of anthropology and chair of the Department of Social Sciences at Indian River Community College in Ft. Pierce, Florida. He is the author of two books on anthropological theory: *Science, Reason, and Anthropology* and *The Human Enterprise.* He is coauthor of several others, including *Encyclopedia of Cultural Anthropology, Anthropology of Religion,* and *Emics and Etics.*

NORMAN LEVITT is professor of mathematics at Rutgers University and coauthor (with Paul R. Gross) of *Higher Superstition: The Academic Left and Its Quarrels with Science.*

SCOTT O. LILIENFELD is assistant professor of psychology at Emory University in Atlanta. His research interests include the causes and diagnosis of personality disorders and pseudoscientific practices in psychology.

LEE LOEVINGER is a lawyer in the firm of Hogan & Hartson of Washington, D.C. He is a member of Sigma Xi, a former chair of the American Bar Association's Science and Technology Section of of the National Conference of Lawyers and Scientists, and writes frequently on aspects of science. He was formerly an associate justice of the Minnesota Supreme Court, an Assistant Attorney General of the United States, and member of the Federal Communications Commission.

ELIZABETH F. LOFTUS is professor of psychology and adjunct professor of law at the University of Washington in Seattle and president of the American Psychological Society, 1998–99. She is coauthor of *The Myth of Repressed Memory* (1994).

TIMOTHY E. MOORE is a psychology professor at Glendon College, York University, Toronto. He has been writing reviews and conducting research on subliminal influence for 15 years. He has acted as a consultant to both government and the private sector on issues of subliminal manipulation. In 1990 he testified as an expert witness for the defense during the Judas Priest trial.

JAMES A. MULICK is a professor of psychology at the Ohio State University.

JOE NICKELL is senior research fellow of CSICOP and the *Skeptical Inquirer's* "Investigative Files" columnist. A former magician, detective, and academic, he is author, coauthor, or editor of 16 books.

BERNARD ORTIZ de MONTELLANO, an organic chemist, is professor of anthropology at Wayne State University. He is a long-time advocate for minorities in science and a founding member of the Society for the Advancement of Chicanos and Native Americans in Science. He is the author of *Aztec Medicine, Nutrition, and Health* (1990).

ANTHONY R. PRATKANIS is a social psychologist who studies social influence processes. He is coauthor of *Age of Propaganda: The Everyday Use and Abuse of Persuasion* and is currently professor of psychology at the University of California, Santa Cruz.

CARL SAGAN was the David Duncan Professor of Astronomy and Space Science and director of the Laboratory for Planetary Studies at Cornell University. His last books were *The Demon-Haunted World: Science As a Candle in the Dark* and *Billions and Billions: Thoughts on Life and Death at the Brink of the Millennium*. This chapter is based on his keynote address at the 1994 CSICOP conference in Seattle.

KENNETH SAVITSKY is an assistant professor of psychology at Williams College.

GLENN T. SEABORG is associate director-at-large of the Lawrence Berkeley National Laboratory, University Professor at the University of California, and Chairman of the Lawrence Hall of Science. He received the Nobel Prize in chemistry in 1951. On August 30, 1997, the International Union of Pure and Applied Chemistry officially named element 106 *seaborgium*, after him.

ELIE A. SHNEOUR is director of the Biosystems Research Institute in San Diego and a neurochemist who studied the relationship between brain function and cognitive potential.

MATTHEW SMITH is a research assistant at the University of Hertfordshire.

ROY STEMMAN is a writer on the paranormal and editor of *Reincarnation* magazine.

VICTOR J. STENGER is professor of physics at the University of Hawaii. He does research in neutrino psychics and astrophysics. He is the author of *Not by Design* (1988), *Physics and Psychics* (1990), and *The Unconscious Quantum* (1995).

JAMES R. STEWART is with the department of social behavior, University of South Dakota, Vermillion, South Dakota.

JOHN H. TAYLOR is in the anthropology program at the University of Texas at Arlington.

JEFFREY S. VICTOR is professor of sociology at the SUNY Jamestown Community College. He is author of *Satanic Panic* (1993), winner of the H.L. Mencken Award for best book of 1994.

DONALD WEST is a criminologist at Cambridge University.

JEFF WISEMAN is a freelance writer who assisted in the experiments described in "Eyewitness Testimony and the Paranormal."

RICHARD WISEMAN is the Perrott-Warrick Senior Research Fellow at the University of Hertfordshire and author of *Deception and Self-Deception: Investigating Psychics* (1997).

NAME INDEX